Classical, Semi-classical and Quantum Noise

Leon Cohen • H. Vincent Poor • Marlan O. Scully
Editors

Classical, Semi-classical and Quantum Noise

Springer

Editors
Leon Cohen
Hunter College and Graduate Center
City University of New York
Park Ave. 695
10065 New York
USA
leon.cohen@hunter.cuny.edu

H. Vincent Poor
School of Engineering and Applied Science
Princeton University
Olden Street
08544 Princeton, New Jersey
USA
poor@princeton.edu

Marlan O. Scully
Princeton University
Princeton, NJ 08544

Texas A&M University
College Station, TX 77843
USA
mscully@princeton.edu

ISBN 978-1-4419-6623-0 e-ISBN 978-1-4419-6624-7
DOI 10.1007/978-1-4419-6624-7
Springer New York Dordrecht Heidelberg London

Library of Congress Control Number: 2011943092

Printed on acid-free paper

Springer is part of Springer Science+Business Media (www.springer.com)

Preface

David Middleton was a towering figure of the Twentieth Century engineering and science. He was the originator of many fundamental ideas and methods, and was one of the founders of statistical communication theory. To honor David's contributions, a symposium, "The Middleton Meeting on Classical, Semiclassical, and Quantum Noise," was held on November 2 and 3, 2007, at Princeton University. David was present at the symposium and gave one of the talks. This book is an outcome of that meeting.

David's seminal contributions span more than six decades and have had a major impact on many fields, including radar, sonar, communications, noise theory, and probability theory, among many others. His work was a unique blend of fundamental theory and applications. During the second world war, a young David Middleton working with John H. Van Fleck (who would go on to receive the Nobel Prize in Physics) devised the notion of the matched filter, which is the most basic method used in detecting signals in noise. At the same time, and working independently, D.O. North developed the same idea. In 1960, after making many pioneering contributions, David published a book that has become a classic in the field: "An Introduction to Statistical Communication Theory." This is a monumental book that had, and continues to have, a profound impact on researchers in many fields. It became the basic book for learning stochastic processes and communication theory for thousands of engineers and scientists over the intervening years.

Over the past 10 years or so, David was writing a new book entitled "Elements of Non-Gaussian Space-Time Statistical Communication Theory, Propagation, Noise, and Signal Processing in the Canonical Channel." Sadly, David passed away on November 16, 2008, at the age of 88 years, before this second masterwork could be completed. However, the material that he had already completed represents a major contribution in itself, and will be published posthumously in the near future. In the final chapter of the current volume, we include the Introduction to his new book. It is a wonderful presentation of the field of communication, and of his own views on the field that he developed over the past 60 years.

We wish to express our deep appreciation to the authors who contributed to this volume, which is a very fitting tribute to our late friend and colleague David Middleton.

Leon Cohen
H. Vincent Poor
Marlan O. Scully

Contents

Contributors

Julian J. Bussgang Student of David Middleton, Founder and President of Signatron, Inc., Lexington, MA, USA, Life Fellow of the IEEE

James V. Candy Lawrence Livermore National Laboratory, Livermore, CA, USA

Leon Cohen Department of Physics, Hunter College of the City University of New York, New York, USA

Joseph H. Eberly Rochester Theory Center and Department of Physics and Astronomy, University of Rochester, Rochester, NY, USA

Salim El Rouayheb ECE Department, Texas A&M University, College Station, TX, USA

Lorenzo Galleani Politecnico di Torino, Corso Duca degli Abruzzi, Torino, Italy

Costas N. Georghiades ECE Department, Texas A&M University, College Station, TX, USA

Godfrey Gumbs Physics Department, Hunter College of the City University of New York, New York, NY, USA

Steven Kay Department of Electrical, Computer, and Biomedical Engineering, University of Rhode Island, Kingston, RI, USA

Laszlo B. Kish Department of Electrical and Computer Engineering, Texas A&M University, College Station, TX, USA

Patrick J. Loughlin Department of Bioengineering & ECE, University of Pittsburgh, Pittsburgh, PA, USA

Eugeniy Mikhailov College of William and Mary, Williamsburg, VA, USA

Athina P. Petropulu ECE Department, Rutgers The State University of New Jersey, Piscataway, NJ, USA

H. Vincent Poor School of Engineering and Applied Science, Princeton University, Princeton, NJ, USA

Muralidhar Rangaswamy Air Force Research Laboratory Sensors Directorate, Hanscom AFB, MA, USA

Yuri V. Rostovtsev Department of Physics, University of North Texas, Denton, TX, USA Or Department of Physics, University of North Texas, 1155 Union Circle #311427, Denton, TX, USA

Wolfgang P. Schleich Institut für Quantenphysik, Universität Ulm, Ulm, Germany

Marlan O. Scully Institute for Quantum Studies and Department of Physics, Texas A&M University, College Station, TX, USA

Applied Physics and Materials Science Group, Engineering Quad, Princeton University, Princeton, NJ, USA

Jeffrey H. Shapiro Research Laboratory of Electronics, Massachusetts Institute of Technology, Cambridge, MA, USA

Yanhua Shih Department of Physics, University of Maryland, Baltimore County, Baltimore, MD, USA

Michael F. Shlesinger United States Naval Academy, Physics Department, Annapolis MD, USA

Office of Naval Research, Code 30, Arlington, VA, USA

Andrii S. Sizhuk Institute for Quantum Studies and Department of Physics, Texas A&M University, College Station, TX, USA

Qingqing Sun Department of Physics and Institute of Quantum Studies, Texas A&M University, College Station, TX, USA

Anatoly A. Svidzinsky Institute for Quantum Studies and Department of Physics, Texas A&M University, College Station, TX, USA

George R. Welch Texas A&M University, College Station, TX, USA

Sabine Wölk Institut für Quantenphysik, Universität Ulm, Ulm, Germany

Xueshi Yang Marvell Semiconductor Inc., Santa Clara, CA, USA

Ting Yu Department of Physics and Engineering Physics, Stevens Institute of Technology, Hoboken, NJ, USA

M. Suhail Zubairy Department of Physics and Institute of Quantum Studies, Texas A&M University, College Station, TX, USA

Chapter 1
David Middleton

Julian J. Bussgang

David Middleton, who had a remarkable career spanning seven decades, was a major theoretician, an outstanding academician, and a practical contributor to defense organizations and military contracting companies. He is best known for his seminal work in establishing the field of statistical communication theory. His pioneering theoretical work on modeling communication channels and systems contributed to the modern growth of data and wireless communications. His publications include two books and numerous papers on different aspects of signal processing, radar, communications, electromagnetic interference, and underwater signal handling.

David Middleton was born on April 19, 1920 in New York City to Scudder and Lucile (Davidson) Middleton. His preparatory schooling was at Deerfield Academy in Massachusetts, where his aptitude in mathematics soon became apparent to his teachers. He graduated in 1938. He then attended Harvard College, where he rowed crew and played lacrosse. He was elected to Phi Beta Kappa, and in 1942 graduated *summa cum laude* with a bachelor's degree in physics.

The United States entered World War II in late 1941, during Middleton's senior year. Thus, after receiving his bachelor's degree, Middleton, in addition to continuing as a teaching fellow in electronics at Harvard, became an instructor in an accelerated, basic course on electromagnetic field theory and radar for Army and Navy Signal Corps officers, before they went on for further radar training at the much larger Radiation Laboratory at MIT.

He joined Harvard's Radio Research Laboratory (RRL), operating under the supervision of the Office of Scientific Research and Development, to work on the so-called Project G-19. There, he was assigned to the "Theoretical Group" under the direction of Professor Frederick E. Terman of Stanford University. This group worked primarily on radar counter-measures: how to effectively carry out electronic jamming and passive jamming with chaff.

J.J. Bussgang (✉)
Student of David Middleton, Founder and President of Signatron, Inc., Lexington, MA, USA, Life Fellow of the IEEE

L. Cohen et al. (eds.), *Classical, Semi-classical and Quantum Noise*,
DOI 10.1007/978-1-4419-6624-7_1, © Springer Science+Business Media, LLC 2012

At RRL, Middleton was a special research assistant (1942–45) to Professor John H. Van Vleck, who later received the Nobel Prize in Physics (1977). [1] His research with Van Vleck led Middleton to become involved in the mathematical issues of describing "random noise" analytically, which became the foundation of his future work on the statistical analysis of signal processing. In 1943, their work on the theory of random noise led them to a meeting with Stephen O. Rice of Bell Laboratories, another pioneer in the field of noise modeling and signal processing. Following this meeting, Middleton and Rice developed a special professional relationship. [2]

Middleton continued to study physics and electronics as Van Vleck's doctoral student at Harvard and received his A.M. degree in 1945 and his Ph.D. in physics in 1947.

Simultaneously, but independently of D. O. North (RCA, Princeton), [3] Middleton and Van Vleck developed the fundamental matched-filter concept critical to detecting signals in noise, and jointly publishing an early paper on the performance of radar receivers. Their 1946 paper, drawing on earlier classified reports, was the first open-literature publication of the matched-filter principle, which has become the most fundamental principle of communications receiver design. The term "matched filter" was coined by them in their 1944 Harvard Radio Research Laboratory (RRL) report, a forerunner of their later unclassified paper of the same title. [4] Middleton went on to publish a fundamental paper on the effects of a video filter on the detection of pulsed signals in noise. [5]

Prof. Van Vleck, although basically a physicist, was also the author of a theorem, first published in a wartime report, on the autocorrelation function of clipped noise. [6] Later, he and Middleton published a joint paper on the spectrum of clipped noise. [7] Clipped noise analysis proved to be particularly important to underwater sound applications, where large amplitude variations are commonly encountered and cause processing difficulties.

It is one of Middleton's early, remarkable accomplishments, lesser known among those in communication theory that, as a graduate student, after taking a course with Professor Ronold W. P. King, he published important joint papers with Prof. King on the second-order theory for determining antenna impedance. [8] Unrelated to his future signal-processing work, this fundamental theory and method are used to this day.

Middleton remained at Harvard to do post-doctoral studies with Prof. Leon Brillouin and in 1949 was appointed assistant professor of applied physics in the Department of Engineering and Applied Physics, continuing his research on the statistical modeling of noise, communication channels, and the processing of signals for radar and radio communications. He developed and taught new courses in statistical communication theory, theory of signals in noise, and related probability methods.

Middleton had several doctoral students at Harvard (including: J. J. Bussgang, [9] G. E. Fellows, [10] H. W. Fuller, [11] J. A. Mullen, [12] and D. Van Meter [13]) to whom he suggested excellent thesis topics and with whom he worked closely on fundamental communication issues. The work with Van Meter introduced Statistical Decision Theory into communication analysis. The work with Bussgang introduced Wald's Statistical Sequential Detection into signal detection.

Middleton left Harvard in 1955 and pursued his work as a consulting physicist to industry, university laboratories, and the U. S. Government. His work expanded greatly to include electromagnetic compatibility, with particular attention to non-Gaussian noise models, [14] interference, digitized data, and signal processing for man-made and natural electromagnetic and acoustic environments. His mathematical and analytical work, initially related to radar, found important applications in telecommunications, sonar, and seismology.

In 1960, Middleton published his groundbreaking book *An Introduction to Statistical Communication Theory*, which became the fundamental text for teaching stochastic processes and communication theory. [15] He published more than 170 papers and a second book, *Topics in Communication Theory* (1965). [16] Prof. Leon Cohen of City University of New York wrote in his paper on "the history of noise," "Middleton's book *Introduction to Statistical Communication Theory* is one of those texts that is so extraordinary for its clarity and depth that one marvels at it and its author. It is probably the greatest book ever written on noise, probability theory, and stochastic processes.... The classic book on noise, written with style and elegance, covers a panoramic view unmatched by any other publication." [17] This book was translated into many other languages, including Russian.

His governmental clients included Department of Defense, Department of Transportation, Institute of Defense Analysis, Institute for Telecommunication Sciences (U.S. Dept. of Commerce), NASA, Naval Underwater Systems Center (NUSC), National Oceanic & Atmospheric Administration (NOAA), Naval Research & Development (NRAD–San Diego), NSA, Lawrence Livermore National Laboratories, Office of Naval Research, Office of Telecommunications Policy (Executive Office of the President), USAF Cambridge Research Center, etc. From 1970 to 1977, Middleton served as a member of the U.S. Naval Advisory Research Committee, and from 1988 to 1991, on the Scientific Advisory Board of the Supercomputing Research Center, Institute of Defense Analysis.

His many nongovernmental clients included: Applied Physics Laboratories (University of Washington), Applied Research Laboratories (University of Texas), COMSAT, General Dynamics Information Technical Institute, General Electric, GTE-Sylvania, Honeywell, Karmanos Cancer Institute, Kildare Corp., MIT Lincoln Laboratory, Radiation Laboratory (Johns Hopkins University), RAND Corp., Raytheon, [18] Science Applications International Corporation (SAIC), and Sperry-Rand. As a busy consultant, with clients all across the country and abroad, he carried everywhere a memorable, leather briefcase, bulging with all his working material, as he traveled from place to place.

Middleton continued teaching as adjunct professor at Columbia University (1960–61), Rensselaer Polytechnic Institute (1961–70), Rice University (1979–89), University of Denver (1972–75), University of Texas (1960–71), and University of Rhode Island (1966–2004). He had several doctoral students writing excellent theses at these different places (e.g., A. Spaulding [19]). He maintained extended and friendly relationships with his former students long after they graduated.

Middleton received many honors and recognition for his work. In 1958, he was elected fellow of the IEEE "for contributions to the theory of noise in electronic systems." He was also elected as a fellow of the American Physical Society (1951), the Acoustical Society of America (1978), and the New York Academy of Sciences (1991) and as a member of the American Association for the Advancement of Science (1959). In 1998, he was elected as a member of the National Academy of Engineering.

His paper with W. H. Huggins received first prize at the 1955 National Electronic Conference, [20] as did another paper at the 1979 EMC Society of the IEEE International Conference in Rotterdam. [21] One of his papers received the IEEE Electromagnetic Compatibility Society Transactions Prize, [22] and two other papers the EMC Society's 1979 Prize. [23] Together with A. D. Spaulding, he was a recipient of the 1978 Outstanding Authorship Award of the Institute for Telecommunications Sciences (U.S. Dept. of Commerce) and of the ITS 1978 Special Achievement Award. [24]

He was listed in *Who's Who In America* (Marquis), *Who's Who in the World* (Marquis), *Who in Engineering* (AAES), *Leaders in Electronics* (McGraw Hill), *American Men and Women of Science* (Gale), *Who's Who in Technology Today*, (J. Dick Publishing), *Who's Who in Frontier of Science & Technology* (Marquis), *World Who's Who in Science* (Marquis-Who's Who), and *Who's Who in Optical Science and Engineering* (Marquis-Who's Who).

Recognized internationally, Middleton was highly regarded in the Soviet Union. In the 1970s, at the height of the Cold War, he traveled to the Soviet Union for scientific exchanges and to lecture. He served as the technical editor for the English version (1978) of V. V. Ol'shevskii's *Statistical Methods in Sonar*. [25] Mark A. Bykhovsky, the head of the Radio Research and Development Institute (NIIR) in Moscow, profiled Middleton in his book as the founder of statistical communication theory and included Middleton's biography along with those of other distinguished scientists in the field such as Wiener, Shannon, Rice, and Kotelnikov. [26] Middleton's relationships with Russian scientists were particularly important to him, because his maternal grandfather had immigrated to the United States from Russia in the late 19th century.

Middleton's interest in nonlinear signal processing and non-Gaussian noise had begun early in his career [27] and continued virtually to his death. In the last years of his life, he worked on the sequel to his first book, entitled *Elements of Non-Gaussian Statistical Communication Theory: A Space-Time Treatment,* to be published by the IEEE Press. [28] The new book was intended to summarize his work of over sixty-five years in statistical communication theory as well as present new results from his more recent research by adding time analysis to earlier methods.

Over the years, Middleton was quite involved with IRE and IEEE, and with several of the IEEE societies. Therefore, Michael Geselowitz, director of the IEEE History Center, recorded, as part of the oral history project, Middleton's recollections of his academic and consulting careers, the comments on his research approaches to communication theory, and his views on the impact of computers and the Internet. [29] In August 2008, three months before he died, IEEE and Princeton University hosted a symposium in honor of his long and distinguished career and numerous contributions to the field of statistical information theory.

In 1945, while still a graduate student, Middleton had married Nadea Butler. They first lived in Cambridge and then moved to Concord, MA, where they raised four children in a beautiful home they built on the Assabet River. He and Nadea were divorced in 1971. In 1971, he married Joan (Bartlett) Reed, and moved from Concord, MA, to New York City but continued to maintain an office in Cambridge, MA through the 1970s.

Throughout his life, he was a tall, upright man, determined and focused, with a smile, enormous energy, and a sharp wit. He enjoyed classical music and played Beethoven sonatas, Schubert pieces, and Chopin scherzos on the piano with great gusto and nuance. Precise in speech, courtly in manner, he was also a quick wit and prone to making outrageous puns. He seemed to work incessantly, even during his vacations on Cape Cod, where he enjoyed swimming, playing tennis, and taking walks on the beach.

He had a lifelong interest in art, inherited from a poet father, a violin-playing mother, and a playwright uncle. In an era before computer graphics, he spent hours creating his own drawings, diagrams and representation of equations to illustrate his work. His drafting talents were also put to use to entertain his young family with drawings of imaginative scenes and cartoons of animals; cats were his particular favorites. In addition to his large collection of scientific texts, he also collected books on history, biography, philosophy, and other challenging topics. Dr. Middleton's technical papers have been donated to M.I.T., and his technical library, to the University of Maryland.

David Middleton died on November 16, 2008 at age 88, at Lenox Hospital in New York. He is survived by his wife of thirty-seven years, Joan Bartlett Middleton; the four children by his first wife: Susan of Ashfield, MA, Leslie of Charlottesville, VA, Blake of New York City, and George of Carlisle, MA; three stepsons: Christopher Reed of Philmont, NY, Andrew Reed of NYC, and Henry Reed of Paris, France; and three grandchildren. [30]

References

1. D. Middleton, "Some personal reminiscences: Communication theory and a Nobel prize," *IEEE Commun. Soc. Mag.*, vol. 16, no. 4, July 1978, 9–11.
2. D. Middleton, "S. O. Rice and the theory of random noise: some personal recollections," *IEEE Trans. on Inform. Theory,* IT-34, no. 6, Nov. 1988, 1367–1373.
3. D. O. North, "An analysis of the factors which determine signal/noise discrimination in pulsed carrier systems," RCA Labs, Princeton, NJ, *Rep. PTR-6C*, 1943.
4. J. H. Van Vleck and D. Middleton, "A theoretical comparison of the visual, aural, and meter reception of pulsed signals in the presence of noise," *J. Appl. Phys.*, vol. 17, Nov. 1946, 940–971; D. Middleton, same title, Ph.D. Physics Research Thesis, Harvard University, 1947.

5. D. Middleton, "The effect of a video filter on the detection of pulsed signals in noise," *J. Appl. Phys.*, vol. 21, Aug. 1950, 734–740.
6. J. H. Van Vleck. "The spectrum of clipped noise," Harvard *RRL Report 51*, July 21, 1943.
7. J. H. Van Vleck and D. Middleton, "The spectrum of clipped noise," *Proc. IEEE,* vol. 54, Jan. 1966, 2–19.
8. R. King and D. Middleton, "The cylindrical antenna; current and impedance," *Quart. Appl. Math.* 3, 1946, 302–335; D. Middleton and R. King, "The cylindrical antenna, a comparison of theories," *J. Appl. Phys.*, vol. 17, 1946, 273–284.
9. J. Bussgang, D. Middleton, "Optimum sequential detection of signals in noise," *IRE Transactions on Information Theory*, vol. 1, Issue 3, Dec. 1955, 5–18.
10. G. E. Fellows, "Experimental study of intensity spectra after half-wave detection of signals in noise," Harvard University, Ph.D. Thesis 1955.
11. H. W. Fuller, "Experimental study of signals and noise in a frequency modulation receiver," Harvard University, Ph.D. Thesis 1956.
12. J. A. Mullen and D. Middleton, "The rectification of nonGaussian noise," *Quart. J. Appl. Math.*, vol. 15, 1958, 395–419.
13. D. Van Meter, D. Middleton, "Modern statistical approaches to reception in communication theory," *IRE Professional Group on Information Theory*, vol. 4, Issue 4, Sept. 1954, 119–145; "On optimum multiple-alternative detection of signals in noise," *IRE Trans. on Information Theory*, vol. 1, no. 2, Sept. 1955, 1–9.
14. D. Middleton, "Canonical non-Gaussian noise models: their implications for measurement and for prediction of receiver performance," *IEEE Trans. EMC-21*, Issue 3, Aug. 1979, 209–220.
15. *An Introduction to Statistical Communication Theory*, McGraw-Hill, 1960–72; reissued Peninsula 1987–96; IEEE Classic Reissue, IEEE Press, 1996–.
16. *Topics in Communication Theory*, McGraw-Hill 1965–1972; reissued Peninsula, 1987.
17. L. Cohen, "The history of noise [on the 100th anniversary of its birth]," *IEEE Signal Processing Magazine*, vol. 22, Issue 6, Nov. 2005, 38 & 44.
18. R. Esposito, D. Middleton, and J. A. Mullen, "Advantages of amplitude and phase adaptivity in the detection of signals subject to slow Rayleigh fading," *IEEE Trans. Inform. Theory*, **IT-11**(4), 473–482 (October 1965); D. Middleton and R. Esposito, "Simultaneous optimum detection and estimation of signals in noise," *IEEE Trans. Inform. Theory*, **IT-14**(3), May 1968, 434–444.
19. A. D. Spaulding, "Optimum reception in an impulsive interference environment," Ph.D., Dep. Elec. Eng., Univ. Denver, Denver, Colo., May 1975; A. D. Spaulding and D. Middleton, "Optimum reception in an impulsive environment," *IEEE Trans. Comm*, "Part I: Coherent Detection," **COM-25** (9), Sept. 1977, 910–923; "... Part II: Incoherent Detection," *ibid.*, 924–934.
20. W. H. Huggins, D. Middleton, "A comparison of the phase and amplitude principles in signal detection," *Proc. Natl. Electronics Conf.*, vol. 11, Oct. 1955, 304.
21. D. Middleton, "Canonical non-Gaussian noise models: their implications for measurement and prediction of receiver performance," *Proc. 3rd Int'l Symp.*, paper 71, May 1979, 365–376; republished, See note 14.
22. D. Middleton, "Statistical-Physical Models of Electromagnetic Interference," *IEEE Trans. EMC-19*, Issue 3, Part I, Aug. 1977, 106–127.
23. D. Middleton, "Procedures for determining the parameters of the first-order canonical model of Class A and Class B electromagnetic interference models," *IEEE Trans. EMC-21*, no. 3, Aug. 1979, 190–208.
24. See note 19.
25. Acous. Inst., Acad. of Sciences, U.S.S.R., "Studies in Soviet Sciences" (Consultants Bureau and Plenum Pub Corp., New York and London, 1978).
26. M. A. Bykhovsky, **круги памяти**, *Circles of Memory, Historical Essays on XXth Century Progress in the Field of Radio Communications and Broadcasting*, International Center of Technical & Scientific Info., Moscow, 2001.

27. D. Middleton, "Some general results on the theory of noise through non-linear devices," *Qtly. Appl. Math.*, V (4), 1948, 445–498.
28. D. Middleton, *Non-Gaussian Statistical Communication Theory*, Wiley–IEEE Press, 2012.
29. Interviews #388 and 473; www.ieeeghn.org/wiki/index.php/David_Middleton_Oral_History_(2000) and www.ieeeghn.org/wiki/index.php/David_Middleton_Oral_History_(2007).
30. The author wishes to express his appreciation for the assistance received from the children of David Middleton and from Dr. J. A. Mullen, one of Middleton's former students.

Chapter 2
Sequential Bayesian Detection: A Model-Based Approach

James V. Candy

2.1 Introduction

Sequential detection is a methodology developed essentially by Wald [1] in the late 1940s providing an alternative to the classical batch methods evolving from the basic Neyman–Pearson theory of the 1930s [2, 3]. From the detection theoretical viewpoint, the risk (or error) associated with a decision typically decreases as the number of measurements increases. Sequential detection enables a decision to be made more rapidly (in most cases) employing fewer measurements while maintaining the same level of risk. Thus, the aspiration is to reduce the decision time while maintaining the risk for a fixed sample size. Its significance was truly brought to the forefront with the evolution of the digital computer and the fundamental idea of acquiring and processing data in a sequential manner. The seminal work of Middleton [2,4–7] as well as the development of sequential processing techniques [8–13] during the 1960s provided the necessary foundation for the sequential processor/detector that is applied in a routine manner today [7–11, 13].

This chapter investigates the idea of a sequential processor from the statistical perspective and relates its operation to that of the classical sequential Neyman–Pearson detector developed by Wald and termed the sequential probability ratio test (*SPRT*). Here, the investigation of simple binary hypothesis testing evolves as a common thread that leads to Wald's *SPRT*. We start with the batch detector developing its sequential variant and then show how the *SPRT* easily follows from the results of Neyman–Pearson theory [1–3, 9–11].

In Sect. 2.2 we develop the sequential detector and compare it to the *SPRT* indicating the similarities and differences. In Sect. 2.3, we discuss how the Gauss–Markov linear processor can be applied to the model-based problem followed by the nonlinear case in Sect. 2.4.

J.V. Candy (✉)
Lawrence Livermore National Laboratory P.O. 808, L-151 Livermore, CA 94551, USA
e-mail: candy1@llnl.gov

L. Cohen et al. (eds.), *Classical, Semi-classical and Quantum Noise*,
DOI 10.1007/978-1-4419-6624-7_2, © Springer Science+Business Media, LLC 2012

2.2 Sequential Detection

In this section, we briefly develop sequential detection theory which will be expanded to include the model-based approach in the section to follow. We start with the "batch" solution and then develop the sequential approach.

Let us assume that we have a set of measurements defined by $\mathrm{M}_{t_k} := \{m(t_k), \cdots, m(t_1), m(t_0)\}$ and we would like to decide between two hypothesis, H_0 and H_1 respectively. We choose the Neyman–Pearson criterion to develop our detector [10]. Recall that the Neyman–Pearson theorem states that a detector is optimal if it maximizes the probability of detection, P_{DET} (or minimizes the miss probability) for any false alarm rate less than a pre-specified value, say P^*_{FA}. The theorem follows directly from a constrained optimization problem formulation using Lagrange multipliers (see [3, 10] for details), which yields the solution based on the ratio of likelihoods, $\Pr(M_{t_k}|H_i)$; $i = 0, 1$ (for the binary case). That is, to maximize P_{DET} for a fixed value of P^*_{FA}, we have the likelihood ratio, $\mathcal{L}(t_k)$ or equivalently the sufficient statistic defined by the joint density functions

$$\mathcal{L}(t_k) := \mathcal{L}\left(\mathrm{M}_{t_k}\right) = \frac{\Pr\left(\mathrm{M}_{t_k}|H_1\right)}{\Pr\left(\mathrm{M}_{t_k}|H_0\right)} \underset{H_0}{\overset{H_1}{\gtrless}} \mathrm{T} \tag{2.1}$$

for $\Pr(\cdot)$ the respective joint probabilities under each hypothesis. Here, the threshold T is found from integrating the false alarm density for the prespecified value

$$P_{\mathrm{FA}} = \int_{\mathrm{T}}^{\infty} \Pr\left(\mathrm{M}_{t_k}|H_0\right) \mathrm{d}P = P^*_{\mathrm{FA}}. \tag{2.2}$$

Thus, based on the Neyman–Pearson criterion, the optimal detector can be found for "batch" mode as

$$\mathcal{L}(t_k) = \frac{\Pr\left(\mathrm{M}_{t_k}|H_1\right)}{\Pr\left(\mathrm{M}_{t_k}|H_0\right)} = \frac{\Pr\left(\mathbf{m}(t_k), \cdots, \mathbf{m}(t_0)|H_1\right)}{\Pr\left(\mathbf{m}(t_k), \cdots, \mathbf{m}(t_0)|H_0\right)}. \tag{2.3}$$

It follows from the chain rule of probability theory [10] that

$$\begin{aligned}\Pr\left(\mathrm{M}_{t_k}|H_i\right) &= \prod_{\ell=0}^{t_k} \Pr\left(\mathbf{m}(t_{k-\ell})|\mathrm{M}_{t_{k-\ell-1}}; H_i\right) \\ &= \Pr\left(\mathbf{m}(t_k)|\mathrm{M}_{t_{k-1}}; H_i\right) \times \Pr\left(\mathbf{m}(t_{k-1})|\mathrm{M}_{t_{k-2}}; H_i\right) \\ &\quad \times \cdots \times \Pr\left(\mathbf{m}(1)|\mathbf{m}(0); H_i\right) \times \Pr\left(\mathbf{m}(0); H_i\right)\end{aligned} \tag{2.4}$$

which can be expressed succinctly using Bayes rule as

$$\Pr\left(\mathrm{M}_{t_k}|H_i\right) = \Pr\left(\mathbf{m}(t_k), \mathrm{M}_{t_{k-1}}|H_i\right) = \Pr\left(\mathbf{m}(t_k)|\mathrm{M}_{t_{k-1}}; H_i\right)\Pr\left(\mathrm{M}_{t_{k-1}}|H_i\right). \tag{2.5}$$

Substituting this expression into (2.1), we obtain

$$\begin{aligned}\mathcal{L}(t_k) &= \left[\frac{\Pr\left(\mathrm{M}_{t_{k-1}}|H_1\right)}{\Pr\left(\mathrm{M}_{t_{k-1}}|H_0\right)}\right]\frac{\Pr\left(\mathbf{m}(t_k)|\mathrm{M}_{t_{k-1}}; H_1\right)}{\Pr\left(\mathbf{m}(t_k)|\mathrm{M}_{t_{k-1}}; H_0\right)}\\ &= \mathcal{L}(t_{k-1})\frac{\Pr\left(\mathbf{m}(t_k)|\mathrm{M}_{t_{k-1}}; H_1\right)}{\Pr\left(\mathbf{m}(t_k)|\mathrm{M}_{t_{k-1}}; H_0\right)},\end{aligned} \tag{2.6}$$

which is precisely the sequential form of the likelihood ratio. It is also clear that not just the likelihood function can be used but any monotonic function of the likelihood can also be used as well [10]. Taking natural logarithms of both sides of the equation, and defining $\Lambda(t_k) := \ln\mathcal{L}(t_k)$, we obtain the sequential log-likelihood ratio as

$$\Lambda(t_k) = \Lambda(t_{k-1}) + \ln\Pr\left(\mathbf{m}(t_k)|\mathrm{M}_{t_{k-1}}; H_1\right) - \ln\Pr\left(\mathbf{m}(t_k)|\mathrm{M}_{t_{k-1}}; H_0\right). \tag{2.7}$$

To construct the sequential probability ratio test (*SPRT*) or equivalently the sequential likelihood ratio test (*SLRT*) based on the Neyman–Pearson criterion, we must define two time-varying thresholds $\mathrm{T}_0(t_k)$ and $\mathrm{T}_1(t_k)$. If the likelihood ratio at time t_k is greater than threshold $\mathrm{T}_1(t_k)$, we accept hypothesis H_1. If it is less than $\mathrm{T}_0(t_k)$, we accept H_0, but if its value lies between the thresholds, we *continue* to take another sample. The sequential test differs from the fixed sample size batch test and is capable of handling nonstationary statistics.

To implement the *SLRT* at the t_k^{th}-stage, we must know the prior probabilities or equivalently the prior likelihood ratio, $\mathcal{L}(t_{k-1})$ (or $\Lambda(t_{k-1})$) containing all of the past information and the corresponding stage conditional probabilities $\Pr(\mathbf{m}(t_k)|M_{t_{k-1}}; H_i)$ to sequentially update the likelihood. The t_k^{th} stage thresholds are constructed following Neyman–Pearson theory in terms of the detection and false alarm probabilities as

$$\mathrm{T}_0(t_k) = \frac{P_{\text{DET}}}{P_{\text{FA}}} = \frac{1-P_{\text{MISS}}}{P_{\text{FA}}}, \quad \text{and} \quad \mathrm{T}_1(t_k) = \frac{1-P_{\text{DET}}}{1-P_{\text{FA}}} = \frac{P_{\text{MISS}}}{1-P_{\text{FA}}}. \tag{2.8}$$

In the case of the log-likelihood, the thresholds are:

$$\begin{aligned}&\Lambda(t_k) \geq \ln\mathrm{T}_1(t_k) && [\text{Accept } H_1]\\ &\ln\mathrm{T}_0(t_k) < \Lambda(t_k) < \ln\mathrm{T}_1(t_k) && [\text{Continue}]\\ &\Lambda(t_k) \leq \ln\mathrm{T}_0(t_k) && [\text{Accept } H_0].\end{aligned} \tag{2.9}$$

This completes the fundamental concepts for the construction of the sequential detection approach, next we investigate the development of the model-based sequential detector.

2.3 Model-Based Sequential Detection: Linear Case

In this section, we develop the model-based approach to sequential detection employing the *SLRT* as the mechanism to implement the model-based designs starting with the linear time-varying (nonstationary) problem (see [13, 14] for more details). We begin with the development of the generic Gauss–Markov signal model defined by its *state* (signal) *vector*, $\mathbf{s}(t_k)$, where the linear state-space process model is given by

$$\mathbf{s}(t_k) = A(t_{k-1})\mathbf{s}(t_{k-1}) + B(t_{k-1})\mathbf{u}(t_{k-1}) + \mathbf{w}(t_{k-1}) \tag{2.10}$$

with corresponding measurement model as

$$\mathbf{m}(t_k) = C(t_k)\mathbf{s}(t_k) + \mathbf{v}(t_k), \tag{2.11}$$

where $\mathbf{s}$, $\mathbf{w}$ are the N_s-dimensional signal and process noise vectors, with $\mathbf{m}$, $\mathbf{v}$ the N_m-dimensional measurement and noise vectors along with $\mathbf{u}$ the N_u-dimensional known input vector. Both $\mathbf{w}$ and $\mathbf{v}$ are zero-mean Gaussian with respective covariances, $R_{\mathbf{ww}}(t_{k-1})$ and $R_{\mathbf{vv}}(t_k)$. The corresponding system, input and measurement matrices are appropriately dimensioned and given by $A(t_{k-1})$, $B(t_{k-1})$ and $C(t_k)$, respectively. The initial state vector is Gaussian with $\mathbf{s}(0) \sim \mathcal{N}(\bar{\mathbf{s}}(0), P_{\mathbf{ss}}(0))$.

Because the underlying distributions are Gaussian, we know that the optimal solution to the signal estimation problem is given by the model-based processor (Kalman filter) providing the predicted conditional mean estimate, $\hat{\mathbf{s}}(t_k|t_{k-1}) := E\{\mathbf{s}(t_k)|M_{t_{k-1}}\}$[1] with corresponding predicted conditional (error) covariance, $\tilde{P}(t_k|t_{k-1})$ [12–15].

With this signal model in hand, we can now define the binary problem to decide whether the measurement contains the signal or just noise alone, that is, we are testing the hypotheses that:

$$\begin{aligned} H_0 &: \quad \mathbf{m}(t_k) = \mathbf{v}(t_k) && \text{[NOISE]} \\ H_1 &: \quad \mathbf{m}(t_k) = C(t_k)\mathbf{s}(t_k) + \mathbf{v}(t_k) && \text{[SIGNAL+NOISE]} \end{aligned}$$

[1] This notation is defined in terms of predicted *conditional means* and *covariances* by: $\hat{\mathbf{s}}(t_k|t_{k-1})$ and $\tilde{P}(t_k|t_{k-1}) := \text{cov}(\tilde{\mathbf{s}}(t_k|t_{k-1}))$ for the *predicted state estimation error*, $\tilde{\mathbf{s}}(t_k|t_{k-1}) := \mathbf{s}(t_k) - \hat{\mathbf{s}}(t_k|t_{k-1})$.

From the underlying Gauss–Markov assumption, the sequential likelihood-ratio solution is specified by the ratio of conditional Gaussian posterior distributions, $\Pr(\mathbf{m}(t_k)|M_{t_{k-1}};H_i);\ i=0,1$. That is,

$$\Pr(\mathbf{m}(t_k)|M_{t_{k-1}};H_0) \sim \mathcal{N}(\mathbf{m}(t_k):\mathbf{0},R_{\mathbf{vv}}(t_k))$$
$$\Pr(\mathbf{m}(t_k)|M_{t_{k-1}};H_1) \sim \mathcal{N}(\mathbf{m}(t_k):C(t_k)\hat{\mathbf{s}}(t_k|t_{k-1}),R_{\mathbf{ee}}(t_k)),$$

where the *innovations*, $\mathbf{e}(t_k)$ and its corresponding covariance, $R_{\mathbf{ee}}(t_k)$ are obtained as outputs of the model-based processor (Kalman filter) specified by:

$$\begin{aligned}\mathbf{e}(t_k) &= \mathbf{m}(t_k)-\hat{\mathbf{m}}(t_k|t_{k-1}) = \mathbf{m}(t_k)-C(t_k)\hat{\mathbf{s}}(t_k|t_{k-1})\\ R_{\mathbf{ee}}(t_k) &= C(t_k)\tilde{P}(t_k|t_{k-1})C'(t_k)+R_{\mathbf{vv}}(t_k),\end{aligned} \tag{2.12}$$

Thus, under the null hypothesis we have that

$$\Pr(\mathbf{m}(t_k)|M_{t_{k-1}};H_0) = \frac{1}{(2\pi)^{N_m/2}|R_{\mathbf{vv}}(t_k)|^{1/2}}\exp\left\{-\frac{1}{2}\mathbf{m}'(t_k)R_{\mathbf{vv}}^{-1}(t_k)\mathbf{m}(t_k)\right\} \tag{2.13}$$

while the conditional probability under alternate hypothesis is given by

$$\begin{aligned}\Pr(\mathbf{m}(t_k)\mid M_{t_{k-1}};H_1) &= \frac{1}{(2\pi)^{N_m/2}|R_{\mathbf{ee}}(t_k)|^{1/2}}\\ &\quad\times\exp\left\{-\frac{1}{2}(\mathbf{m}(t_k)-\hat{\mathbf{m}}(t_k|t_{k-1}))'R_{ee}^{-1}(t_k)(\mathbf{m}(t_k)-\hat{\mathbf{m}}(t_k|t_{k-1}))\right\}\end{aligned} \tag{2.14}$$

or simply

$$\Pr(\mathbf{m}(t_k)|M_{t_{k-1}};H_1) = \frac{1}{(2\pi)^{N_m/2}|R_{\mathbf{ee}}(t_k)|^{1/2}}\exp\left\{-\frac{1}{2}\mathbf{e}'(t_k)R_{\mathbf{ee}}^{-1}(t_k)\mathbf{e}(t_k)\right\}. \tag{2.15}$$

Moving all *known* terms to the threshold, the required sequential log-likelihood of (2.7) for the linear Gauss–Markov signal model becomes

$$\begin{aligned}\Lambda(t_k) &= \Lambda(t_{k-1})-\frac{1}{2}\mathbf{e}'(t_k)R_{\mathbf{ee}}^{-1}(t_k)\mathbf{e}(t_k)+\frac{1}{2}\mathbf{m}'(t_k)R_{\mathbf{vv}}^{-1}(t_k)\mathbf{m}(t_k)\ \underset{H_0}{\overset{H_1}{\gtrless}}\ \mathcal{T}(t_k)\\ \mathcal{T}(t_k) &= \ln \mathrm{T}(t_k)-\ln\left(\frac{1}{(2\pi)^{N_m/2}|R_{\mathbf{ee}}(t_k)|^{1/2}}\right)+\ln\left(\frac{1}{(2\pi)^{N_m/2}|R_{\mathbf{vv}}(t_k)|^{1/2}}\right)\end{aligned} \tag{2.16}$$

All that remains is to specify the predicted and corrected means and covariances to implement the sequential detector. These are available as part of the model-based algorithm given by:

$$\begin{aligned}
\hat{\mathbf{s}}(t_k|t_{k-1}) &= A(t_{k-1})\hat{\mathbf{s}}(t_{k-1}|t_{k-1}) + B(t_{k-1})\mathbf{u}(t_{k-1}) && \text{[Prediction]}\\
\tilde{P}(t_k|t_{k-1}) &= A(t_{k-1})\tilde{P}(t_{k-1}|t_{k-1})A'(t_{k-1}) + R_{\mathbf{ww}}(t_{k-1}) && \text{[Prediction Cov.]}\\
\hat{\mathbf{s}}(t_k|t_k) &= \hat{\mathbf{s}}(t_k|t_{k-1}) + K(t_k)\mathbf{e}(t_k) && \text{[Correction]}\\
\tilde{P}(t_k|t_k) &= [\mathbf{I} - K(t_k)C(t_k)]\,\tilde{P}(t_k|t_{k-1}) && \text{[Correction Cov.]}
\end{aligned}$$

for $K(t_k)$ the corresponding gain (see [14] for more details), next we consider the nonlinear case.

2.4 Model-Based Sequential Detection: Nonlinear Case

In this section, we use the development of the linear case as a road map and develop the sequential model-based detector for the nonlinear case. We start with the nonlinear (approximate) Gauss–Markov model given by:

$$\begin{aligned}
\mathbf{s}(t_k) &= \mathbf{a}\,[\mathbf{s}(t_{k-1})] + \mathbf{b}\,[\mathbf{u}(t_{k-1})] + \mathbf{w}(t_{k-1})\\
\mathbf{m}(t_k) &= \mathbf{c}\,[\mathbf{s}(t_k)] + \mathbf{v}(t_k),
\end{aligned} \tag{2.17}$$

where (as before) $\mathbf{s}$, $\mathbf{w}$ are the N_s-dimensional signal and process noise vectors with $\mathbf{m}$, $\mathbf{v}$ the N_m-dimensional measurement and noise vectors along with $\mathbf{u}$ the N_u-dimensional known input vector. Both $\mathbf{w}$ and $\mathbf{v}$ are zero-mean Gaussian with respective covariances, $R_{\mathbf{ww}}(t_{k-1})$ and $R_{\mathbf{vv}}(t_k)$. The corresponding *nonlinear* system, input and measurement vector functions are appropriately dimensioned and given by $\mathbf{a}[\cdot]$, $\mathbf{b}[\cdot]$ and $\mathbf{c}[\cdot]$, respectively. The initial state vector is Gaussian with $\mathbf{s}(0) \sim \mathcal{N}(\overline{\mathbf{s}}(0), P_{\mathbf{ss}}(0))$.

Because the underlying distributions are assumed approximately Gaussian, we know that one possible solution to the signal estimation problem is given by the model-based processor (extended Kalman filter[2]) providing the predicted conditional mean estimate, $\hat{\mathbf{s}}(t_k|t_{k-1})$, with corresponding predicted conditional (error) covariance, $\tilde{P}(t_k|t_{k-1})$ [14].

[2] It is well known that some of the modern variants currently available offer alternatives such as the unscented Kalman filter or particle filter [14, 16–26] that are better than the *EKF* but we choose this formulation since it easily tracks the linear case developed previously.

With this nonlinear signal model in hand, we can now define the binary problem to decide whether the measurement contains the signal or just noise alone, that is, we are testing the hypotheses that:

$$H_0: \quad \mathbf{m}(t_k) = \mathbf{v}(t_k) \qquad \text{[NOISE]}$$
$$H_1: \quad \mathbf{m}(t_k) = \mathbf{c}\left[\mathbf{s}(t_k)\right] + \mathbf{v}(t_k) \qquad \text{[SIGNAL+NOISE]}.$$

Before we proceed with the sequential detection development for nonlinear, non-Gaussian problems, we have chosen to employ the first-order Taylor series representation to approximate the nonlinear vector functions. The development evolves quite naturally from a linearized model-based processor (linearized Kalman filter) [14]. These approximations when expanded about a reference $\mathbf{s} = \mathbf{s}^*$ take the general form

$$\mathbf{c}\left[\mathbf{s}(t_k)\right] \approx \mathbf{c}\left[\mathbf{s}^*(t_k)\right] + \mathbf{C}\left[\mathbf{s}^*(t_k)\right] \times \left(\mathbf{s}(t_k) - \mathbf{s}^*(t_k)\right) + \text{H.O.T.} \tag{2.18}$$

with $\mathbf{C}[\mathbf{s}^*(t_k)] := \frac{\partial}{\partial \mathbf{s}}\mathbf{c}[\mathbf{s}]\big|_{\mathbf{s}=\mathbf{s}^*(t_k)}$. Each of the nonlinear system, input and measurement functions are approximated in this manner using the Taylor series. For the ad-hoc nonlinear *EKF* processor, the most current estimate available, $\hat{\mathbf{s}}(t_k|*)$, is used as a reference during the prediction and correction steps for the linearization, that is, the state estimates are

$$\hat{\mathbf{s}}(t_k|t_{k-1}) = \mathbf{a}\left[\hat{\mathbf{s}}(t_{k-1}|t_{k-1})\right] + \mathbf{b}\left[\mathbf{u}(t_{k-1})\right]$$
$$\hat{\mathbf{s}}(t_k|t_k) = \hat{\mathbf{s}}(t_k|t_{k-1}) + \mathbf{K}\left[\hat{\mathbf{s}}(t_k|t_{k-1})\right]\mathbf{e}(t_k), \tag{2.19}$$

where the Jacobians appear in the covariance equations.

With this in mind and from the underlying *approximate* Gauss–Markov assumption, the sequential likelihood-ratio solution is again specified by the ratio of conditional Gaussian posterior distributions, $\Pr(\mathbf{m}(t_k)|M_{t_{k-1}}; H_i);\ i = 0, 1$ as before. That is,

$$\Pr(\mathbf{m}(t_k)|M_{t_{k-1}}; H_0) \sim \mathcal{N}\left(\mathbf{m}(t_k) : \mathbf{0}, R_{\mathbf{vv}}(t_k)\right)$$
$$\Pr(\mathbf{m}(t_k)|M_{t_{k-1}}; H_1) \sim \mathcal{N}\left(\mathbf{m}(t_k) : \mathbf{c}\left[\hat{\mathbf{s}}(t_k|t_{k-1})\right], R_{\mathbf{ee}}(t_k)\right).$$

where the *innovations*, $\mathbf{e}(t_k)$ and its corresponding covariance, $R_{\mathbf{ee}}(t_k)$ are obtained as outputs of the model-based processor (extended Kalman filter) specified by:

$$\mathbf{e}(t_k) = \mathbf{m}(t_k) - \hat{\mathbf{m}}(t_k|t_{k-1}) = \mathbf{m}(t_k) - \mathbf{c}\left[\hat{\mathbf{s}}(t_k|t_{k-1})\right]$$
$$R_{\mathbf{ee}}(t_k) = \mathbf{C}\left[\hat{\mathbf{s}}(t_k|t_{k-1})\right]\tilde{P}(t_k|t_{k-1})\mathbf{C}'\left[\hat{\mathbf{s}}(t_k|t_{k-1})\right] + R_{\mathbf{vv}}(t_k). \tag{2.20}$$

Thus, under the null hypothesis we have that

$$\Pr(\mathbf{m}(t_k)|M_{t_{k-1}};H_0) = \frac{1}{(2\pi)^{N_m/2}|R_{\mathbf{vv}}(t_k)|^{1/2}} \exp\left\{-\frac{1}{2}\mathbf{m}'(t_k)R_{\mathbf{vv}}^{-1}(t_k)\mathbf{m}(t_k)\right\} \tag{2.21}$$

while the conditional probability under alternate hypothesis is given by

$$\Pr(\mathbf{m}(t_k)|M_{t_{k-1}};H_1) = \frac{1}{(2\pi)^{N_m/2}|R_{\mathbf{ee}}(t_k)|^{1/2}} \exp\left\{-\frac{1}{2}\mathbf{e}'(t_k)R_{\mathbf{ee}}^{-1}(t_k)\mathbf{e}(t_k)\right\}. \tag{2.22}$$

Again moving all known terms to the threshold, the required sequential log-likelihood of (2.7) for the nonlinear Gauss–Markov signal model becomes

$$\Lambda(t_k) = \Lambda(t_{k-1}) - \frac{1}{2}\mathbf{e}'(t_k)R_{\mathbf{ee}}^{-1}(t_k)\mathbf{e}(t_k) + \frac{1}{2}\mathbf{m}'(t_k)R_{\mathbf{vv}}^{-1}(t_k)\mathbf{m}(t_k) \underset{H_0}{\overset{H_1}{\gtrless}} \mathcal{T}(t_k)$$

$$\mathcal{T}(t_k) = \ln \mathrm{T}(t_k) - \ln\left(\frac{1}{(2\pi)^{N_m/2}|R_{\mathbf{ee}}(t_k)|^{1/2}}\right) + \ln\left(\frac{1}{(2\pi)^{N_m/2}|R_{\mathbf{vv}}(t_k)|^{1/2}}\right). \tag{2.23}$$

All that remains is to specify the predicted and corrected covariances to implement the sequential detector, since the state estimates we have given in (2.19) above. These are available as part of the nonlinear model-based algorithm (*EKF*) given by:

$$\tilde{P}(t_k|t_{k-1}) = \mathbf{A}\left[\hat{\mathbf{s}}(t_k|t_{k-1})\right]\tilde{P}(t_{k-1}|t_{k-1})\mathbf{A}'\left[\hat{\mathbf{s}}(t_k|t_{k-1})\right] + R_{\mathbf{ww}}(t_{k-1}) \quad \text{[Prediction Cov.]}$$

$$\tilde{P}(t_k|t_k) = \left[\mathbf{I} - \mathbf{K}\left[\hat{\mathbf{s}}(t_k|t_{k-1})\right]\mathbf{C}\left[\hat{\mathbf{s}}(t_k|t_{k-1})\right]\right]\tilde{P}(t_k|t_{k-1}) \quad \text{[Correction Cov.]}$$

for $\mathbf{K}\left[\hat{\mathbf{s}}(t_k|t_{k-1})\right]$ the corresponding gain (see [14] for more details). This completes the development.

2.5 Summary

In this chapter, we have motivated and summarized the development of sequential detection theory from the original Neyman–Pearson theory following the Wald approach [1]. We then showed how a model-based approach can be incorporated into the sequential paradigm. We developed the Gauss–Markov (state-space) representation of both linear and nonlinear systems capable of capturing the dynamics of many physics-based problems [14] and showed how they can be embedded into the sequential theoretical framework. The solution leads to the implementation of a linear model-based processor (linear Kalman filter) as well as the nonlinear model-based scheme (extended Kalman filter).

References

1. Wald A (1973) Sequential analysis. Dover Publications, New York
2. Middleton D (1960) Introduction to statistical communication theory. McGraw-Hill, New York
3. Van Trees H (1968) Detection, estimation and modulation theory, Pt. 1. Wiley, New York
4. Middleton D, Esposito R (1968) Simultaneous optimum detection and estimation of signals in noise. IEEE Trans Inform Theory IT-14(3):434–444
5. Middleton D, Esposito R (1970) New results in the theory of simultaneous optimum detection and estimation of signals in noise. Problemy Peredachi Informatsii 6(3):3–20
6. Sullivan E, Middleton D (1993) Estimation and detection issues in matched-field processing. IEEE Trans Oceanic Eng 18(3):156–167
7. Candy J, Middleton D (1995) Joint detection-estimation: a model-based solution, **LLNL Report**, UCRL-JC-120219
8. Basseville M, Nikiforov I (1993) Detection by abrupt changes: theory and application. Prentice-Hall, New Jersey
9. Scharf L (1991) Statistical signal processing: detection, estimation, and time series analysis. Addison-Wesley, Massachusetts
10. Kay S (1998) Fundamentals of statistical signal processing: detection theory. Prentice-Hall, New Jersey
11. Therrian C (1991) Random signal processing: detection theory. Prentice-Hall, New Jersey
12. Jazwinski A (1970) Stochastic processes and filtering theory. Academic Press, New York
13. Sage A, Melsa J (1971) Estimation theory with applications to communcations and control. McGraw-Hill, New York
14. Candy JV (2006) Model-based signal processing. Wiley/IEEE Press, New Jersey
15. Anderson B, Moore J (1979) Optimum filtering. Prentice-Hall, New Jersey
16. Julier SJ, Uhlmann JK, Durrant-Whyte HF (2000) "A new method for the nonlinear transformation of means and covariances in filters and estimators." IEEE Trans Autom Contr 45(3):477–482
17. Haykin S, de Freitas N (2004) "Special Issue: Sequential state estimation: from Kalman filters to particle filters." Proc IEEE 92(3):399–574
18. Godsill S, Djuric P (2002) "Special issue: Monte Carlo methods for statistical signal processing." IEEE Trans Signal Proc 50:173–499
19. Ristic B, Arulampalam S, Gordon N (2004) Beyond the Kalman filter: particle filters for tracking applications. Artech House, Boston
20. Candy JV (2009) Bayesian signal processing: classical, modern and particle filtering. Wiley/IEEE Press, New Jersey
21. Gordon N, Salmond D, Smith AFM (1993) Novel approach to nonlinear/non-Gaussian Bayesian state estimation. IEE Proc F 140:107–113
22. Arulampalam M, Maskell S, Gordon N, Clapp T (2002) "A tutorial on particle filters for online nonlinear/non-gaussian Bayesian tracking." IEEE Trans Signal Proc 50(2):174–188
23. Djuric P, Kotecha J, Zhang J, Huang Y, Ghirmai T, Bugallo M, Miguez J (2003) "Particle filtering." IEEE Signal Proc Mag 20(5):19–38

24. Doucet A, Wang X (2005) "Monte Carlo methods for signal processing." IEEE Signal Proc Mag 24(5):152–170
25. Cappe O, Godsill S, Moulines E (2007) "An overview of existing methods and recent advances in sequential Monte Carlo." Proc IEEE 95(5):899–924
26. Candy J (2007) "Bootstrap particle filtering." IEEE Signal Proc Mag 24(4):73–85

Chapter 3
The Propagation of Noise Fields in a Dispersive Medium

Leon Cohen

3.1 Introduction

Our aim is to show how a noise field propagates in a medium that has dispersion and attenuation. For example, suppose noise with known statistical properties is generated at a particular spatial point and we ask for its statistical properties at other spatial points at a later time. We show that phase space methods are particularly effective and intuitive to study such problems. We deal with fields and stochastic processes that are not necessarily stationary either locally or globally. In the next section, we review deterministic wave propagation and we also introduce a new method for studying propagation by way of the Wigner distribution. Subsequently, we discuss the issue as to how to handle nonstationary stochastic processes using the Wigner spectrum. After these issues we address the propagation of noise fields in a dispersive deterministic medium. By a deterministic medium we mean that the medium has no random aspects and that the random aspects come in only in the initial generation of the noise field.

We divide our considerations into two cases depending on the initial conditions. For a pulse, denoted by $u(x,t)$, we differentiate the two cases as:

$$\text{Initially given}: u(x,t_0) \qquad \text{(Case A)} \tag{3.1}$$

$$: u(x_0,t) \qquad \text{(Case B)} \tag{3.2}$$

In both cases, one wants to determine $u(x,t)$. In Case A, one disturbs the medium at time t_0 and are given $u(x,t_0)$ for all x. For example, a string is plucked at time zero and then let go. In Case B, we stay at one position, here taken as x_0, and generate

L. Cohen (✉)
Department of Physics, Hunter College of the City University of New York, USA
e-mail: Leon.Cohen@hunter.cuny.edu

L. Cohen et al. (eds.), *Classical, Semi-classical and Quantum Noise*,
DOI 10.1007/978-1-4419-6624-7_3, © Springer Science+Business Media, LLC 2012

a wave (signal) as a function of time. Examples of this case are active radar, sonar, or fiber optics. Since we deal with linear wave propagation, there is no loss of generality if we take $x_0 = 0$ in Case B and $t_0 = 0$ in Case A.

3.1.1 Notation

Since we are dealing with two separate cases, we will attempt to be somewhat consistent with the notation so that one can transpose the results from one case into another in a direct manner. There are two Fourier transforms that will come up. For the spatial Fourier transform, we write[1]

$$S(k,t) = \frac{1}{\sqrt{2\pi}} \int u(x,t)\, \mathrm{e}^{-ikx}\, \mathrm{d}x \tag{3.3}$$

with the inverse transform,

$$u(x,t) = \frac{1}{\sqrt{2\pi}} \int S(k,t)\, \mathrm{e}^{ikx}\, \mathrm{d}k. \tag{3.4}$$

Similarly, we will take Fourier transform with respect to the time variables and hence write

$$F(\omega,x) = \frac{1}{\sqrt{2\pi}} \int u(x,t)\, \mathrm{e}^{-\mathrm{i}\omega t}\, \mathrm{d}t \tag{3.5}$$

$$u(x,t) = \frac{1}{\sqrt{2\pi}} \int F(\omega,x)\, \mathrm{e}^{\mathrm{i}\omega t}\, \mathrm{d}\omega. \tag{3.6}$$

3.2 Case A: Initially Given: $u(x,t_0)$

We deal with wave equations with constant coefficients

$$\sum_{n=0}^{N} a_n \frac{\partial^n u}{\partial t^n} = \sum_{n=0}^{M} b_n \frac{\partial^n u}{\partial x^n}. \tag{3.7}$$

As a standard to solve the equation, one substitutes $\mathrm{e}^{ikx-\mathrm{i}\omega t}$ into (3.7) to give [11–16]

[1] All integrals go from $-\infty$ to ∞ unless otherwise noted.

$$\sum_{n=0}^{N} a_n(-\mathrm{i}\omega)^n = \sum_{n=0}^{M} b_n(\mathrm{i}k)^n. \tag{3.8}$$

This defines the dispersion relation and one solves for ω in terms of k

$$\omega(k) = \omega(k) = \omega_{\mathrm{R}}(k) + \mathrm{i}\omega_{\mathrm{I}}(k), \tag{3.9}$$

where $\omega_{\mathrm{R}}(k)$ and $\omega_{\mathrm{I}}(k)$ are the real and imaginary parts, respectively. The group velocity is defined by

$$v_g = \omega_{\mathrm{R}}'(k). \tag{3.10}$$

There could be many solutions to the dispersion relation and each solution is called a mode. The general solution for each mode is [11–16],

$$u(x,t) = \frac{1}{\sqrt{2\pi}} \int S(k,0)\, \mathrm{e}^{\mathrm{i}kx-\mathrm{i}\omega(k)t}\, \mathrm{d}k, \tag{3.11}$$

where $S(k,0)$ is the initial spatial spectrum which is obtained from $u(x,0)$ by way of

$$S(k,0) = \frac{1}{\sqrt{2\pi}} \int u(x,0)\, \mathrm{e}^{-\mathrm{i}kx}\, \mathrm{d}x. \tag{3.12}$$

If one defines the time dependent spectrum by

$$S(k,t) = S(k,0)\, \mathrm{e}^{-\mathrm{i}\omega(k)t}, \tag{3.13}$$

then $u(x,t)$ and $S(k,t)$ form Fourier transform pairs between x and k for all time

$$u(x,t) = \frac{1}{\sqrt{2\pi}} \int S(k,t)\, \mathrm{e}^{\mathrm{i}kx}\, \mathrm{d}k \tag{3.14}$$

$$S(k,t) = \frac{1}{\sqrt{2\pi}} \int u(x,t)\, \mathrm{e}^{-\mathrm{i}kx}\, \mathrm{d}x. \tag{3.15}$$

One can combine (3.11) and (3.12) to give

$$u(x,t) = \frac{1}{2\pi} \iint u(x',0)\, \mathrm{e}^{-\mathrm{i}k(x'-x)-\mathrm{i}\omega(k)t}\, \mathrm{d}k\mathrm{d}x', \tag{3.16}$$

which we can write as

$$u(x,t) = \int u(x',0)\, G(x'-x,t)\mathrm{d}x', \tag{3.17}$$

where

$$G(x'-x,t) = \frac{1}{2\pi} \int \mathrm{e}^{-\mathrm{i}k(x'-x)-\mathrm{i}\omega(k)t}\, \mathrm{d}k. \tag{3.18}$$

As mentioned, the dispersion relation can be complex and that indicates whether we have damping or not. Using (3.13,) we have that

$$|S(k,t)| = \left|S(k,0)\,\mathrm{e}^{-\mathrm{i}\omega(k)t}\right| = |S(k,0)|\,\mathrm{e}^{\omega_{\mathrm{I}}(k)t} \tag{3.19}$$

and we see that the spectrum is a characterization of the damping vs. no damping case. In particular,

$$\text{if } |S(k,t)| = |S(k,0)| \qquad \text{no damping} \tag{3.20}$$

$$|S(k,t)| = |S(k,0)|\,\mathrm{e}^{\omega_{\mathrm{I}}(k)t} \qquad \text{damping.} \tag{3.21}$$

From (3.16), we have that

$$N(t) = \int |u(x,t)|^2\,\mathrm{d}x = \int |S(k,0)\,|^2 \mathrm{e}^{2\omega_{\mathrm{I}}(k)t}\,\mathrm{d}k \tag{3.22}$$

which shows that the normalization changes in time if we have damping which implies that energy is being added or removed. That is, if the dispersion relation has an imaginary part then

$$\int |u(x,t)|^2\,\mathrm{d}x \neq \int |u(x,0)|^2\,\mathrm{d}x. \tag{3.23}$$

If we define a new pulse by

$$u_N(x,t) = \frac{1}{\sqrt{N(t)}}u(x,t) \tag{3.24}$$

$$S_N(k,t) = \frac{1}{\sqrt{N(t)}}S(k,t) \tag{3.25}$$

then

$$\int |u_N(x,t)|^2\,\mathrm{d}x = \int |u_N(x,0)|^2\,\mathrm{d}x \tag{3.26}$$

regardless of whether there is damping or not.

3.2.1 Relation Between Damped and Undamped Waves

Suppose we have a dispersion relation that is real and another one with the same real part but with an additional complex part. Further, suppose the initial pulse is the same. We call the solution to the two cases $u(x,t)$ and $f(x,t)$ respectively,

$$u(x,t) = \frac{1}{\sqrt{2\pi}} \int S(k,0)\, e^{ikx - i\omega_R(k)t + \omega_I(k)t}\, dk \qquad \text{damping} \tag{3.27}$$

$$f(x,t) = \frac{1}{\sqrt{2\pi}} \int S(k,0)\, e^{ikx - i\omega_R(k)t}\, dk \qquad \text{no damping.} \tag{3.28}$$

We now show that one can derive an explicit relation between $u(x,t)$ and $f(x,t)$. Defining the corresponding spectra of each by

$$S_u(k,t) = S(k,0)\, e^{-i\omega_R(k)t + \omega_I(k)t} \tag{3.29}$$

$$S_f(k,t) = S(k,0)\, e^{-i\omega_R(k)t} \tag{3.30}$$

we have that

$$S_u(k,t) = S_f(k,t)\, e^{\omega_I(k)t} \tag{3.31}$$

and therefore, taking the Fourier transform of both sides one obtains

$$u(x,t) = \frac{1}{\sqrt{2\pi}} \int S_f(k,t)\, e^{\omega_I(k)t}\, e^{ikx}\, dk. \tag{3.32}$$

Substituting for $S_f(k,t)$ we have

$$u(x,t) = \frac{1}{2\pi} \iint f(x',t)\, e^{-ik(x'-x)}\, e^{\omega_I(k)t}\, dk\, dx' \tag{3.33}$$

or

$$u(x,t) = \int f(x',t)\, L(x'-x,t) dx' \tag{3.34}$$

where

$$L(x'-x,t) = \frac{1}{2\pi} \int e^{-ik(x'-x)}\, e^{\omega_I(k)t}\, dk \tag{3.35}$$

Similarly,

$$f(x,t) = \frac{1}{\sqrt{2\pi}} \int S_u(k,t)\, e^{-\omega_I(k)t}\, e^{ikx}\, dk \tag{3.36}$$

$$= \frac{1}{2\pi} \iint u(x',t)\, e^{-ik(x'-x)}\, e^{-\omega_I(k)t}\, dk\, dx' \tag{3.37}$$

or

$$f(x,t) = \int u(x',t)\, L(x'-x,-t) dx'. \tag{3.38}$$

These equations allow one to transform from one case to the other and explicitly shows the effects of damping.

3.2.2 Wave Propagation in Phase Space

We now define the appropriate phase space for Case A. The interpretation of $|u(x,t)|^2$ and $|S(k,t)|^2$ are:

$$|u(x,t)|^2 = \text{intensity (or energy) per unit } x \text{ at time } t, \tag{3.39}$$

$$|S(k,t)|^2 = \text{spectral intensity (or energy) per unit } k \text{ at time } t \tag{3.40}$$

We use the spatial-wavenumber Wigner distribution to define the phase space

$$W(x,k,t) = \frac{1}{2\pi}\int u^*(x - \tfrac{1}{2}\tau, t)\, u(x + \tfrac{1}{2}\tau, t)\, \mathrm{e}^{-\mathrm{i}\tau k} \mathrm{d}\tau \tag{3.41}$$

which in terms of the spatial spectrum is

$$W(x,k,t) = \frac{1}{2\pi}\int S^*(k + \tfrac{1}{2}\theta, t)\, S(k - \tfrac{1}{2}\theta, t)\, \mathrm{e}^{-\mathrm{i}\theta x} \mathrm{d}\theta \tag{3.42}$$

This of course is what Wigner did in his original paper [23] for the quantum mechanical wave function but we are doing it for a general wave [3,5,20,21]. This Wigner distribution satisfies the marginal requirements [3,4]

$$\int W(x,k,t)\, \mathrm{d}k = |u(x,t)|^2 \tag{3.43}$$

$$\int W(x,k,t)\, \mathrm{d}x = |S(k,t)|^2 \tag{3.44}$$

We now obtain the equation of evolution for $W(x,k,t)$. Substituting (3.13) into (3.42) we have

$$W(x,k,t) = \frac{1}{2\pi}\int S^*(k + \theta/2, 0)\; S(k - \theta/2, 0)\; \mathrm{e}^{-\mathrm{i}\theta x} \mathrm{e}^{\mathrm{i}[\omega^*(k+\theta/2) - \omega(k-\theta/2)]t} \mathrm{d}\theta. \tag{3.45}$$

However, at $t = 0$

$$W(x,k,0) = \frac{1}{2\pi}\int S^*(k + \theta/2, 0)\; S(k - \theta/2, 0)\; \mathrm{e}^{-\mathrm{i}\theta x} \mathrm{d}\theta \tag{3.46}$$

and therefore

$$S^*(k + \theta/2, 0)\; S(k - \theta/2, 0) = \int W(x,k,0)\, \mathrm{e}^{\mathrm{i}\theta x} \mathrm{d}x. \tag{3.47}$$

Inserting (3.47) into (3.16) results in

$$W(x,k,t) = \frac{1}{2\pi} \iint W(x',k,0)\, \mathrm{e}^{\mathrm{i}\theta(x'-x)} \mathrm{e}^{\mathrm{i}[\omega^*(k+\theta/2)-\omega(k-\theta/2)]t} \mathrm{d}\theta \mathrm{d}x'. \tag{3.48}$$

This equation is exact and relates the Wigner distribution at time t with the initial Wigner distribution. It is sometimes convenient to write it in terms of a Greens function

$$W(x,k,t) = \frac{1}{2\pi} \iint W(x',k,0) G_W(x'-x,t) \mathrm{d}x' \tag{3.49}$$

with

$$G_W(x'-x,t) = \frac{1}{2\pi} \int \mathrm{e}^{\mathrm{i}\theta(x'-x)} \mathrm{e}^{\mathrm{i}[\omega^*(k+\theta/2)-\omega(k-\theta/2)]t} \mathrm{d}\theta. \tag{3.50}$$

Just as we related the propagation of a pulse when there is attenuation and not we do so here for the Wigner distribution. Consider first the Wigner distribution with no damping

$$W_f(x,k,t) = \frac{1}{2\pi} \int S_f^*(k+\tfrac{1}{2}\theta,t)\, S_f(k-\tfrac{1}{2}\theta,t)\, \mathrm{e}^{-\mathrm{i}\theta x} \mathrm{d}\theta, \tag{3.51}$$

which gives

$$S_f^*(k+\tfrac{1}{2}\theta,t)\, S_f(k-\tfrac{1}{2}\theta,t) = \int W_f(x,k,t)\, \mathrm{e}^{\mathrm{i}\theta x} \mathrm{d}x. \tag{3.52}$$

Now, the Wigner distribution with damping is

$$W(x,k,t) = \frac{1}{2\pi} \int S^*(k+\tfrac{1}{2}\theta,t)\, S(k-\tfrac{1}{2}\theta,t)\, \mathrm{e}^{-\mathrm{i}\theta x} \mathrm{d}\theta \tag{3.53}$$

$$= \frac{1}{2\pi} \int S_f^*(k+\tfrac{1}{2}\theta,t)\, S_f(k-\tfrac{1}{2}\theta,t)\, \mathrm{e}^{[\omega_{\mathrm{I}}(k+\theta/2)-\omega_{\mathrm{I}}(k-\theta/2)]t} \mathrm{e}^{-\mathrm{i}\theta x} \mathrm{d}\theta \tag{3.54}$$

and therefore

$$W(x,k,t) = \frac{1}{2\pi} \int W_f(x',k,t)\, \mathrm{e}^{[\omega_{\mathrm{I}}(k+\theta/2)-\omega_{\mathrm{I}}(k-\theta/2)]t} \mathrm{e}^{-\mathrm{i}\theta(x-x')} \mathrm{d}\theta\, \mathrm{d}x'. \tag{3.55}$$

3.2.3 *Differential Equation for a Mode*

To obtain the equation of motion for the Wigner distribution, we first obtain the differential equation for $u(x,t)$. Differentiating (3.11) with respect to time, one has

$$i\frac{\partial}{\partial t} u(x,t) = \frac{1}{\sqrt{2\pi}} \int \omega(k) S(k,t_0)\, \mathrm{e}^{\mathrm{i}kx-\mathrm{i}\omega(k)t}\, \mathrm{d}k. \tag{3.56}$$

This is an integral equation that can be converted into a differential equation

$$\mathrm{i}\frac{\partial}{\partial t}u(x,t) = \omega\left(\frac{1}{i}\frac{\partial}{\partial x}\right)u(x,t). \qquad (3.57)$$

In addition, differentiating (3.13), the differential equation for the spatial spectrum, $S(k,t)$, is

$$\mathrm{i}\frac{\partial}{\partial t}S(k,t) = \omega(k)S(k,t). \qquad (3.58)$$

Using the method of Galleani and Cohen [7, 8], one can now derive the Wigner distribution corresponding to (3.57). It is

$$\mathrm{i}\frac{\partial}{\partial t}W(x,k,t) = \left[\omega\left(k+\frac{1}{2i}\frac{\partial}{\partial x}\right) - \omega^*\left(k-\frac{1}{2i}\frac{\partial}{\partial x}\right)\right]W(x,k,t). \qquad (3.59)$$

By expanding $\omega(k)$ in a power series, we have

$$\frac{\partial}{\partial t}W(x,k,t) \sim \left[2\omega_{\mathrm{I}}(k) - v(k)\frac{\partial}{\partial x} - \frac{1}{4}\omega_{\mathrm{I}}^{(2)}(k)\frac{\partial^2}{\partial x^2} + \frac{1}{24}\omega_{\mathrm{R}}^{(3)}(k)\left(\frac{\partial}{\partial x}\right)^3 \cdots\right]W(x,k,t), \qquad (3.60)$$

where $v(k)$ is the group velocity as per (3.10)

3.2.4 *Approximation*

Loughlin and Cohen have shown that an excellent approximation to the Wigner distribution as given by (3.55) is [3–5, 20, 21]

$$W_a(x,k,t) \approx \mathrm{e}^{2\omega_{\mathrm{I}}(k)t}W(x - v_g(k)t), k, 0). \qquad (3.61)$$

This approximation shows that, other than to calculate the Wigner distribution of the initial wave, one just substitutes $x - v_g(k)t$ for x in the initial Wigner distribution and multiplies by $\mathrm{e}^{2\omega_{\mathrm{I}}(k)t}$. It also gives considerable insight because it shows how the Wigner distribution approximately evolves, namely that each point in phase space evolves with constant velocity where the velocity is the group velocity.

We now give two derivations of (3.61). In (3.60) we truncate the series after the first derivative and call the approximate Wigner distribution $W_a(x,k,t)$. Then,

$$\frac{\partial}{\partial t}W_a(x,k,t) = \left[2\omega_{\mathrm{I}}(k) - v(k)\frac{\partial}{\partial x}\right]W_a(x,k,t). \qquad (3.62)$$

The solution to this equation is (3.61). Alternatively, in (3.55) we expand the terms in the exponent in a power series in θ

$$\omega_{\mathrm{R}}(k+\theta/2)-\omega_{\mathrm{R}}(k-\theta/2)=\sum_{n=0}^{\infty}\frac{\omega_{\mathrm{R}}^{(2n+1)}(k)}{(2n+1)!}\frac{\theta^{2n+1}}{2^{2n}}\sim\omega_{\mathrm{R}}'(k)\theta+\frac{1}{24}\omega_{\mathrm{R}}^{(3)}(k)\theta^{3}\ldots \tag{3.63}$$

$$\omega_{\mathrm{I}}(k+\theta/2)+\omega_{\mathrm{I}}(k-\theta/2)=\sum_{n=0}^{\infty}\frac{\omega_{\mathrm{I}}^{(2n)}(k)}{(2n)!}\frac{\theta^{2n}}{2^{2n-1}}\sim 2\omega_{\mathrm{I}}(k)+\frac{1}{8}\omega_{\mathrm{I}}^{(2)}(k)\theta^{2}\ldots, \tag{3.64}$$

where $\omega^{(n)}(k)$ signifies the nth derivative with respect to k. If we keep only the first term in each series and substitute into (3.55) we have that

$$W(x,k,t)\approx\frac{1}{2\pi}\iint W(x',k,0)\mathrm{e}^{2\omega_{\mathrm{I}}(k)t}\,\mathrm{e}^{\mathrm{i}\theta(x'-x)}\mathrm{e}^{\mathrm{i}v_g(k)\theta t}\mathrm{d}\theta\mathrm{d}x' \tag{3.65}$$

$$=\mathrm{e}^{2\omega_{\mathrm{I}}(k)t}\int W(x',k,0)\mathrm{e}^{2\omega_{\mathrm{I}}(k)t}\,\delta(x'-x+v_g(k)t)\mathrm{d}x'= \tag{3.66}$$

which gives (3.61).

3.3 Case B: Initially Given: $u(x_0,t)$

One substitute $\mathrm{e}^{\mathrm{i}\omega t-\mathrm{i}kx}$ to obtain

$$\sum_{n=0}^{N}a_n(\mathrm{i}\omega)^n=\sum_{n=0}^{M}b_n(-\mathrm{i}k)^n \tag{3.67}$$

and solves for k in terms of ω

$$k=k(\omega)=k_{\mathrm{R}}(\omega)+\mathrm{i}k_{\mathrm{I}}(\omega). \tag{3.68}$$

Of course, for this case we could have kept $\mathrm{e}^{\mathrm{i}kx-\mathrm{i}\omega t}$ instead $\mathrm{e}^{\mathrm{i}\omega t-\mathrm{i}kx}$ but this would produce a number of equations that would have awkward minus signs. By using $\mathrm{e}^{ikx-\mathrm{i}\omega t}$, the transcription from one case to another is straightforward. The "group slowness", or the "unit transit time" is defined by

$$\tau_g(\omega)=k_{\mathrm{R}}'(\omega). \tag{3.69}$$

The general solution for each mode is

$$u(x,t)=\frac{1}{\sqrt{2\pi}}\int F(\omega,0)\,\mathrm{e}^{\mathrm{i}\omega t-\mathrm{i}k(\omega)x}\,\mathrm{d}\omega, \tag{3.70}$$

where

$$F(\omega,0) = \frac{1}{\sqrt{2\pi}} \int u(0,t)\, e^{-i\omega t}\, dt. \tag{3.71}$$

Defining the x-dependent spectrum by

$$F(\omega,x) = F(\omega,0)\, e^{-ik(\omega)x}, \tag{3.72}$$

one has that $u(x,t)$ and $F(\omega,x)$ form Fourier transform pairs between t and ω as

$$u(x,t) = \frac{1}{\sqrt{2\pi}} \int F(\omega,x)\, e^{i\omega t}\, d\omega \tag{3.73}$$

$$F(\omega,x) = \frac{1}{\sqrt{2\pi}} \int u(x,t)\, e^{-i\omega t}\, dt. \tag{3.74}$$

Using (3.70) and (3.71), we have

$$u(x,t) = \frac{1}{2\pi} \iint u(x',0)\; e^{-i\omega(t'-t)-ik(\omega)x}\, d\omega dt' \tag{3.75}$$

or

$$u(x,t) = \int u(0,t')\; G_2(t'-t,x) dt', \tag{3.76}$$

where

$$G_2(t'-t,x) = \frac{1}{2\pi} \int e^{-i\omega(t'-t)-ik(\omega)x}\, d\omega. \tag{3.77}$$

As in Case A, the spectrum characterizes of the damping vs. no damping case,

$$\text{if } |F(\omega,x)| = |F(\omega,0)| \qquad \text{no damping} \tag{3.78}$$

$$|F(\omega,x)| = |F(\omega,0|\, e^{k_1(\omega)x} \qquad \text{damping}\,. \tag{3.79}$$

The normalization is

$$N(x) = \int |u(x,t)|^2\, dt = \int |F(\omega,0\,|^2 e^{2k_1(\omega)x}\, d\omega \tag{3.80}$$

and as before if we define a new pulse by

$$u_N(x,t) = \frac{1}{\sqrt{N(x)}} u(x,t) \tag{3.81}$$

$$F_N(\omega,x) = \frac{1}{\sqrt{N(x)}} F(\omega,x), \tag{3.82}$$

then

$$\int |u_N(x,t)|^2\, dt = \int |u_N(x,0)|^2\, dt \tag{3.83}$$

regardless of whether there is damping or not.

3.3.1 Relation Between Damped and Undamped Waves

We define

$$u(x,t) = \frac{1}{\sqrt{2\pi}} \int F(\omega,0)\ \mathrm{e}^{\mathrm{i}\omega t - \mathrm{i}k_{\mathrm{R}}(\omega)x + k_{\mathrm{I}}(\omega)x}\, \mathrm{d}\omega \qquad \text{damping} \tag{3.84}$$

$$f(x,t) = \frac{1}{\sqrt{2\pi}} \int F(\omega,0)\ \mathrm{e}^{\mathrm{i}\omega t - \mathrm{i}k_{\mathrm{R}}(\omega)x} \mathrm{d}\omega \qquad \text{no damping} \tag{3.85}$$

and the same derivation as for Case A leads to

$$F_u(\omega,x) = F_f(\omega,x)\mathrm{e}^{k_{\mathrm{I}}(\omega)x} \tag{3.86}$$

and

$$u(x,t) = \frac{1}{2\pi} \iint f(x,t')\ \mathrm{e}^{-\mathrm{i}\omega(t'-t)}\ \mathrm{e}^{k_{\mathrm{I}}(\omega)x}\ \mathrm{d}\omega \mathrm{d}t'. \tag{3.87}$$

Also,

$$f(x,t) = \frac{1}{2\pi} \iint u(x,t')\ \mathrm{e}^{-\mathrm{i}\omega(t'-t)}\ \mathrm{e}^{-k_{\mathrm{I}}(\omega)x}\ \mathrm{d}\omega \mathrm{d}t'. \tag{3.88}$$

3.3.2 Phase Space

For this case, we use the time-frequency Wigner distribution at position x,

$$W(t,\omega,x) = \frac{1}{2\pi} \int u^*(x, t - \tfrac{1}{2}\tau)\, u(x, t + \tfrac{1}{2}\tau)\, \mathrm{e}^{-\mathrm{i}\tau\omega}\, \mathrm{d}\tau, \tag{3.89}$$

which can be expressed in terms of the spectrum as [4]

$$W(t,\omega,x) = \frac{1}{2\pi} \int F^*\left(\omega + \tfrac{1}{2}\omega', x\right)\ F\left(\omega - \tfrac{1}{2}\omega', x\right)\ \mathrm{e}^{-\mathrm{i}\omega' t}\, \mathrm{d}\omega'. \tag{3.90}$$

This defines the time-frequency phase-space at position x. The interpretation of $|\, u(x,t)\, |^2$ and $|\, F(\omega,x)\, |^2$ is:

$$|\, u(x,t)\, |^2 = \text{intensity (or energy) per unit time } t, \text{ at position } x \tag{3.91}$$

$$|\, F(\omega,x)\, |^2 = \text{spectral intensity (or energy) per unit frequency at position } x \tag{3.92}$$

and the time-frequency Wigner distribution satisfies

$$\int W(t,\omega,x)\, \mathrm{d}\omega = |\, u(x,t)\, |^2 \tag{3.93}$$

$$\int W(t,\omega,x)\,dt = |\,F(\omega,x)\,|^2. \tag{3.94}$$

The same derivation that led to (3.55) leads to

$$W(t,\omega,x) = \frac{1}{2\pi}\iint W(t',\omega,0)e^{i\omega'(t'-t)}\, e^{i[k^*(\omega+\omega'/2)-k(\omega-\omega'/2)]x}\, dt'd\omega', \tag{3.95}$$

which relates the Wigner distribution at position x to the Wigner distribution at position $x = 0$.

3.3.3 Differential Equation for a Mode

Differentiating (3.70) with respect to position

$$i\frac{\partial}{\partial x}u(x,t) = \frac{1}{\sqrt{2\pi}}\int k(\omega)F(\omega,0)\,e^{i\omega t - ik(\omega)x}\,d\omega \tag{3.96}$$

and converting to a differential equations we have

$$i\frac{\partial}{\partial x}u(x,t) = k\left(\frac{1}{i}\frac{\partial}{\partial t}\right)u(x,t), \tag{3.97}$$

In addition, differentiating (3.72)

$$i\frac{\partial}{\partial x}F(\omega,x) = k(\omega)F(\omega,x) \tag{3.98}$$

and the differential equations for the Wigner distribution is

$$i\frac{\partial}{\partial x}W(t,\omega,x) = \left[k\left(\omega + \frac{1}{2i}\frac{\partial}{\partial t}\right) - k^*\left(\omega - \frac{1}{2i}\frac{\partial}{\partial t}\right)\right]W(t,\omega,x). \tag{3.99}$$

As we derived the relationship between the Wigner distribution for the damped and undamped wave in Case A, a similar calculation for this case gives

$$W(t,\omega,x) = \frac{1}{2\pi}\int W_f(t',\omega,x)\,e^{[k_1(\omega+\theta/2)-k_1(\omega-\theta/2)]x}\,e^{-i\theta(t-t')}d\theta dt', \tag{3.100}$$

where $W_f(t',\omega,x)$ is the Wigner distribution for the undamped case.

Also, the derivation that led to (3.61) for the approximation now leads to

$$W(t,\omega,x) \approx e^{2k_1(\omega)x}\,W(t-\tau_g(\omega)\,x,\omega,0). \tag{3.101}$$

3.4 Nonstationary Noise Processes in Phase-Space

In this section, we discuss how noise can be formulated using phase-space methods. The advantage is that one can study nonstationary noise in a direct manner. To deal with nonstationary noise, we need a generalization of the standard power spectrum. Suppose we have a random function $z(t)$, one can think of particular realization and substitute into the Wigner distribution, (3.89) and then take the ensemble average of it [17–19]

$$\overline{W}(t,\omega) = \frac{1}{2\pi}\int E\left[z^*\left(t-\tfrac{1}{2}\tau\right)z\left(t+\tfrac{1}{2}\tau\right)\right]\mathrm{e}^{-\mathrm{i}\tau\omega}\mathrm{d}\tau. \tag{3.102}$$

$\overline{W}_z(t,\omega)$ is called the *Wigner spectrum* ,which can be thought of as the instantaneous spectrum of a random process $z(t)$. The marginal conditions become

$$\int \overline{W}(t,\omega)\,\mathrm{d}\omega = E\left[\,|z(t)\,|^2\right] \tag{3.103}$$

$$\int \overline{W}(t,\omega)\,\mathrm{d}t = E\left[\,|\,Z(\omega)\,|^2\,\right]. \tag{3.104}$$

As standard we define the autocorrelation function by way of

$$R(t_1,t_2) = E[z(t_1)z^*(t_2)] \tag{3.105}$$

and hence the Wigner spectrum can be written as

$$\overline{W}_z(t,\omega) = \frac{1}{2\pi}\int R(t+\tau/2,t-\tau/2)\mathrm{e}^{-\mathrm{i}\tau\omega}\mathrm{d}\tau. \tag{3.106}$$

Taking the inverse,

$$R(t+\tau/2,t-\tau/2) = \int \overline{W}(t,\omega,x)\mathrm{e}^{\mathrm{i}\tau\omega}\mathrm{d}\omega, \tag{3.107}$$

also letting $t_2 = t-\tau/2$ and $t_1 = t+\tau/2$, we also have

$$R(t_1,t_2) = \int \overline{W}\left(\frac{t_1+t_2}{2},\omega\right)\mathrm{e}^{-\mathrm{i}(t_2-t_1)\omega}\mathrm{d}\omega. \tag{3.108}$$

3.4.1 Noise Types in Terms of the Wigner Spectrum

Galleani and Lopresti have shown how various noise types are formulated in terms of the Wigner spectrum [6]. We briefly summarize their results and then discuss how these various noise types propagate.

3.4.1.1 Stationary Processes

A process is stationary if the autocorrelation function depends on the difference of the two times,

$$R(t_1, t_2) = R(t_1 - t_2). \tag{3.109}$$

The Wigner spectrum is therefore

$$\overline{W}(t,\omega) = \frac{1}{2\pi}\int R(t+\tau/2, t-\tau/2)\mathrm{e}^{-\mathrm{i}\tau\omega}\mathrm{d}\tau \tag{3.110}$$

$$= \frac{1}{2\pi}\int R(\tau)\mathrm{e}^{-\mathrm{i}\tau\omega}\mathrm{d}\tau, \tag{3.111}$$

which is, as expected, independent of time and equal to the standard power spectrum.

3.4.1.2 White Noise

For white noise

$$R(t_1, t_2) = N_0\delta(t_1 - t_2) \tag{3.112}$$

and hence the Wigner spectrum is

$$\overline{W}(t,\omega) = \frac{1}{2\pi}\int R(t+\tau/2, t-\tau/2)\mathrm{e}^{-\mathrm{i}\tau\omega}\mathrm{d}\tau \tag{3.113}$$

$$= \frac{N_0}{2\pi}\int \delta(\tau)\mathrm{e}^{-\mathrm{i}\tau\omega}\mathrm{d}\tau, \tag{3.114}$$

which gives

$$\overline{W}(t,\omega) = \frac{1}{2\pi}N_0. \tag{3.115}$$

Thus, for white noise, the Wigner spectrum is independent of time and frequency.

3.4.1.3 Cyclostationary Noise

Cyclostationary noise is when the autocorrelation function is periodic in both variables

$$R(t_1 + T, t_2 + T) = R(t_1, t_2), \tag{3.116}$$

where T is the period. The Wigner distribution for such a process is

$$\overline{W}_{\mathrm{CS}}(t,\omega) = \frac{1}{2\pi}\int R(t+\tau/2, t-\tau/2)\mathrm{e}^{-\mathrm{i}\tau\omega}\mathrm{d}\tau \tag{3.117}$$

$$= \frac{1}{2\pi} \int R(t + T + \tau/2, t + T - \tau/2) e^{-i\tau\omega} d\tau, \qquad (3.118)$$

which gives that

$$\overline{W}(t, \omega) = \overline{W}(t + T, \omega) \qquad (3.119)$$

and thus we see that the Wigner spectrum is periodic in time.

3.4.1.4 Local Stationary

By a locally stationary process one means that the process is approximately stationary properties in an interval of time. We have shown that a random process is locally stationary when its time-varying spectrum is approximately factorable into its marginal at the time of interest [9, 10]

$$\overline{W}(t_0, \omega) \sim \overline{W}_t(t_0) \overline{W}_\omega(\omega) \text{ in the neighborhood of } t_0 \qquad (3.120)$$

Also, we have shown that this result is equivalent to the criteria formulated by Silverman [22].

3.4.1.5 Modulated White-Noise

Modulated white noise is defined by

$$z(t) = \sigma(t) f(t), \qquad (3.121)$$

where $f(t)$ is white noise with zero mean and $\sigma(t)$ is a deterministic function. The autocorrelation function for $f(t)$ is

$$R_f(t_1, t_2) = N_0 \delta(t_1 - t_2) \qquad (3.122)$$

and Wigner spectrum is then

$$\overline{W}(t, \omega) = \frac{1}{2\pi} \int R(t + \tau/2, t - \tau/2) e^{-i\tau\omega} d\tau \qquad (3.123)$$

$$= \frac{N_0}{2\pi} \int \sigma(t + \tau/2, t - \tau/2) \delta(\tau) e^{-i\tau\omega} d\tau, \qquad (3.124)$$

which gives

$$\overline{W}(t, \omega) = \frac{N_0}{2\pi} \sigma^2(t). \qquad (3.125)$$

3.5 Propagation of Noise

Our aim in this section is to obtain the propagation equation for the Wigner spectrum and the nonstationary autocorrelation function. In this section, we deal with the case where the noise is generated at a particular position and we wish to determine the statistics of the noise at another position as the noise propagates. Of course, we assume that we know the statistics at the point of generation. By taking the ensemble average of (3.95) we have

$$\overline{W}(t,\omega,x) = \frac{1}{2\pi}\iint \overline{W}(t',\omega,0)\mathrm{e}^{\mathrm{i}\omega'(t'-t)}\,\mathrm{e}^{\mathrm{i}[k^*(\omega+\omega'/2)-k(\omega-\omega'/2)]x}\,\mathrm{d}t'\mathrm{d}\omega'. \tag{3.126}$$

This in principle solves the problem. To obtain the equation for the autocorrelation function, we use

$$R(t+\tau/2,t-\tau/2,x) = \int \overline{W}(t,\omega,x)\mathrm{e}^{\mathrm{i}\tau\omega}\mathrm{d}\omega \tag{3.127}$$

and after some algebra we obtain that

$$R(t+\tau/2,t-\tau/2,x) = \left(\frac{1}{2\pi}\right)^2 \iint R(t'+\tau'/2,t'-\tau'/2,0)\mathrm{e}^{\mathrm{i}\omega(\tau-\tau')}\mathrm{e}^{\mathrm{i}\omega'(t'-t)}$$
$$\mathrm{e}^{\mathrm{i}[k^*(\omega+\omega'/2)-k(\omega-\omega'/2)]x}\mathrm{d}\tau'\mathrm{d}t'\mathrm{d}\omega'\mathrm{d}\omega \tag{3.128}$$

and

$$R(t_1,t_2,x) = \left(\frac{1}{2\pi}\right)^2 \iint\iint R(t_1',t_2',0)\mathrm{e}^{\mathrm{i}\omega(t_1-t_2-t_1'+t_2')}\mathrm{e}^{\mathrm{i}\omega'(t_1'+t_2'-t_1-t_2)/2}$$
$$\mathrm{e}^{\mathrm{i}[k^*(\omega+\omega'/2)-k(\omega-\omega'/2)]x}\mathrm{d}\omega'\mathrm{d}\omega dt_1 dt_2. \tag{3.129}$$

We now consider the noises that were discussed in Sect. 3.4.1 and ascertain how they propagate.

3.5.1 No Dispersion

The dispersionless case is obtained when

$$k(\omega) = \frac{1}{c}\omega \tag{3.130}$$

which gives

$$k^* (\omega + \omega'/2) - k (\omega - \omega'/2) = \frac{\omega'}{c}, \tag{3.131}$$

and using (3.126) we have

$$\overline{W}(t, \omega, x) = \overline{W}(t - x/c, \omega, 0). \tag{3.132}$$

This shows that when we there is no dispersion the statistics at point x are the same as at the point of origin except that time is delayed by the amount of time it takes to travel from zero to the observation point x. Similarly, using (3.128) and (3.129) for the autocorrelation function, we have

$$R(t + \tau/2, t - \tau/2, x) = R(t - x/c + \tau/2, t - x/c - \tau/2, 0) \tag{3.133}$$

and

$$R(t_1, t_2, x) = R_0(t_1 - x/c, t_2 - x/c, 0). \tag{3.134}$$

3.5.2 *White Noise*

For white noise, the Wigner spectrum at the point of origin is

$$\overline{W}_{\mathrm{WN}}(t, \omega) = \frac{N_0}{2\pi}. \tag{3.135}$$

The Wigner spectrum at position x is then

$$\overline{W}(t, \omega, x) = \frac{1}{2\pi} \iint \overline{W}(t', \omega, x_0) e^{i\omega'(t'-t)} e^{i[k^*(\omega+\omega'/2)-k(\omega-\omega'/2)]x} \, dt' d\omega' \tag{3.136}$$

$$= \frac{1}{2\pi} \iint \frac{N_0}{2\pi} e^{i\omega'(t'-t)} e^{i[k^*(\omega+\omega'/2)-k(\omega-\omega'/2)]x} \, dt' d\omega' \tag{3.137}$$

or

$$\overline{W}(t, \omega, x) = \frac{N_0}{2\pi} e^{i[k^*(\omega)-k(\omega)]x}. \tag{3.138}$$

If the dispersion relation is real, then

$$\overline{W}(t, \omega, x) = \frac{N_0}{2\pi} = \overline{W}(t, \omega, 0) \tag{3.139}$$

and hence we have white noise at other points. However if we have a complex dispersion relation, then

$$\overline{W}(t,\omega,x) = \frac{N_0}{2\pi}\,\mathrm{e}^{2ik_1(\omega)x}. \tag{3.140}$$

3.5.3 *Stationary Process*

From (3.111) we have that for a stationary process

$$\overline{W}(t,\omega,0) = \overline{W}_S(\omega,0) \tag{3.141}$$

and hence

$$\overline{W}(t,\omega,x) = \frac{1}{2\pi}\iint \overline{W}(t',\omega,x_0)\mathrm{e}^{\mathrm{i}\omega'(t'-t)}\,\mathrm{e}^{\mathrm{i}[k^*(\omega+\omega'/2)-k(\omega-\omega'/2)]x}\,\mathrm{d}t'\mathrm{d}\omega' \tag{3.142}$$

$$= \frac{1}{2\pi}\iint \overline{W}_S(\omega,x_0)\mathrm{e}^{\mathrm{i}\omega'(t'-t)}\,\mathrm{e}^{\mathrm{i}[k^*(\omega+\omega'/2)-k(\omega-\omega'/2)]x}\,\mathrm{d}t'\mathrm{d}\omega', \tag{3.143}$$

which gives

$$\overline{W}(t,\omega,x) = \overline{W}_S(\omega,0)\,\mathrm{e}^{\mathrm{i}[k^*(\omega)-k(\omega)]x}. \tag{3.144}$$

This shows that the process remains time independent and hence stationary. However, it could be attenuated. If the dispersion relation is real, then

$$\overline{W}(t,\omega,x) = \overline{W}_S(\omega,0), \tag{3.145}$$

but if the dispersion relation is complex, then we have

$$\overline{W}(t,\omega,x) = \overline{W}_S(\omega,0)\mathrm{e}^{2ik_1(\omega)x}. \tag{3.146}$$

3.5.4 *Local Stationarity*

From (3.120), the condition for local stationary at the time point t_0 is

$$\overline{W}(t_0,\omega) \sim \overline{W}_t(t_0)\overline{W}_\omega(\omega) \text{ in the neighborhood of } t_0. \tag{3.147}$$

The Wigner spectrum at position x is then

$$\overline{W}(t,\omega,x) = \frac{1}{2\pi}\iint \overline{W}(t',\omega,x_0)\mathrm{e}^{\mathrm{i}\omega'(t'-t)}\,\mathrm{e}^{\mathrm{i}[k^*(\omega+\omega'/2)-k(\omega-\omega'/2)]x}\,\mathrm{d}t'\mathrm{d}\omega' \tag{3.148}$$

$$= \frac{1}{2\pi}\iint \overline{W}_t(t',0)\overline{W}_\omega(\omega,0)\mathrm{e}^{\mathrm{i}\omega'(t'-t)}\,\mathrm{e}^{\mathrm{i}[k^*(\omega+\omega'/2)-k(\omega-\omega'/2)]x}\,\mathrm{d}t'\mathrm{d}\omega' \tag{3.149}$$

$$= \frac{1}{2\pi}\overline{W}_\omega(\omega,0)\iint \overline{W}_t(t',0)\mathrm{e}^{\mathrm{i}\omega'(t'-t)}\,\mathrm{e}^{\mathrm{i}[k^*(\omega+\omega'/2)-k(\omega-\omega'/2)]x}\,\mathrm{d}t'\mathrm{d}\omega'. \tag{3.150}$$

In a dispersionless medium we have

$$\overline{W}(t,\omega,x) = \overline{W}_\omega(\omega,0)\overline{W}_t(t-x/c,0), \tag{3.151}$$

which shows that it remains locally stationary. But if, for example,

$$k(\omega) = \frac{1}{c}\omega + \gamma\omega^2/2, \tag{3.152}$$

then

$$k^*\left(\omega+\omega'/2\right) - k\left(\omega-\omega'/2\right) = \omega'(1/c+\gamma\omega) \tag{3.153}$$

and we obtain that

$$\overline{W}(t,\omega,x) = \overline{W}_\omega(\omega,0)\overline{W}_t(t-(1/c+\gamma\omega)x,0). \tag{3.154}$$

This is not a factorable and hence the distribution does not remain locally stationary.

3.5.5 Modulated White Noise

As per (3.121), we take

$$\overline{W}(t,\omega,0) = \frac{N_0}{2\pi}\sigma^2(t) \tag{3.155}$$

and substitute into (3.126)

$$\overline{W}(t,\omega,x) = \frac{1}{2\pi}\iint \overline{W}(t',\omega,0)\mathrm{e}^{\mathrm{i}\omega'(t'-t)}\,\mathrm{e}^{\mathrm{i}[k^*(\omega+\omega'/2)-k(\omega-\omega'/2)]x}\,\mathrm{d}t'\mathrm{d}\omega' \tag{3.156}$$

$$= \frac{1}{2\pi}\frac{N_0}{2\pi}\iint \sigma^2(t')\mathrm{e}^{\mathrm{i}\omega'(t'-t)}\,\mathrm{e}^{\mathrm{i}[k^*(\omega+\omega'/2)-k(\omega-\omega'/2)]x}\,\mathrm{d}t'\mathrm{d}\omega'. \tag{3.157}$$

Hence, it does not remain modulated white noise. As an example consider

$$k(\omega) = \frac{1}{c}\omega + \gamma\omega^2/2, \tag{3.158}$$

and using (3.157) we obtain

$$\overline{W}(t,\omega,x) = \frac{N_0}{2\pi}\sigma^2(t - x/c - \gamma\omega x). \tag{3.159}$$

3.5.6 Cyclostationary Noise

As per (3.115) cyclostationary noise is when the Wigner spectrum is

$$\overline{W}_{\mathrm{CS}}(t,\omega,0) = \overline{W}_{\mathrm{CS}}(t+T,\omega,0). \tag{3.160}$$

Substituting into (3.126) gives

$$\overline{W}(t,\omega,x) = \overline{W}(t-T,\omega,0), \tag{3.161}$$

which shows that the process remains cyclostationary except the time is displaced by the period.

3.6 Propagation of Spatial Noise Processes

In this section, we consider the propagation of spatial noise propagation. That is, suppose that at time $t = 0$ we are given a noise field with spatial statistical characterization, how will that noise field evolve? In this case, we have a spatial random function $z(x)$, and define the spatial Wigner spectrum by

$$\overline{W}(x,k) = \frac{1}{2\pi}\int E[z^*(x - \tfrac{1}{2}\tau)z(x + \tfrac{1}{2}\tau)]\mathrm{e}^{-\mathrm{i}\tau k}\mathrm{d}\tau. \tag{3.162}$$

The spatial autocorrelation is defined by

$$R(x_1,x_2) = E[x(t_1)x^*(t_2)] \tag{3.163}$$

and the spatial Wigner spectrum can be written as

$$\overline{W}(x,k) = \frac{1}{2\pi}\int R(x+\tau/2, x-\tau/2)\mathrm{e}^{-\mathrm{i}\tau k}\mathrm{d}\tau. \tag{3.164}$$

We also have

$$R(x+\tau/2, x-\tau/2) = \int \overline{W}_z(x,k)\mathrm{e}^{-\mathrm{i}\tau k}\mathrm{d}k \tag{3.165}$$

and

$$R(x_1,x_2) = \int \overline{W}\left(\frac{x_1+x_2}{2}, k\right)\mathrm{e}^{-\mathrm{i}(x_2-x_1)k}\mathrm{d}k. \tag{3.166}$$

3.6.1 Propagation of Noise

Using (3.48) we have

$$\overline{W}(x,k,t) = \frac{1}{2\pi}\iint \overline{W}(x',k,0)\,\mathrm{e}^{\mathrm{i}\theta(x'-x)}\mathrm{e}^{\mathrm{i}[\omega^*(k+\theta/2)-\omega(k-\theta/2)]t}\mathrm{d}\theta\mathrm{d}x' \tag{3.167}$$

and also

$$R(x+\tau/2, x-\tau/2, t) = \int \overline{W}(x,k,t)\mathrm{e}^{\mathrm{i}\tau k}\mathrm{d}k \tag{3.168}$$

$$R(x+\tau/2, x-\tau/2, t) = \left(\frac{1}{2\pi}\right)^2\iint R(x'+\tau'/2, x'-\tau'/2, 0)\mathrm{e}^{\mathrm{i}k(\tau-\tau')}\mathrm{e}^{\mathrm{i}k'(x'-x)}$$
$$\mathrm{e}^{\mathrm{i}[\omega^*(k+\theta/2)-\omega(k-\theta/2)]t}\,\mathrm{d}\tau'\mathrm{d}x'\mathrm{d}k'\mathrm{d}k \tag{3.169}$$

$$R(x_1,x_2,x) = \left(\frac{1}{2\pi}\right)^2\iint R(x_1',x_2',0)\mathrm{e}^{\mathrm{i}k(x_1-x_2-x_1'+x_2')}$$
$$\mathrm{e}^{\mathrm{i}k'(x_1'+x_2'-x_1-x_2)/2}\mathrm{e}^{\mathrm{i}[\omega^*(k+\theta/2)-\omega(k-\theta/2)]t}\,\mathrm{d}x\mathrm{d}x'\mathrm{d}k'\mathrm{d}k. \tag{3.170}$$

Consider now some special cases. If the medium has no dispersion, then

$$\omega(k) = ck, \tag{3.171}$$

and substituting this into (3.167) we obtain

$$\overline{W}(x,k,t) = \overline{W}(x-ct,k,0) \tag{3.172}$$

$$R(x+\tau/2, x-\tau/2, x) = R(x-ct+\tau/2, x-ct-\tau/2, 0) \tag{3.173}$$

and

$$R(x_1,x_2,t) = R(x_1-ct, x_2-ct, 0) \tag{3.174}$$

3.6.1.1 Spatial White Noise

For spatial white noise

$$R(x_1, x_2) = N_0\delta(x_1 - x_2), \tag{3.175}$$

one obtains

$$\overline{W}(x,k,t) = \frac{N_0}{2\pi}\,\mathrm{e}^{\mathrm{i}[\omega^*(k)-\omega(k)]t}. \tag{3.176}$$

If the dispersion relation is real, then

$$\overline{W}(x,k,t) = \frac{N_0}{2\pi} = \overline{W}(x,k,0). \tag{3.177}$$

However, if we have a complex dispersion relation, then

$$\overline{W}(x,k,t) = \frac{N_0}{2\pi}\,\mathrm{e}^{2\mathrm{i}\omega_1(k)t}. \tag{3.178}$$

3.6.2 *Spatially Stationary Process*

A process is stationary if the autocorrelation function depends on the difference of the two spatial points. Using Eq. (3.167), we obtain

$$\overline{W}(x,k,t) = \overline{W}_S(k,0)\mathrm{e}^{\mathrm{i}[\omega^*(k)-\omega(k)]t}. \tag{3.179}$$

This shows the process remains spatially stationary if the dispersion relation is real

$$\overline{W}(x,k,t) = \overline{W}_S(k,0). \tag{3.180}$$

However, if the dispersion relation is complex, then we have

$$\overline{W}(x,k,t) = \overline{W}_S(\omega,0)\mathrm{e}^{2\mathrm{i}\omega_1(k)t}. \tag{3.181}$$

3.6.3 *Spatially Local Stationary*

The condition for local stationary at the a spatial point x_0 is

$$\overline{W}(x_0,k) \sim \overline{W}_x(x_0)\overline{W}_k(k) \text{ in the neighborhood of } x_0. \tag{3.182}$$

The Wigner spectrum at position x is

$$\overline{W}(x,k,t) = \frac{1}{2\pi}\overline{W}_k(k,0)\iint \overline{W}_x(x',0)\mathrm{e}^{\mathrm{i}k'(x'-x)}\,\mathrm{e}^{\mathrm{i}[\omega^*(k+k'/2)-\omega(k-k'/2)]t}\,\mathrm{d}x'\mathrm{d}k'. \tag{3.183}$$

In a dispersionless medium using (3.167) we have

$$\overline{W}(x,k,t) = \overline{W}_k(k,0)\overline{W}_x(x-ct,0), \tag{3.184}$$

which shows that it remains spatially locally stationary. However, if for example

$$\omega(k) = ck + \gamma k^2/2, \tag{3.185}$$

then (3.167) gives

$$\overline{W}(x,k,t) = \overline{W}_k(k,0)\overline{W}_x(x-(c+\gamma k)t,0). \tag{3.186}$$

This is not a factorable and hence the distribution does not remain locally spatially stationary.

3.6.4 *Spatially Modulated White Noise*

Spatially modulated white noise is when

$$z(x) = \sigma(x) f(x), \tag{3.187}$$

where $f(x)$ is white noise with zero mean

$$R_f(x_1,x_2) = N_0\delta(x_1-x_2) \tag{3.188}$$

and $\sigma(x)$ is a deterministic function. The Wigner spatial spectrum is

$$\overline{W}(x,k,0) = \frac{N_0}{2\pi}\sigma^2(x). \tag{3.189}$$

Substituting into (3.167), we have

$$\overline{W}(x,k,t) = \frac{1}{2\pi}\frac{N_0}{2\pi}\iint \sigma^2(x')\mathrm{e}^{ik'(x'-x)}\mathrm{e}^{i[\omega^*(k+\theta/2)-\omega(k-\theta/2)]t} \tag{3.190}$$

and hence in general it does not remain modulated white noise. For example, if

$$\omega(k) = ck + \gamma k^2/2 \tag{3.191}$$

gives

$$\overline{W}(x,k,t) = \frac{N_0}{2\pi}\sigma^2(x-ct-\gamma kt). \tag{3.192}$$

3.6.5 Spatial Cyclostationary Noise

Spatial cyclostationary noise is when the spatial autocorrelation function is periodic in both variables

$$R(x_1 + X, x_2 + X) = R(x_1, x_2) \tag{3.193}$$

and this condition implies that

$$\overline{W}_{CS}(x, k, 0) = \overline{W}_{CS}(x + X, k, 0). \tag{3.194}$$

Substituting into (3.167) gives

$$\overline{W}(x, k, t) = \overline{W}(x - X, k, 0) \tag{3.195}$$

and hence it remains cyclostationary but displaced by a distance X.

References

1. Cohen L (1989) Time-frequency distributions – A review. Proc IEEE 77:941–981
2. Cohen L (1995) Time-frequency analysis. Prentice-Hall, NJ
3. Cohen L (2001) Wigner distribution and pulse propagation. Proc SPIE 4474:20–24
4. Cohen L, Loughlin P (2007) Dispersion its effects, and compensation. In: Sadjadi F (ed) Physics of automatic target recognition. Springer, Berlin
5. Cohen L, Loughlin P, Okopal G (2008) Exact and approximate moments of a propagating pulse. J Mod Optic 55:3349–3358
6. Galleani L, Lo Presti L, Fosson S (2003) The Wigner spectrum and its application to stochastic processes. IEEE-EURASIP NSIP 03, 8–11 June 2003, Grado, Italy
7. Galleani L, Cohen L (2002) The Wigner distribution for classical systems. Phys Lett A 302:149–155
8. Galleani L, Cohen L (2003) Time-frequency Wigner distribution approach to differential equations. In: Barner K, Arce G (eds.) Nonlinear signal and image processing: theory, methods, and applications. CRC Press, Boca Raton
9. Galleani L, Cohen L, Suter B (2006) Locally stationary noise and random processes. In: Gabriel Cristóbal (ed.) Information optics. AIP, New York, pp 514–519
10. Galleani L, Cohen L, Suter B (2007) Local stationarity and time-frequency distributions. Proc SPIE 6313:63130Z-1-5
11. Jackson JD (1992) Classical electrodynamics. Wiley, New York
12. Lighthill J (1978) Waves in fluids. Cambridge University Press, Cambridge
13. Morse PH and Ingard KU (1968) Theoretical acoustics. McGraw-Hill, New York
14. Tolstoy I, Clay CS (1966) Ocean acoustics: theory and experiment in underwater sound. Acoustical Society of America, New York
15. Whitham GB (1974) Linear and nonlinear waves. Wiley, New York
16. Graff K (1975) Wave motion in elastic solids. Oxford University Press, Oxford
17. Mark WD (1970) Spectral analysis of the convolution and filtering of non-stationary stochastic processes. J Sound Vib 11:19–63
18. Martin W (1982) Time-frequency analysis of random signals. In: Proc ICASSP '82, pp 1325—1328

19. Martin W, Flandrin P (1985) Wigner–Ville spectral analysis of nonstationary processes. IEEE Trans Acoust Speech Signal Process 33:1461–1470
20. Loughlin P, Cohen L (2005) A Wigner approximation method for wave propagation. J Acoust Soc Am 118(3):1268–1271
21. Loughlin P, Cohen L (2004) Phase-space approach to wave propagation with dispersion and damping. Proc SPIE 5559:221–231
22. Silverman RA (1957) Locally stationary random processes. IRE Trans Inf Theory 3:182–187
23. Wigner EP (1932) On the quantum correction for thermodynamic equilibrium. Phys Rev 40:749–759

Chapter 4
How Does Noise Affect a Quantum State?

Joseph H. Eberly and Ting Yu

4.1 Background: Nomenclature

Noise is usually registered indirectly as decay of correlation. Quantum mechanics has a unique form of correlation labelled entanglement. Recent research has focused on the question whether entanglement is subject to a unique form of noise-induced decorrelation. The answer has been found to be positive and the consequences are currently being explored.

If the correlated systems are at sites remotely located from each other, meaning that their wave functions do not overlap, their entanglement is called nonlocal. Because of their remoteness, nonlocally entangled systems need not interact with each other and in principle can be separately locally manipulated, allowing the entanglement to be transported and exploited. Local entanglement is much more commonly found, e.g. in states of strongly interacting and/or mutually overlapping electrons in stable atomic orbits or in condensed-matter lattices. Such entanglement, mandated by the Pauli Principle, is currently considered less interesting than nonlocal entanglement.

Einstein labelled as "spooky" the nonlocal distribution of quantum pair-correlation between noninteracting entities, and he felt that its existence was a sufficient ground for questioning the completeness of quantum mechanics [1]. Following tests of inequalities conceived by Bell [2] that were carried out in experiments initiated by Clauser and Freedman, and confirmed as well as significantly extended by

J.H. Eberly (✉)
Rochester Theory Center and Department of Physics and Astronomy, University of Rochester, Rochester, New York 14627-0171, USA
e-mail: eberly@pas.rochester.edu

T. Yu
Department of Physics and Engineering Physics, Stevens Institute of Technology, Hoboken, New Jersey 07030, USA
e-mail: ting.yu@stevens.edu

L. Cohen et al. (eds.), *Classical, Semi-classical and Quantum Noise*,
DOI 10.1007/978-1-4419-6624-7_4,

Aspect et al. [3–8], there is now general support of the view that entanglement permits greater correlation than any local and realistic classical theory of information can provide. This higher degree of correlation is accepted to be a resource that can be exploited for new approaches to both computation and communication.

The nature of quantum entanglement is to a remarkable extent still mysterious. For example, no proven measure of the degree of entanglement is available for a mixed state of more than two quantum systems at a time. Even then, measures are missing unless the evolution takes place in no more than a 6-dimensional state space $(spin - 1/2) \otimes (spin - 1)$. Thus, most calculations have been designed for entanglement of two qubits, meaning any pair of two-state systems analogous to spin-1/2 particles, for which the most common measure is termed concurrence [9]. In any event, qubits serve as the building blocks for most applications being contemplated for entanglement. One can think of a qubit pair as a memory unit in a network, storing quantum information in their correlation for later recall. Since entanglement is an information measure, we can dispense with physical interactions between the two units, and focus entirely on the way their information content is disrupted by noise.

4.2 Background: Evolution

In the presence of noise, entanglement correlation is subject to degradation, as any other type of correlation. However, the degradation of entanglement has a feature not previously found in any Markovian-noise relaxation theories we are aware of: calculations of the time dependence of entanglement, undertaken within the last ten years for mathematical models [10–13] and physical processes [14–18], have exposed the existence of the ESD effect. The new label ESD can be interpreted as Early-Stage Decoherence or Early-Stage Disentanglement, or most popularly as Entanglement Sudden Death, but in all cases ESD refers to the complete disappearance of entanglement after only a finite time, under the influence of weak local noise. That is, ESD does not follow the standard "half-life" rule of decay, in which a constant fraction of remaining correlation is lost in every successive equal time interval. Half-life and non-half-life decays are illustrated in Fig. 4.1 for the physical example of vacuum noise, which is responsible for the disentangling effects of spontaneous emission. A review of pre-ESD entanglement evolution, with speculations about multiparty entanglement, has been given [19].

The dynamic evolution of the degree of entanglement contained in any two-qubit joint state (pure or mixed) is measured by Wootters' concurrence [9], which varies from $C = 1$ for a maximally entangled state to $C = 0$ for a separable state. Figure 4.1 shows two quite different routes of decay from $C = 2/3$ to $C = 0$. Concurrence may be calculated explicitly from the density matrix ρ for qubits A and B:

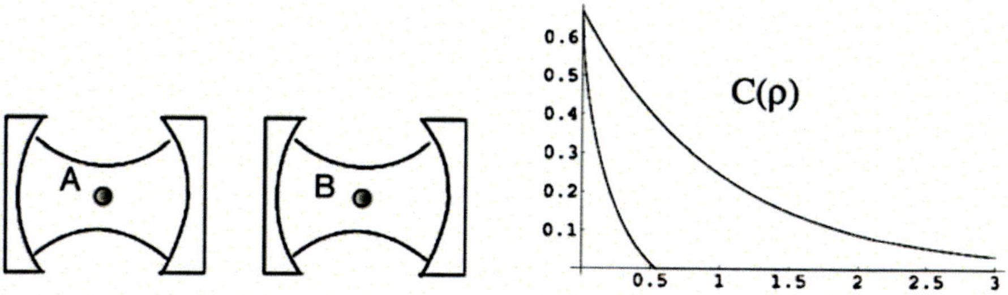

Fig. 4.1 *Left figures* show two qubits as two-level atoms A and B confined to separate overdamped cavities into which the atoms can emit a photon when excited. *Right figure* shows distinct routes to zero of AB entanglement, one following the standard exponential decay of atomic probability and the other showing sudden death decay, i.e., reaching zero in a finite time and remaining zero (from Yu and Eberly [14])

$$C(\rho) = \max\left(0,\ \sqrt{\lambda_1} - \sqrt{\lambda_2} - \sqrt{\lambda_3} - \sqrt{\lambda_4}\right)$$
$$= \max\left(0,\ Q(t)\right), \tag{4.1}$$

where $Q(t)$ is a useful abbreviation for the given combination of λ_i, which are the eigenvalues in decreasing order of the auxiliary matrix

$$\zeta = \rho^{AB}\left(\sigma_y^A \otimes \sigma_y^B\right)\rho^{AB*}\left(\sigma_y^A \otimes \sigma_y^B\right), \tag{4.2}$$

where ρ^* denotes the complex conjugation of ρ in the standard ($\pm$) basis and σ_y is the usual Pauli matrix.

In the following, we will describe the evolution of entanglement under noise-induced relaxation of a class of bipartite density matrices having the "X" form (containing only diagonal and antidiagonal elements) [20]:

$$\rho^{AB} = \begin{pmatrix} a & 0 & 0 & w \\ 0 & b & z & 0 \\ 0 & z^* & c & 0 \\ w^* & 0 & 0 & d \end{pmatrix}, \tag{4.3}$$

where $a+b+c+d = 1$. This class of mixed state arises naturally in a wide variety of physical situations including pure Bell states as well as the well-known Werner mixed state as special cases. Its notable feature for us is that it contains two-particle coherence (all of the anti-diagonal elements) without any single-particle coherence (i.e., no off-diagonal elements in the reduced density matrices of the separate units).

Methods that have been used to track evolution induced by noise include master equations, stochastic Schrödinger equations, and Kraus operator-sum representations. Until the discovery of the ESD effect, it was common to assume that local coherence times for single qubits (e.g., conventional T_1 and T_2 times) also serve as measures of joint-state (entanglement) coherence times, but after the discovery of ESD, this is no longer realistic because T_1 and T_2 times are always associated with normal half-life type decay.

Following earlier studies of the different effects of "global" noise (i.e., affecting two qubits in the same way) and independent "local" noises [21,22], we calculated a (finite) ESD time for local vacuum noise: $\tau_{esd} = \ln(1 + 1/\sqrt{2})\tau_0$, where τ_0 is the usual spontaneous lifetime. As Fig. 4.1 shows, we demonstrated that initial states of equal concurrence may have quite different evolution trajectories and some of them may avoid ESD. This result has been extended from vacuum noise to more common environmental noise types, including classical amplitude and phase noise [23].

4.3 The Biggest Difference

Many consequences of the ESD effect, and analogs of it, are now predicted, and two laboratory observations of it have appeared [24, 25]. Most have been generated by very broadband, effectively Markovian, noise but an expanded effort into non-Markovian territory has begun [26–29]. Here, we want to draw particular attention to what appears to be the biggest difference between entanglement decay and the wide range of decay processes previously recognized in physics, staying within the Markov-based regime.

It is widely observed in physics that relaxation is an additive effect under weak noises. That is, in situations where nonlinear feedback effects are absent the correlation relaxation rate of a system exposed to a collection of weak noises is the sum of the relaxation rates associated with the noises separably. More explicitly, we have considered [30] the Markov decay of a category of two-qubit density matrices with a single positive parameter λ for simplicity:

$$\rho_\lambda^{AB} = \frac{1}{9}\begin{pmatrix} 1 & 0 & 0 & 0 \\ 0 & 4 & \lambda & 0 \\ 0 & \lambda & 4 & 0 \\ 0 & 0 & 0 & 0 \end{pmatrix}. \tag{4.4}$$

In the notation of the generic X matrix in (4.3), this matrix has $a = 1/9,\ b = c = 4/9,\ z = \lambda/9,\ w = d = 0$. There is non-zero entanglement initially: $C_\lambda(0) = 2\lambda/9$, and its Markovian phase-noise and amplitude-noise time dependences can be found from solutions of the associated master equations, with phase and amplitude decay rates given by γ and Γ, as explained in [14, 30].

The result for phase noise is:

$$C_\lambda(0) \to C_\lambda(t) = (2\lambda/9)\exp[-\gamma t],$$

so entanglement decays smoothly exponentially to zero under phase noise. For amplitude noise (e.g., quantum vacuum noise), a direct calculation for the one-parameter 4×4 above gives the following time dependences for the density matrix elements:

$$z = \frac{\lambda}{9} \to z(t) = \frac{\lambda}{9}\exp[-\Gamma t], \tag{4.5}$$

$$a = \frac{1}{9} \to a(t) = \frac{1}{9}\exp[-2\Gamma t], \tag{4.6}$$

$$d = 0 \to d(t) = \frac{1}{9}\omega^4 + \frac{8}{9}\omega^2, \tag{4.7}$$

where $\omega = \sqrt{1 - \exp[-\Gamma t]}$. The concurrence is given by

$$C_\lambda(0) \to C_\lambda(t) = \frac{2}{9}\left[\lambda - \sqrt{\omega^4 + 8\omega^2}\right]\exp[-\Gamma t].$$

Therefore, for any value of the parameter λ between 3 and 4, our bipartite entanglement under amplitude noise also decays smoothly and exponentially to zero. With these two exercises in hand, we conclude that for this category of mixed two-system states, and for the stated range for λ, the entanglement decays exponentially to zero in the presence of either weak amplitude noise or weak phase noise.

Now, we consider the issue of additivity. We will do this in two stages, first considering the response of the density matrix itself to the simultaneous presence of both phase and amplitude noise, and then the response of the concurrence. The elements $a = 1/9$ and $d = 0$ are not altered by phase noise (phase decay rate is $\gamma = 0$), but undergo decay at the rate Γ in response to amplitude noise. The off-diagonal element $z = \lambda/9$ decays at the sum of the separate phase- and amplitude-noise rates:

$$z(0) = \frac{\lambda}{9} \to z(t) = \frac{\lambda}{9}\exp[-(\gamma + \Gamma)t]. \tag{4.8}$$

Thus, the presence of more than one noise source has the expected effect on the density matrix – the decay rates for the elements are additive, with the appropriate rate of phase noise relaxation being added to the separate rate of amplitude noise relaxation for all density matrix elements.

However when the evolution of quantum entanglement is considered, the consequences are strikingly different, as shown in the three parts of Fig. 4.2. The regions for which ESD occurs are identified by gray shading. ESD is absent from the regions with white backgrounds. The squares encompass all possible physical values for a and z in the joint density matrix (4.3). The short horizontal line inside the squares at $a = 1/9$ near the bottom covers the range of z values allowed in (4.4), i.e., covering the specific example whose time dependences are calculated for matrix (4.4). Remarkably, for that continuous range of allowed values of

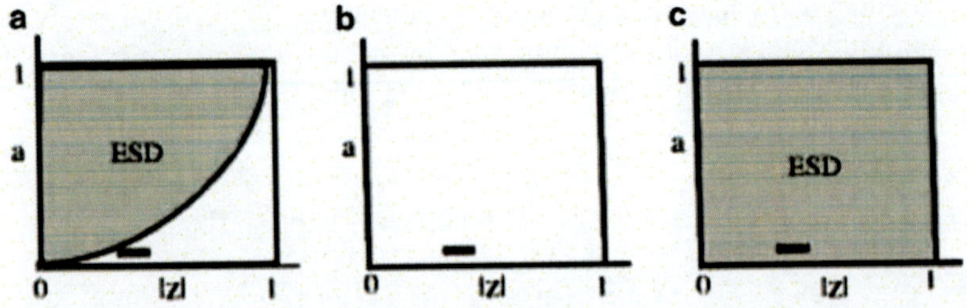

Fig. 4.2 The squares show the physically allowed region of the parameters a and z in (4.3) with $d = 0$. Each square shows by *gray shading* the a-z zone where ESD occurs (adapted from [30]). The shading differs depending on whether we are taking account of (**a**) amplitude noise, (**b**) phase noise, or (**c**) combined phase and amplitude noise. The consequences are interpreted in the text

z (given by $4 \geq \lambda \geq 3$) the two-party system is completely free from ESD, but only if subjected to a single noise (either of the first two squares). As soon as both phase and amplitude noises are applied together (third square), ESD cannot be avoided by any choice of initial conditions, even those for which neither phase noise nor amplitude noise causes ESD.

4.4 Summary

The conclusions we have presented raise the issue whether ESD creates new barriers to achievement of quantum information processes in which entanglement will play a central role. This has to be taken seriously, we believe, at least until a method is discovered to estimate (or calculate) the time at which ESD will strike. So far no generic guidelines are known for this [31].

One must also say that methods for avoiding or recovering from decorrelation are known. There are two main types, employing redundancy or symmetries, both with classical antecedents. Quantum error correction (see [32]) employs algorithms that engage redundancy in a sufficiently sophisticated way to recover lost entanglement, so that action in advance of the (so far unknown) ESD time can be effective. The known methods are very demanding in resource usage, and must be employed before ESD, because they cannot recover entanglement that has been completely lost. A recent report [33] suggests that even action in advance of ESD may not be sufficient, demonstrating an example in which application of error correction has the possibility to induce instead of repair the ESD effect.

A method for delaying rather than correcting entanglement decay engages so-called decoherence-free subspaces. That is, when sufficient symmetries in Hilbert space can be found, it may be possible to "hide" entanglement's evolutionary path from decorrelating effects (see [34]). This is a quantum analog of classical methods that exploit symmetries to keep correlations protected from noise. However, situations such as shown on the left of Fig. 4.1, where two atoms are located in separate cavities, can be regarded as a paradigm for two-party relaxation with no symmetries at all because each atom is subjected to an independent local noise. Unfortunately, practical situations involving remotely entangled qubits are not only the most desirable, but also the most likely to have completely independent noise interactions, and thus have no symmetries to exploit.

What is worse, of course, is that these rather unfavorable prospects are made all the more likely when the results presented here are taken into account. It is unrealistic in any practical sense to hope that only one source of noise will attack a correlation of interest. Our main point here has been to focus on results obtained for ESD under multiple noises. It appears nearly inevitable that ESD should be expected.

Acknowledgements JHE and TY acknowledge grant support from the US Army Research Office (48422-PH) and the National Science Foundation (PHY06-01804 and PHY07-58016).

References

1. Einstein A, Podolsky B, Rosen N (1935) Phys Rev 47:779
2. Bell JS (1964) Physics 1:195
3. Clauser JF (1969) Bull Am Phys Soc 14:578
4. Clauser JF, Holt RA, Horne MA, Shimony A (1969) Phys Rev Lett 23:880
5. Freedman SJ, Clauser JF (1972) Phys Rev Lett 28:938
6. Fry ES, Thompson RC (1976) Phys Rev Lett 37:465
7. Aspect A, Grangier P, Roger G (1981) Phys Rev Lett 47:460
8. Aspect A, Dalibard J, Roger G (1982) Phys Rev Lett 49:1804
9. Wootters WK (1998) Phys Rev Lett 80:2245
10. For example, see Rajagopal AK, Rendell RW (2001) Phys Rev A 63:022116
11. Zyczkowski K, Horodecki P, Horodecki M, Horodecki R (2001) Phys Rev A 65:012101
12. Daffer S, Wodkiewicz K, McIver JK (2003) Phys Rev A 67:062312
13. Diosi L (2003) In: Benatti F, Floreanini R (eds) Irreversible quantum dynamics. Springer, Berlin, pp 157–163
14. Yu T, Eberly JH (2004) Phys Rev Lett 93:140404
15. Dodd PJ, Halliwell JJ (2004) Phys Rev A 69:052105
16. Dodd PJ (2004) Phys Rev A 69:052106
17. Paz JP, Roncaglia AJ (2008) Phys Rev Lett 100:220401
18. Chou C-H, Yu T, Hu BL (2008) Phys Rev E 77:011112
19. Mintert F, Carvalho ARR, Kus M, Buchleitner A (2005) Phys Rep 415:207
20. Yu T, Eberly JH (2007) J Quant Comp Inf 7:495
21. Yu T, Eberly JH (2002) Phys Rev B 66:193306
22. Yu T, Eberly JH (2003) Phys Rev B 68:165322
23. Yu T, Eberly JH (2006) Opt Commun 264:393–397
24. Almeida MP, de Melo F, Hor-Meyll M, Salles A, Walborn SP, Souto Ribeiro PH, Davidovich L (2007) Science 316:579. See also Eberly JH, Yu T (2007) Science 316:555
25. Laurat J, Choi KS, Deng H, Chou CW, Kimble HJ (2007) Phys Rev Lett 99:180504
26. For example, see Ban M, Kitajima S, Shibata F (2005) J Phys A 38:7161
27. Ban M (2006) J Phys A 39:1927–1943
28. Bellomo B, Lo Franco R, Compagno G (2007) Phys Rev Lett 99:160502
29. Yu T, Eberly JH (2010) Opt Commun 283:676–680

30. Yu T, Eberly JH (2006) Phys Rev Lett 97:140403
31. Yu T, Eberly JH (2009) Science 323:598–601
32. See Calderbank AR, Shor PW (1996) Phys Rev A 54:1098, Bennett CH, DiVincenzo DP, Smolin JA, Wootters WK (1996) Phys Rev A 54:3824–3851, and Steane A (1996) Phys Rev A 54:4741
33. Sainz I, Bjork G (2008) Phys Rev A 77:052307
34. Lidar DA, Whaley KB (2003) In: Benatti F, Floreanini R (eds) Irreversible quantum dynamics. Springer, Berlin, pp 83–120

Chapter 5
Graph Theoretic Methods in Coding Theory

Salim El Rouayheb and Costas N. Georghiades

5.1 Introduction

Let $\Sigma_q = \{0, 1, \ldots, q-1\}$ be an alphabet of order q. A q-ary (unrestricted) code C of length n and size $|C|$ is a subset of Σ_q^n containing $|C|$ elements called codewords. The Hamming weight $wt(c)$ of a codeword c is the number of its nonzero entries. A constant-weight code is a code where all the codewords have the same Hamming weight. The Hamming distance $d(c, c')$ between two codewords c and c' is the number of positions where they have different entries. The minimum Hamming distance of a code C is the largest integer Δ such that $\forall c, c' \in C, d(c, c') \geq \Delta$.

Let $A_q(n, d)$ be the maximum size of a q-ary code of length n and minimum Hamming distance d [1, Chap. 17]. $A(n, d, w)$ is defined similarly for binary codes with constant weight w. Finding the values of $A_q(n, d)$ and $A(n, d, w)$ is a fundamental problem in "classical" coding theory [1,2]. This problem is considered to be very difficult and was in fact described in [3], as "a hopeless task". For this reason, much of the research done has focused on bounding these quantities. Note that the dual problem of finding the maximal order of a set of codewords satisfying an upper bound on their pairwise Hamming distance (anticodes) is well studied in extremal combinatorics. Surprisingly enough, it has a closed form solution [3–5].

Inthis paper, we showcase the basic interplay between graph theory and coding theory. Many known bounds on $A_q(n, d)$ and $A(n, d, w)$ follow directly from basic properties of graphs, such as relations among the clique, independence and chromatic numbers of graphs. Other can be proven using deeper algebraic results. For example, using a property of vertex transitive graphs, an inequality relating the maximal size of codes and that of anticodes can be found, leading thus to several bounds on $A_q(n, d)$ and $A(n, d, w)$.

S. El Rouayheb (✉) • C.N. Georghiades
ECE Department,Texas A&M University, College Station, TX 77843, USA
e-mail: rouayheb@ieee.org; c-georghiades@tamu.edu

L. Cohen et al. (eds.), *Classical, Semi-classical and Quantum Noise*,
DOI 10.1007/978-1-4419-6624-7_5, © Springer Science+Business Media, LLC 2012

This paper is organized as follows. In Sect. 5.2, we briefly introduce some of the needed background in graph theory. In Sect. 5.3, we use the tools introduced in the previous section to derive bounds on the maximum size of unrestricted codes. In Sect. 5.4, we focus on constant weight codes and derive some bounds and inequalities on their maximal size.

5.2 Graph Theory Background

We start by giving a brief summary of some graph theoretical concepts and results that will be needed in this paper. For more details, we refer the interested reader to [6] and [7].

5.2.1 *Basic Notation and Results*

A *graph* is a pair $G = (V, E)$ of sets such that the elements of E are subsets of order two of V. The elements of V are the *vertices* of the graph G and those of E are its *edges*. For any graph X, we let $V(X)$ denote its vertex set and $E(G)$ its edge set.

Two vertices u and v of G ($u, v \in V$) are *adjacent* if $\{u, v\}$ is an edge of G ($\{u, v\} \in E$), and we write $u \sim v$. If all the vertices of G are pairwise adjacent, then G is *complete*. A complete graph on n vertices is denoted as K^n. Two vertices that are not adjacent are called *independent*. The degree $d(v)$ of a vertex v is the number of vertices adjacent to v. The maximum degree of the graph G is then defined as $\Delta(G) := \max\{d(v); v \in V\}$. The graph G is called connected if for any disjoint partition V_1 and V_2 of its vertex set, i.e. $V_1 \cup V_2 = V$ and $V_1 \cap V_2 = \emptyset$, there exists at least one vertex in V_1 that is adjacent to a vertex in V_2. A graph C_n having $V(C_n) = \{v_1, \ldots, v_n\}$ and $E(C_n) = \{\{v_1, v_2\}, \{v_2, v_3\}, \ldots, \{v_n, v_1\}\}$ is called an odd *cycle* if n is odd, even otherwise.

The complement of a graph G is the graph $\bar{G}$ defined over the same vertex set but where two vertices are adjacent in $\bar{G}$ iff they are not in G. We denote by $\omega(G)$ the *clique number* of a graph G, defined as the largest number of vertices of G that are pairwise adjacent. In contrast, $\alpha(G)$, the *independence number* of G, is the largest number of pairwise independent vertices in G. It can be easily seen that $\alpha(G) = \omega(\bar{G})$.

A coloring of the graph G is an assignment of colors to its vertices such that adjacent vertices are never given the same color. Formally, a graph G has a k-*coloring* if there exists a map $c : V \to \{1, \ldots, k\}$ such that $c(u) \neq c(v)$ whenever the two vertices u and v are adjacent. The smallest integer k such that G has a k-coloring is called the chromatic number of G and denoted by $\chi(G)$. It is easy to show that the chromatic number of G is upper bounded by $\Delta(G) + 1$. This bound can be slightly improved in many cases:

Theorem 5.1 (Brooks (1941)). *If G is a connected graph but neither complete nor an odd cycle, then*

$$\chi(G) \leq \Delta(G).$$

Given a coloring of the graph G, vertices assigned the same color are pairwise independent and their average number is $\frac{|V(G)|}{\chi(G)}$. Therefore, using Brooks Theorem, we obtain the following bound on the independence number of a graph G.

Lemma 5.1. *If G is a connected graph but neither complete nor an odd cycle, then*

$$\alpha(G) \geq \frac{|V(G)|}{\Delta(G)}.$$

The next result, known as Turán theorem [8, Theorem 4.1], is a famous result in extremal graph theory and relates the clique number of a graph to the number of its edges. Define

$$M(n, p) := \frac{p-2}{2(p-1)}n^2 - \frac{r(p-1-r)}{2(p-1)},$$

where r is the remainder of the division of n by $p-1$.

Theorem 5.2 (Turán (1941)). *A graph G on n vertices having more than $M(n, p)$ edges satisfies $\omega(G) \geq p$.*

5.2.2 *Algebraic Graph Theory*

We define here the notions of graph automorphism and homomorphism and describe the class of vertex transitive graphs and state some of their useful properties. For the proofs of the theorems presented here and further related details, [7] can be consulted.

Definition 5.1 (Graph Automorphism). Let $G(V, E)$ be a graph and ϕ a bijection from V to itself. ϕ is called an *automorphism* of G iff

$$\forall u, v \in V, u \sim v \Leftrightarrow \phi(u) \sim \phi(v).$$

The set of all automorphisms of G is a group under composition; it is called the automorphism group of G and it is denoted $\mathrm{Aut}(G)$. For example, the complete graph on n vertices K_n has S_n, the symmetric group of order n, as its automorphism group. A graph is *vertex transitive* if the action of its automorphism group on its vertex set is transitive:

Definition 5.2 (Vertex Transitive Graph). A graph $G(V, E)$ is vertex transitive iff $\forall u, v \in V, \exists \phi \in \mathrm{Aut}(G)$ s.t. $\phi(u) = v$.

The following theorem [7, Lemma 7.2.2] gives a very important property of vertex transitive graphs, which will be instrumental in deriving the results in the coming sections.

Theorem 5.3. *Let $G(V, E)$ be a vertex transitive graph, then*

$$\alpha(G)\omega(G) \leq |V(G)|.$$

Let X and Y be two graphs.

Definition 5.3 (Graph Homomorphism). A mapping $f : V(X) \to V(Y)$ is a homomorphism from G to G' if $\forall x, y \in V\ x \sim y \Rightarrow f(x) \sim f(y)$.

Theorem 5.4. *If Y is vertex transitive and there is a homomorphism from X to Y, then*

$$\frac{\alpha(X)}{|V(X)|} \geq \frac{\alpha(Y)}{|V(Y)|}$$

Proof. An application of Lemma 7.14.2 in [7]. □

5.3 Bounds on Unrestricted Codes

In this section, we start applying some of the previously discussed graph theoretical results to obtain some bounds on the maximal size of codes. First, we define a family of graphs called *Hamming graphs* that will be instrumental in establishing the link between codes and graphs.

Definition 5.4 (Hamming Graph [2]). Given the positive integers n, q and d such that $q > 1$ and $2 \leq d \leq n$, the Hamming graph $H_q(n, d)$, has as vertices all the q-ary sequences of length n, and two vertices are adjacent iff their Hamming distance is larger or equal to d. That is, $V(H_q(n, d)) = \Sigma_q^n$, where $\Sigma_q = \{0, 1, \ldots, q-1\}$, and $u \sim v$ iff $d(u, v) \geq d$.

Notice that a q-ary code of length n and minimum Hamming distance d corresponds to a clique in the graph $H_q(n, d)$. Furthermore, $A_q(n, d)$, the maximum size of such code is the clique number of the corresponding Hamming graph. This is concisely stated in the following easy observation which has interesting consequences.

Observation 5.1 $A_q(n, d) = \omega(H_q(n, d))$.

We now give the first application by showing how Lemma 5.1 immediately implies the Gilbert–Varshamov Bound. Taking the graph G to be the complement of the Hamming graph $H_q(n, d)$, we have, by Observation 5.1, $A_q(n, d) = \alpha(G)$. Furthermore, $\Delta(G) = \Delta(\bar{H}_q(n, d)) = \sum_{i=0}^{d-1} \binom{n}{i}(q-1)^i$. Thus, by Lemma 5.1 we get:

Lemma 5.2 (Gilbert–Varshamov Bound).

$$A_q(n,d) \geq \frac{q^n}{\sum_{i=0}^{d-1}\binom{n}{i}(q-1)^i}.$$

For specific numerical values, this bound can be slightly improved by using Turán's theorem as noted in [13]. Next, we will show that the Hamming graphs are vertex transitive. This property will then be used to derive the well-known Singleton and Hamming bounds.

Lemma 5.3. *The Hamming graph $H_q(n,d)$ is vertex transitive.*

Proof. Take $\Sigma_q = \mathbb{Z}_q$, the integers modulo q. For all $u, v, x \in \Sigma_q^n$, define the function $\phi_{u,v}(x) = x + v - u$. $\phi_{u,v}(x)$ is an automorphism of $H_q(n,d)$. In fact, $d(\phi_{u,v}(x), \phi_{u,v}(y)) = d(x+v-u, y+v-u) = \mathrm{wt}(x+v-u-(y+v-u)) = \mathrm{wt}(x-y) = d(x,y)$. Also, $\phi_{u,v}(x)$ takes u to v. □

Thus, we deduce from Theorem 5.3 and Observation 5.1 the following inequality [14]:

Corollary 5.1. $A_q(n,d)\alpha(H_q(n,d)) \leq q^n$

The independence number $\alpha(H_q(n,d))$ of the Hamming graph $H_q(n,d)$ is actually the maximum number of sequences of length n such that the Hamming distance between any two of them is at most $d-1$. A set of sequences satisfying this property is called an *anticode* with maximum distance $d-1$. Define $N_q(n,s)$ to be the maximum number of q-ary sequences of length n that intersect pairwise, i.e., have the same entries, in at least s positions [4]. It follows that

$$\alpha(H_q(n,d)) = N_q(n,t); \quad \text{with } t = n-d+1. \tag{5.1}$$

By bounding from below the value of $N_q(n,t)$ in two different ways, we get the Singleton and the Hamming Bounds [11].

Lemma 5.4 (Singleton Bound). $A_q(n,d) \leq q^{n-d+1}$

Proof. Consider the set $T(n,t)$ of all q-ary sequences of length n having the same value 0 in the first $t = n-d+1$ entries. Therefore, by definition, $N_q(n,t) \geq |T(n,t)| = q^{n-t}$. Then by (5.1) and Corollary 5.1, $A_q(n,d) \leq \frac{q^n}{q^{n-t}} = q^{n-d+1}$. □

Lemma 5.5 (Hamming Bound).

$$A_q(n,d) \leq \frac{q^n}{\sum_{i=0}^{\lfloor \frac{d-1}{2} \rfloor}\binom{n}{i}(q-1)^i}.$$

Proof. Let $r = \lfloor \frac{d-1}{2} \rfloor$ and consider the set (ball) $B(n,r) = \{x \in \Sigma_q^n; wt(x) \leq r\}$. By the triangle inequality, $\forall x, y \in B(n,r), d(x,y) \leq 2r \leq d-1$. Therefore, $N_q(n,t) \geq |B(n,r|$. But $|B(n,r)| = \sum_{i=0}^{r}\binom{n}{i}(q-1)^i$. The result then follows directly from (5.1). □

The number $N_q(n,t)$ is well studied in extremal combinatorics [4, 5], and a closed form for it is known. Thus, exact expressions of $N_q(n,t)$ can be used to derive better upper bounds on $A_q(n,d)$ [9]. For instance, if $n-t$ is even, $N_2(n,t) = \sum_{i=0}^{\frac{n-t}{2}} \binom{n}{i}$. Thus, in this case, $B(n, \lfloor \frac{d-1}{2} \rfloor)$ is a maximal anticode and no improvement can be made in this case on the Hamming bound. However, when $n-t$ is odd, $N_2(n,t) = 2\sum_{i=0}^{\frac{n-t-1}{2}} \binom{n-1}{i}$ [4, Theorem Kl] and [10]. Therefore, we obtain the following improvement on the Hamming bound for even values of d [11, 12].

Lemma 5.6.

$$A(n,d) \leq \frac{2^{n-1}}{\sum_{i=0}^{\frac{d-2}{2}} \binom{n-1}{i}}, \quad \text{if } d \text{ is even.} \tag{5.2}$$

Using the exact expression of $N_q(n,t)$ given in Theorem 2 in [5] or the Diametric theorem of [4], we get this improved upper bound on $A_q(n,d)$ for nonbinary alphabets.

Lemma 5.7. *For* $q \geq 3$, $t = n - d + 1$ *and* $r = \lfloor \min\{\frac{n-t}{2}, \frac{t-1}{q-2}\} \rfloor$,

$$A_q(n,d) \leq \frac{q^{t+2r}}{\sum_{i=0}^{r} \binom{t+2r}{i}(q-1)^i}. \tag{5.3}$$

Note that for $q \geq t+1$, $N_q(n,t) = q^{n-t}$ [5, Corollary 1], i.e. a maximal anticode would be the trivial set $T(n,t)$ described in the proof of Lemma 5.4. In this case, the bound of (5.3) boils down to the Singleton bound.

For d even and n not much larger than t, the next lemma provides another improvement on the Hamming bound for non-binary alphabets.

Lemma 5.8. *For* d *odd and* $n \leq t + 1 + \frac{\log t}{\log(q-1)}$

$$A_q(n,d) \leq \frac{q^{n-1}}{\sum_{i=0}^{\frac{d-2}{2}} \binom{n-1}{i}(q-1)^i} \tag{5.4}$$

Proof. Under the conditions of this lemma, $N_q(n,t) = q\sum_{i=0}^{\frac{d-2}{2}} \binom{n-1}{i}(q-1)^i$ [4, Eq. (1.7)]. The result then follows from Corollary 5.1. □

By constructing homomorphisms between Hamming graphs with different parameters, we get the following recursive inequalities on $A_q(n,d)$ by Theorem 5.4:

Lemma 5.9.

$$A_q(n,d) \leq \frac{1}{q} A_q(n+1, d+1). \tag{5.5}$$

$$A_q(n,d) \leq q A_q(n-1, d). \tag{5.6}$$

$$A_q(n,d) \le \frac{q^n}{(q-1)^n} A_{q-1}(n,d). \tag{5.7}$$

Proof. Let $\phi_1 : \Sigma_q^{n+1} \to \Sigma_q^{n+1}$ such that $\phi_1((x_1,\dots,x_n,x_{n+1})) = (x_1,\dots,x_n)$. ϕ_1 is a graph homomorphism from $\bar{H}_q(n+1,d+1)$ to $\bar{H}_q(n,d)$. Applying then Theorem 5.4, we get (5.5). Similarly, taking $\phi_2 : \Sigma_q^{n-1} \to \Sigma_q^n$ such that $\phi_1((x_1,\dots,x_{n-1})) = (x_1,\dots,x_n,0)$, we get (5.6). The third inequality is obtained by taking $\phi_3 : \Sigma_{q-1}^n \hookrightarrow \Sigma_q^n$ to be the inclusion map. □

5.4 Bounds on Constant-Weight Codes

Let $A(n,2\delta,w)$ be the maximum possible number of codewords in a *binary* code of length n, constant weight w, and minimum distance 2δ [2,15].

Let $K(n,2\delta,w)$ be the graph whose vertices are all the binary sequences of length n and weight w and where two vertices u,v are adjacent iff $d(u,v) \ge 2\delta$. In analogy with Hamming graphs, we observe here the following:

Observation 5.2 $A(n,2\delta,w) = \omega(K(n,2\delta,w))$.

Let $\binom{[n]}{w}$ denote the set of all subsets of $[n] = \{1,2,\dots,n\}$ of order w. There is a natural bijection ν between $V(K(n,2\delta,w))$ and $\binom{[n]}{w}$, namely, $\forall u \in V(K(n,2\delta,w))$, $\nu(u) = U = \{i; u(i) = 1\}$.

Lemma 5.10. *$\forall p,q \in V(K(n,2\delta,w))$, $p \sim q$ iff $|P \cap Q| \le w-\delta$ where $P = \nu(q)$ and $Q = \nu(q)$.*

Proof. $2\delta \le d(p,q) = |(P \cap \bar{Q}) \cup (\bar{P} \cap Q)| = 2w - 2|P \cap Q|$. □

Lemma 5.11. *$K(n,2\delta,w)$ is vertex transitive.*

Proof. For any two vertices p,q of $K(n,2\delta,w)$, any bijection on $[n]$ such that the image of $P = \nu(p)$ is $Q = \nu(q)$, takes p to q and belongs to $Aut(K(n,2\delta,w))$. □

The first result that follows directly from Lemma 5.11 is the Bassalygo–Elias inequality [15] which relates the maximum size of constant-weight codes to that of unrestricted codes.

Lemma 5.12 (Bassalygo–Elias inequality).

$$A(n,d) \le \frac{2^n}{\binom{n}{w}} A(n,d,w)$$

Proof. Consider the two graphs $Y = \bar{H}(n,d)$ and $X = \bar{K}(n,d,w)$. Y is vertex transitive. Since X is an induced subgraph of Y, the inclusion map is a homomorphism that takes X to Y. The result then follows from applying Theorem 5.4. □

By the same token, we can show the following inequalities. The first two are known as Johnson's bounds [16].

Lemma 5.13.

$$A(n, d, w) \leq \frac{n}{w} A(n-1, d, w-1) \tag{5.8}$$

$$A(n, d, w) \leq \frac{n}{n-w} A(n-1, d, w) \tag{5.9}$$

$$A(n, d, w) \geq \frac{n-w+1}{w} A(n, d, w-1) \tag{5.10}$$

$$A(n, d, w) \geq \frac{w+1}{n-w} A(n, d+2, w+1) \tag{5.11}$$

Proof. We start by giving a proof for the first Johson's bound in (5.8) using the techniques described in this chapter. Consider the two graphs $X = \bar{K}(n-1, d, w-1)$ and $Y = \bar{K}(n, d, w)$ and let ϕ be a map from $V(X)$ to $V(Y)$ such that for any sequence u of length n and weight w, $\phi(u)$ is obtained by appending a '1' to u. It is easy to see that $d(\phi(u), \phi(v)) = d(u, v)$. Therefore, if u and v are connected in X, *i.e.*, $d(u, v) < d$, $\phi(u)$ and $\phi(v)$ are connected in Y. Therefore, ϕ is a graph homomorphism from X to Y and (5.8) follows directly from Theorem 5.4. Equation (5.9) can be obtained similarly by considering the homomorphism from $\bar{K}(n-1, d, w)$ to $\bar{K}(n, d, w)$ that appends a '0' any sequence of length n and weight w.

To prove (5.10), consider the mapping ϕ' from $V(Y)$ to $V(Z)$, where $Z = \bar{K}(n, d, w-1)$, defined as follows: for any $u \in V(Y)$, $\phi'(u)$ is obtained by setting the first occurrence of '1' in u to '0'. Therefore, for any $u, v \in V(Y)$, we have $d(\phi'(u), \phi'(v)) \leq d(u, v)$. Hence, ϕ' is a graph homomorphism from Y to Z and (5.10) is recovered by Theorem 5.4. Equation (5.11) is obtained in a similar way by considering the homomorphism from $\bar{K}(n, d, w)$ to $\bar{K}(n, d+2, w+1)$ that sets the first occurence of a '0' in a sequence to '1'. □

Next, we use the vertex transitivity property of the graphs $K(n, 2\delta, w)$ to rederive a number of upper bounds on $A(n, d, w)$.

Lemma 5.14. *Let* $t = w - \delta + 1$, *then*

$$A(n, 2\delta, w) \leq \frac{\binom{n}{w}}{\binom{n-t}{w-t}} \tag{5.12}$$

Proof. Let $G = K(n, d, w)$. Since G is vertex transitive, we have

$$A(n, 2\delta, w)\alpha(G) \leq |V(G)| = \binom{n}{w}.$$

Define $M(n,w,s)$ as in [3] to be the maximum number of subsets of $[n]$ of order w that intersect pairwise in at least s elements. By Lemma 5.10, $\alpha(G) = M(n,w,t)$. But, $M(n,w,t) \geq \binom{n-t}{w-t}$ (for instance, consider the system of all subsets of $[n]$ of order w that contain the set $\{1,2,\ldots,t\}$). □

The previous bound is the same as the one in Theorem 12 in [15], which was given there with a different proof. One can improve on the bound of Lemma 5.14 by using the exact value of $M(n,w,t)$ [3]. It is known that for $n \geq (w-t+1)(t+1)$, the famous Erdős-ko-Rado theorem [18] holds and $M(n,w,t) = \binom{n-t}{w-t}$ [18, 19]. However, this is not the case for lower values of n.

Lemma 5.15. *Let $t = w - \delta + 1$ and $r = \max\{0, \lceil \frac{\delta(w-\delta)}{n-d} - 1 \rceil\}$, then*

$$A(n,2\delta,w) \leq \frac{\binom{n}{w}}{\sum_{i=t+r}^{w} \binom{t+2r}{i}\binom{n-t-2r}{w-i}}; \tag{5.13}$$

with $\binom{n}{k} = 0$ when $k > n$.

Proof. (sketch) $A(n,d,w) \leq \frac{\binom{n}{w}}{M(n,w,t)}$, then use the exact value of $M(n,w,t)$ given by the main theorem of [3]. □

5.5 Conclusion

We illustrated in this paper the use of graph theoretic techniques to answer a fundamental problem in coding theory, that is determining the maximal size of codes of a certain length and a given minimum Hamming distance. Inequalities involving the independence and clique numbers of general and vertex transitive graphs are shown to lead to many well-known bounds on codes, such as the Hamming, Singleton, Gilbert–Varshamov, and Bassalygo–Elias bounds. Additional interesting applications were omitted here due to space restriction. For instance, advanced results in extremal graph theory were used in [20] to get asymptotic improvement on the Gilbert–Varshamov bound. Furthermore, graph theory has many applications in modern coding theory as in the design of Low Density Parity Check (LDPC) codes and the design of their iterative decoders [21].

References

1. Macwilliams FJ, Sloane NJA (1977) The theory of error-correcting-codes. North-Holland, Amsterdam
2. Sloane NJA (1989) Unsolved problems in graph theory arising from the study of codes. Graph Theory Notes of New York 18:11-20
3. Ahlswede R, Khachatrian LH (1997) The complete intersection theorem for systems of finite sets. Eur J Combinator 18:125–136

4. Ahlswede R, Khachatrian LH (1998) The diametric theorem in Hamming spaces - optimal anticodes. Adv Appl Math 20:429–449
5. Frankl P, Tokushige N (1999) The Erdős-Ko-Rado theorem for integer sequences. Combinatorica 19:55–63
6. Diestel R (2006) Graph theory. Springer, New York
7. Godsil C, Royle G (2001) Algebraic graph theory. Springer, New York
8. van Lint JH, Wilson RM (2001) A course in combinatorics. Cambridge University Press, United Kingdom
9. Ahlswede R (2001) On perfect codes and related concepts. Designs Codes Cryptography 22:221–237
10. Kleitman DJ (1966) On a combinatorial conjecture of Erdős. J Combin Theor 1:209–214
11. El Rouayheb S, Georghiades CN, Soljanin E, Sprintson A (2007) Bounds on codes based on graph theory. Int Symp Inform Theor Nice
12. Matsumoto R, Kurosawa K, Itoh T, Konno T, Uyematsu T (2006) Primal-dual distance bounds of linear codes with applications to cryptography. IEEE Trans Inform Theor 52:4251–4256
13. Tolhuizen L (1997) The generalized Gilbert–Varshamov bound is implied by turàn's theorem. IEEE Trans Inform Theor 43:1605–1606
14. Delsarte P (1973) An algebraic approach to association schemes of coding theory. Phillips J Res 10
15. Agrell E, Vardy A, Zeger K (2000) Upper bounds for constant-weight codes. IEEE Trans Inform Theor 46:2373–2395
16. Johnson SM (1962) A new upper bound for error-correcting codes. IRE Trans Inform Theor IT-8:203–207
17. Levenshtein VI (1974) Upper bound estimates for fixed-weight codes. Probl Inform Trans 7:281–287
18. Erdős P, Ko C, Rado R (1961) Intersection theorems for systems of finite sets. Quart J Math Oxford 12:313–320
19. Wilson RM (1984) The exact bound on the Erdos-ko-Rado Theorem. Combinatorica 4:247–260
20. Jiang T, Vardy A (2000) Asymptotic improvement of the Gilbert–Varshamov bound on the size of binary codes. IEEE Trans Inform Theor 50(8):1655–2395
21. Richardson T, Urbanke R (2008) Modern coding theory. Cambridge University Press, Cambridge

Chapter 6
The Statistics of the Atomic Clock Noise

Lorenzo Galleani

6.1 Introduction

Precise timing has several applications, and their number is steadily increasing. Atomic clocks are the main actors of this phenomenon, a consequence of the fact that they provide the most accurate time measurements currently available. By definition, the accuracy of a clock is its deviation from the Coordinated Universal Time (UTC), the time standard based on the International Atomic Time (TAI), which is obtained by averaging more than 300 atomic clocks throughout the world [1]. The most precise atomic clock presently available guarantees a time error in the determination of the second that is equivalent to ± 1 s in 10 million years.

Because of the performance of atomic clocks, previously unthinkable applications are now a reality, such as satellite navigation. In Global Navigation Satellite Systems (GNSSs), such as GPS and the future European Galileo system, position is determined through the estimation of the distances between the user and the satellites in view [2, 3]. Such distances are obtained by multiplying the speed of light by the time of flight that signals require to travel from the satellites to the user. The position of the user is estimated from the measured distances by using a trilateration method. An error in time implies therefore an error in position. For example, 1 ns of error in time results in a positioning error of 0.3 m. For this reason, GNSSs use atomic clocks onboard the satellites.

Atomic clocks have several other applications. In the metrological field, for instance, they are used to define an increasing number of fundamental constants. An interesting application is also time stamping, which reduces the risk of frauds by precisely dating events, such as banking transactions [4].

L. Galleani (✉)
Politecnico di Torino, Corso Duca degli Abruzzi 24, 10129 Torino, Italy
e-mail: galleani@polito.it

L. Cohen et al. (eds.), *Classical, Semi-classical and Quantum Noise*,
DOI 10.1007/978-1-4419-6624-7_6,

In this chapter, we discuss the statistics of atomic time, and we review the main tools used for its characterization. First, in Sect. 6.2 we introduce the quantities that describe atomic time. Then in Sect. 6.3 we show how frequency analysis is used to characterize clock noise. Finally, in Sect. 6.4 we review the fundamental concept of clock stability. In Sects. 6.3–6.4, we also discuss how frequency and stability analysis can be extended to deal with nonstationary clock behaviors, which are often due to the occurrence of anomalies.

6.2 The Atomic Clock Noise

An atomic clock is made essentially by two parts, namely the oscillator and the counter [1]. In the ideal clock, the oscillator generates a perfect sinusoidal signal

$$u(t) = U_0 \sin(2\pi \nu_0 t), \tag{6.1}$$

where U_0 is the amplitude of the oscillation, and ν_0 is the frequency [5, 6]. If we introduce the *ideal clock reading* $h_0(t)$

$$h_0(t) = t, \tag{6.2}$$

then the clock signal becomes

$$u(t) = U_0 \sin(2\pi \nu_0 h_0(t)). \tag{6.3}$$

The counter accumulates the cycles generated by the oscillator, and it provides an approximation to the ideal clock reading $h_0(t)$. In reality, the sinusoidal signal $u(t)$ is affected by statistical fluctuations, and consequently a more realistic model of the clock signal is given by

$$u(t) = (U_0 + \epsilon(t)) \sin(2\pi \nu_0 t + \varphi(t)). \tag{6.4}$$

The amplitude fluctuations $\epsilon(t)$ are, in general, negligible, and hence the clock model can be simplified to

$$u(t) = U_0 \sin(2\pi \nu_0 t + \varphi(t)), \tag{6.5}$$

which can be written as

$$u(t) = U_0 \sin(2\pi \nu_0 h(t)), \tag{6.6}$$

where we define the *clock reading* $h(t)$ as

$$h(t) = h_0(t) + \frac{\varphi(t)}{2\pi \nu_0}. \tag{6.7}$$

The difference between the clock reading and the ideal clock reading is the *phase deviation*

$$x(t) = h(t) - h_0(t). \tag{6.8}$$

By using (6.7), the phase deviation can be written as

$$x(t) = \frac{\varphi(t)}{2\pi\nu_0}. \tag{6.9}$$

Since $\varphi(t)$ is dimensionless and ν_0 is in hertz, then $x(t)$ has the dimension of seconds.

We note that the oscillation frequency ν_0 of the ideal clock is constant with time, while in the real clock it changes with time due to the statistical fluctuations of the phase $\varphi(t)$. The instantaneous frequency of the oscillation is in fact obtained as

$$\nu(t) = \frac{1}{2\pi}\frac{\mathrm{d}}{\mathrm{d}t}\Phi(t), \tag{6.10}$$

where

$$\Phi(t) = 2\pi\nu_0 h(t). \tag{6.11}$$

The variation of the oscillation frequency with respect to the nominal value ν_0 is a fundamental quantity for atomic clocks. In particular, we define the *normalized frequency deviation* as

$$y(t) = \frac{\nu(t) - \nu_0}{\nu_0}. \tag{6.12}$$

By using (6.7), (6.10), and (6.11) we obtain

$$y(t) = \frac{\mathrm{d}x(t)}{\mathrm{d}t}. \tag{6.13}$$

The experimental observations show that $x(t)$ and $y(t)$ exhibit a noise-like behavior, and they are hence referred to as *clock noise*. Several types of noise are actually present in the phase and frequency deviation, but a dominant role is played by random walk-like noise. If, in fact, we synchronize a clock to a time reference with higher accuracy, we experimentally observe that the clock accumulates a deviation from the reference, which grows with time with a random walk nature. Aside from random walks, other types of noise are experimentally observed in atomic clocks. Five types of noise are in general identified in atomic clocks [7], namely, white phase noise (WPN), flicker phase noise (FPN), white frequency noise (WFN), flicker frequency noise (FFN), and random walk frequency noise (RWFN). In Cesium and Rubidium clocks, white frequency and random walk frequency noise represent the dominant clock noise, while in Maser clocks also flicker frequency noise can be observed. In Fig. 6.1 we show the typical frequency deviation $y(t)$ of a Cesium clock, obtained through numerical simulations performed by using

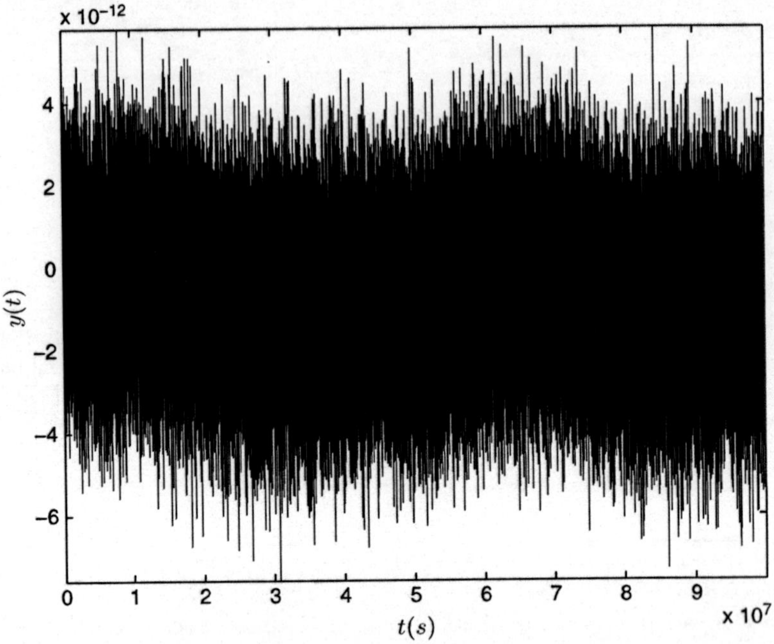

Fig. 6.1 Frequency deviation of a typical Cesium clock. The statistical fluctuations about the mean value are due to the white noise component. The erratic movement of the mean value is a consequence of the random walk component

the two-state model of the atomic clock noise [8]. The instantaneous random fluctuations are due to the white noise component, while the erratic behavior of the mean value is due to the random walk component.

In addition to noise, a deterministic component is also present in the clock phase and frequency deviations. In Cesium and Rubidium clocks, this component is represented by a linear phase drift

$$x_d(t) = x_0 + y_0 t. \tag{6.14}$$

In Maser clocks, a linear frequency component is observed

$$y_d(t) = y_0 + \mathrm{d}t, \tag{6.15}$$

which results in the well known quadratic phase drift

$$x_d(t) = x_0 + y_0 t + \frac{1}{2}\mathrm{d}t^2. \tag{6.16}$$

The constants x_0, y_0, and d can be either estimated from the clock data, or obtained from the specifications given by the clock manufacturer.

6.3 Frequency Analysis

The Fourier spectrum is a fundamental technique for the characterization of clock noise. In the power spectral domain, the five types of clock noise have a power-law representation

$$P_y(f) = kf^{\alpha}, \tag{6.17}$$

where k is a proportionality constant, and the exponent a determines the type of noise. In Table 6.1, we give $P_y(f)$ for the five types of clock noise. The power spectrum is usually plotted in bi-logarithmic scaling, since in this representation the typical clock noises are represented by straight lines. By taking the logarithm of both sides of (6.17), we in fact obtain

$$\log P_y(f) = \log k + \alpha \log f. \tag{6.18}$$

Therefore, the behavior of $\log P_y$ as a function of $\log f$ is a straight line with slope α.

Every type of atomic clock has a characteristic spectral signature, which can be used to establish the performance of the clock and to reveal possible anomalous behaviors. The solid curve shown in Fig. 6.2 represents, for instance, the typical power spectrum of a Cesium clock, obtained by applying Welch's periodogram [9, 10] to a simulated frequency deviation made by $N_1 = 10^4$ samples. The simulations are performed by using the two-state model, with a set of parameters that emphasizes the random walk component of the frequency deviation. It is interesting to see what happens when we estimate the power spectrum on a longer time series. The dashed line in Fig. 6.2 corresponds to the Welch's periodogram evaluated on a time series with $N_2 = 10^6$ samples, where the first $N_1 = 10^4$ samples are the same used before. Since we estimate the power spectrum on a longer time series, we expect the result to be close to the initial power spectrum, although with a lower variance due to the increased number of available samples. In reality, we see that the new spectrum differs from the initial one, and the difference increases as the frequency becomes smaller. The reason of such difference is that the frequency deviation $y(t)$ is a nonstationary random process. The frequency deviation is in fact made by a random walk component plus a white noise. The random walk noise is nonstationary, and its power spectrum is proportional to f^{-2}. The singularity in $f = 0$ makes this random process nonstationary, and consequently its frequency

Table 6.1 Power spectrum $P_y(f)$ and Allan variance $\sigma_y^2(\tau)$ of clock noise

Noise type	Acronym	$P_y(f)$	$\sigma_y^2(\tau)$
White Phase Noise	WPN	f^2	τ^{-2}
Flicker Phase Noise	FPN	f^1	τ^{-2}
White Frequency Noise	WFN	f^0	τ^{-1}
Flicker Frequency Noise	FFN	f^{-1}	τ^0
Random Walk Frequency Noise	RWFN	f^{-2}	τ^1

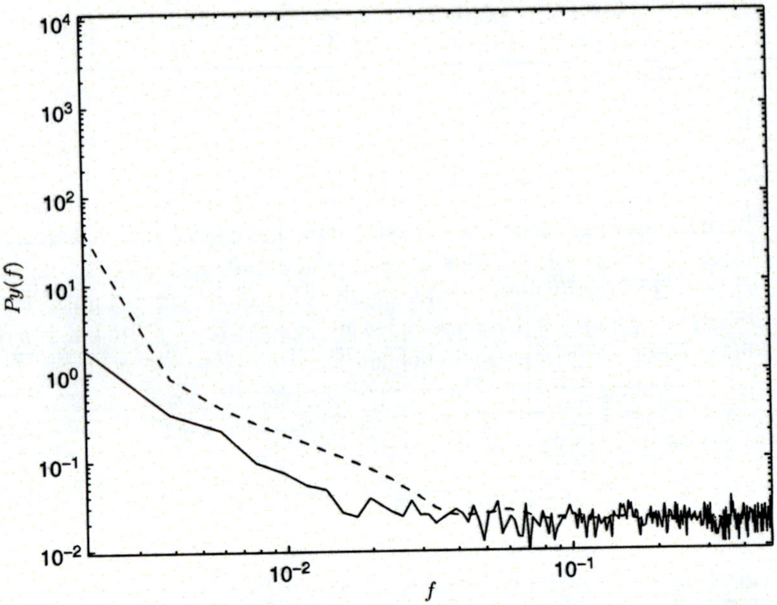

Fig. 6.2 Power spectra of the frequency deviation of a Cesium clock obtained by using Welch's periodogram. The *solid line* represents the power spectrum computed on $N_1 = 10^4$ samples, while the *dashed line* corresponds to a time series with $N_2 = 10^6$ samples (We use dimensionless units)

content varies with time. To represent this time variation, we use time-frequency analysis [11]. We define the time-frequency spectrum of the frequency deviation $y(t)$ through the Wigner spectrum [12]

$$\overline{W}_y(t, f) = \frac{1}{2\pi} \int_{-\infty}^{+\infty} E\left[y^*(t - \tau/2)y(t + \tau/2)\right] e^{-i2\pi f\tau} d\tau, \tag{6.19}$$

which is the expected value of the Wigner distribution [13]. The Wigner spectrum of a random walk frequency noise is given by [14]

$$\overline{W}_y(t, f) = \frac{1}{\pi} \frac{\sin^2 2\pi f t}{(2\pi f)^2} \tag{6.20}$$

for $t \geq 0$. In Fig. 6.3, we show $\overline{W}_y(t, f)$. As it can be noticed, the spectrum increases with time in agreement with the numerical result shown in Fig. 6.2.

Time-frequency analysis can be used to locate possible anomalies in the clock behavior. An anomaly occurs when the clock behavior deviates from the specifications. Therefore, an anomaly produces a change in the clock noise, and consequently in the time-frequency spectrum. Unfortunately, as demonstrated by Fig. 6.2, the time-frequency spectrum changes with time even for the typical clock noises, due to their random walk nature. If a clock behaves according to the specifications,

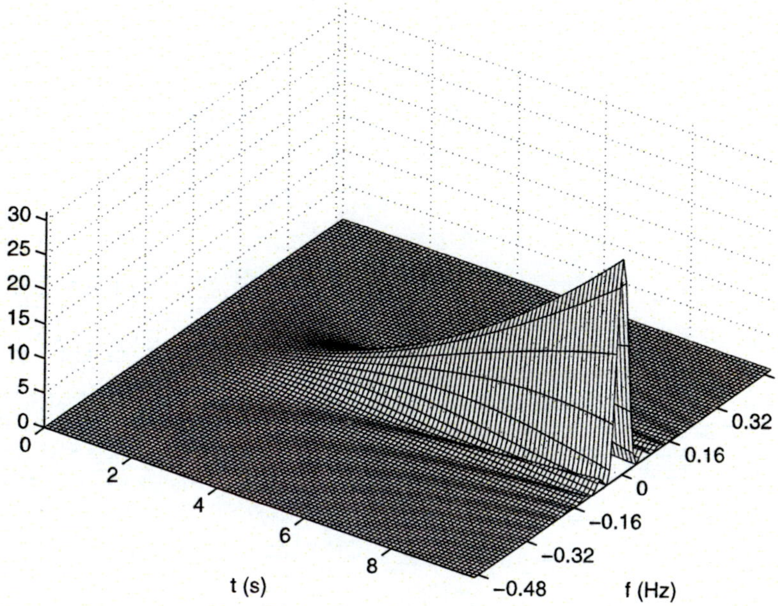

Fig. 6.3 Wigner spectrum of a random walk frequency noise

that is without anomalies, we instead expect its time-frequency spectrum to be constant with time. This problem can be solved by taking the derivative $\Delta y(t)$ of the frequency deviation [15]

$$\Delta y(t) = \frac{\mathrm{d}y(t)}{\mathrm{d}t}. \tag{6.21}$$

The derivative eliminates the singularity in the power spectrum of the clock noise, and it makes the resulting random process $\Delta y(t)$ stationary with time. The time-frequency spectrum $P_{\Delta y}(t, f)$ can be subsequently estimated from the available measurements by using a sliding Welch spectrum. The sliding Welch spectrum is obtained by sliding the Welch's periodogram on the clock data.

In Fig. 6.4, we show the sliding Welch spectrum of the clock noise given in Fig. 6.1. We now see that the spectrum does not increase with time, the only variations being due to the statistical fluctuations of the estimation process. Note that the derivative changes the spectral slopes of the clock noise, since it corresponds to multiplying the original power spectrum by f^2. In Fig. 6.5, we show the sliding Welch spectrum of the sum of a frequency deviation obtained with the identical model used to generate Fig. 6.1, plus a short duration sinusoid. Since the sinusoidal component has a small amplitude, the frequency deviation is practically undistinguishable from Fig. 6.1, and hence we do not show it here. We note that the sliding Welch spectrum reveals the sinusoid, whose approximate time

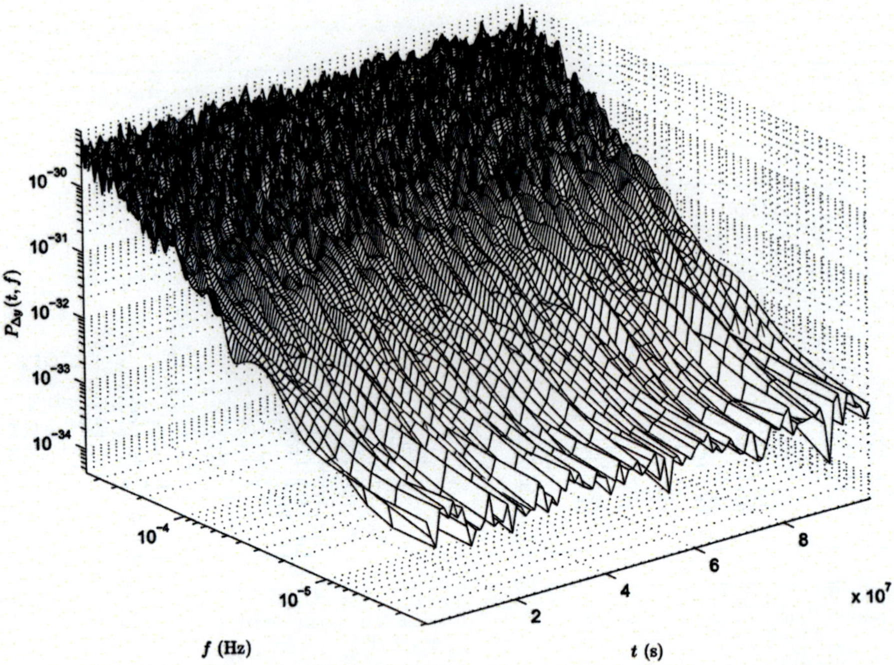

Fig. 6.4 Sliding Welch spectrum $P_{\Delta y}(t, f)$ of the frequency deviation $y(t)$ shown in Fig. 6.1

support can be estimated as $3 \times 10^7\text{s} \leq \text{t} \leq 7 \times 10^7\text{s}$, and whose period roughly equals $1/(10^{-4}\text{Hz}) = 10^4\text{s}$. The presence of sinusoidal components is a typical phenomenon for atomic clocks, especially in space applications [16].

6.4 Stability Analysis

Stability is a key property of atomic clocks. A stable clock has a frequency deviation $y(t)$ which changes slowly with time. To illustrate this concept, we consider a simple example. Suppose that we have an atomic clock, which accumulates a phase deviation of $\Delta x = 10$ ns per day, with respect to an ideal time reference. Its normalized frequency deviation is therefore approximately given by

$$y = \frac{\Delta x}{\Delta t} \simeq \frac{10 \times 10^{-9}\ \text{s}}{10^5\ \text{s}} = 10^{-13}, \tag{6.22}$$

where we use (6.13), and we approximate the $\Delta t = 86,400$ s in a day by $\Delta t \simeq 10^5$ s. The dashed line in Fig. 6.6 represents the frequency deviation, which is constant with time. This clock is considered to be perfectly stable, because, even

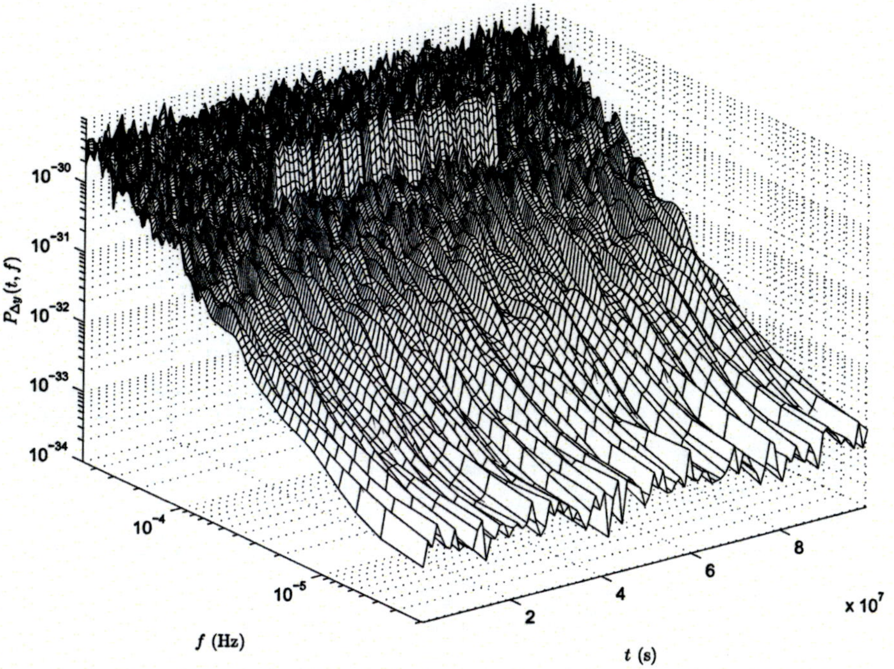

Fig. 6.5 Sliding Welch spectrum $P_{\Delta y}(t, f)$ of the sum of a frequency deviation generated with the same model used for Fig. 6.1, plus a short duration sinusoid

though it steadily accumulates an error with respect to the ideal time, this error is exactly predictable, and it can be removed from the clock time. The process of subtracting the predicted deviation is the *clock correction*.

In reality, as discussed before, the frequency deviation of a clock has a random behavior, represented in Fig. 6.6 by the solid line. The smaller the variability with time of the frequency deviation, the larger the clock stability. The ideal case is represented by the flat dashed line in Fig. 6.6. Therefore, we could define the clock stability through the variance of the frequency deviation. Unfortunately, as discussed in Sect. 6.2, the frequency deviation is a random process that always contains a random walk component. The variance of a random walk increases with time, and eventually diverges when $t \to \infty$. The fact that the variance of the clock noise is a function of time makes this definition of stability impractical.

The *Allan variance* overcomes this problem [17]. The key idea is to evaluate the variance of the derivative of the frequency deviation. The derivative stationarizes the clock noise, and, consequently, the Allan variance is no longer a function of time. The Allan variance is defined as

$$\sigma_y^2(\tau) = \frac{1}{2}\left\langle (\bar{y}(t+\tau) - \bar{y}(t))^2 \right\rangle, \tag{6.23}$$

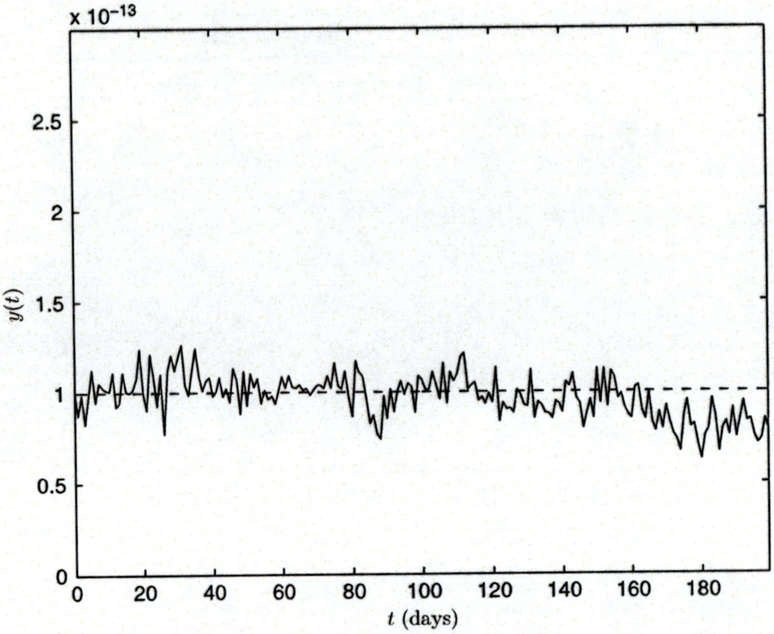

Fig. 6.6 The dashed line represents the frequency deviation of an ideal clock, which is perfectly predictable and, therefore, which can be exactly corrected. The solid line represents instead a real clock, whose frequency deviation has a random walk nature

where τ is the *observation interval*, $\bar{y}(t)$ is the average frequency deviation defined as

$$\bar{y}(t) = \frac{1}{\tau}\int_{t-\tau}^{t} y(t')\mathrm{d}t' \tag{6.24}$$

and $\langle\rangle$ indicates a time average

$$\left\langle(\bar{y}(t+\tau) - \bar{y}(t))^2\right\rangle = \lim_{T\to\infty}\frac{1}{T}\int_{-\frac{T}{2}}^{\frac{T}{2}} (\bar{y}(t+\tau) - \bar{y}(t))^2\,\mathrm{d}t. \tag{6.25}$$

By using (6.13) the Allan variance can be written with respect to the phase deviation

$$\sigma_y^2(\tau) = \frac{1}{2\tau^2}\left\langle(x(t+\tau) - 2x(t) + x(t-\tau))^2\right\rangle. \tag{6.26}$$

In practice, the *Allan deviation* $\sigma_y(t)$ is often used, simply defined as the square root of the Allan variance. The Allan deviation is the standard definition of stability for an atomic clock [18], and it is used by the clock manufacturers to certify the quality of their products. Similarly to the power spectrum, also the Allan variance of the typical clock noises has a power-law behavior, which is shown in Table 6.1.

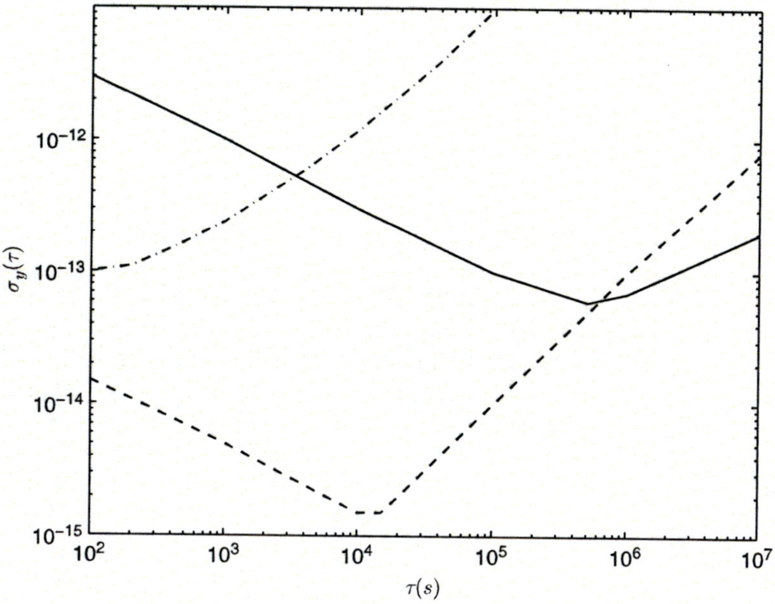

Fig. 6.7 Allan deviations of the most common precise oscillators. The *solid line* represents a Cesium clock, the *dashed line* a Hydrogen Maser, and the *dash dotted line* a quartz clock

Note that white phase noise and flicker phase noise have the same Allan variance. A modified Allan variance can be introduce to discriminate between these two noise types [19].

In Fig. 6.7, we show the Allan deviation of the main atomic clocks, and, for comparison, of a quartz clock. Note that in the bi-logarithmic scaling the atomic clock noise is represented by straight lines. The solid line represents a typical Cesium clock, which is noisy in the short term, that is for small values of the observation interval τ, but highly stable in the long term. The dashed line indicates a Hydrogen Maser, which has the highest stability in the short term, while in the long term shows a lower performance than the Cesium clock. The dash dotted line refers to a quartz oscillator, which has a good stability in the short term, but it is outperformed by atomic clocks in the medium and long term.

Consider now the atomic clock whose frequency deviation is shown in Fig. 6.8. As it can be seen, at the middle of the time series the variance of the clock noise has a sudden increase. Such variation represents an anomaly in the clock behavior, which can be due to several factors, such as to a sudden increase in temperature. If the variance of the clock noise changes with time, the same phenomenon must happen to the clock stability. It is hence necessary to introduce a quantity that represents the behavior of the clock stability as a function of time. The *dynamic Allan variance* (DAVAR) has been recently developed for that reason [20]. The DAVAR is obtained by sliding the Allan variance on the clock data. At every time

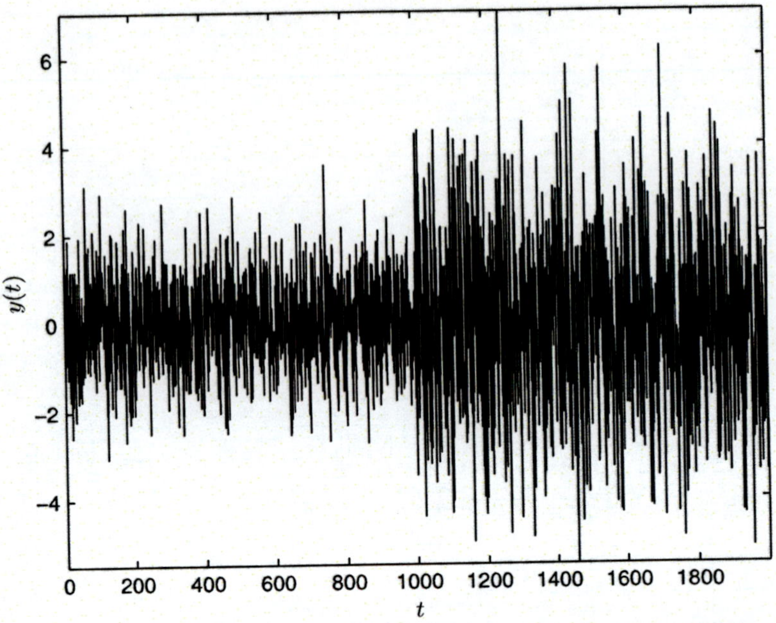

Fig. 6.8 Simulated frequency deviation of an atomic clock. The sudden increase in variance located at $t = 1,000$ represents an anomalous behavior (We use dimensionless units)

instant t the DAVAR $\sigma_y^2(t,\tau)$ is the Allan variance of the frequency deviation $y(t)$ restricted to the interval $(t - T_w/2, t + T_w/2)$, where T_w is the DAVAR window length. By using (6.26), the DAVAR can be computed from the samples $x(t_n)$ of the phase deviation

$$\sigma_y^2(t_n,\tau_k) = \frac{1}{2\tau_k^2}\frac{1}{N_w - 2k}\sum_{m=n-N_w/2+k}^{n+N_w/2-k-1} E\left[(x(t_{m+k}) - 2x(t_m) + x(t_{m-k}))^2\right], \tag{6.27}$$

where $t_n = n\tau_0$, $\tau_k = k\tau_0$, $T_w = N_w\tau_0$, and τ_0 is the sampling time. When N_w is even, the discrete-time observation interval k takes the values $k = 1, 2, \ldots, N_w/2-1$. The operator E is the expectation value, which makes $\sigma_y^2(t,\tau)$ a deterministic quantity. When one realization only of the phase deviation is available, as happens with experimental data, we can estimate the DAVAR by

$$\hat{\sigma}_y^2(t_n,\tau_k) = \frac{1}{2\tau_k^2}\frac{1}{N_w - 2k}\sum_{m=n-N_w/2+k}^{n+N_w/2-k-1} (x(t_{m+k}) - 2x(t_m) + x(t_{m-k}))^2 . \tag{6.28}$$

In practice, we often use the *dynamic Allan deviation* (DADEV) $\sigma_y(t,\tau)$, which is the square root of the DAVAR. In Fig. 6.9, we show the DADEV of the frequency deviation given in Fig. 6.8. We immediately spot the change in variance located at

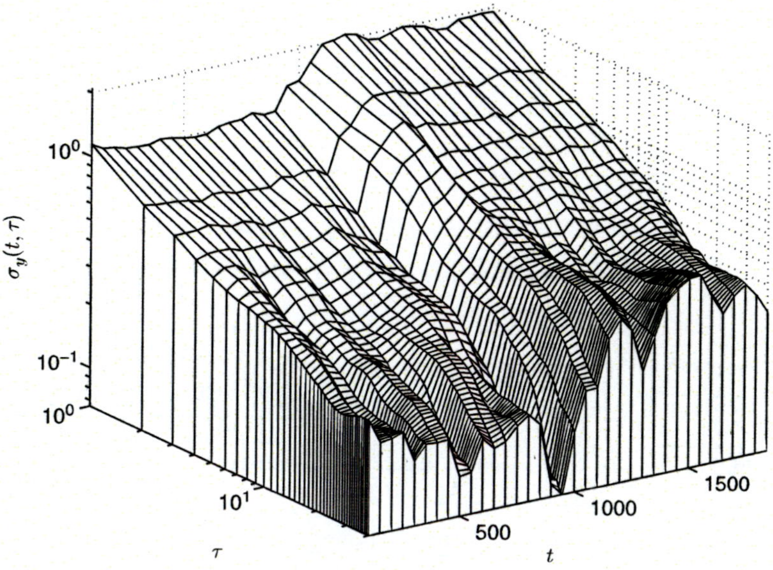

Fig. 6.9 Dynamic Allan deviation of the clock noise shown in Fig. 6.8. The increase in the DADEV surface around $t = 1{,}000$ detects the variance increase of the clock frequency deviation (We use dimensionless units)

$t = 1{,}000$. Before and after the discontinuity, the DADEV surface is stationary with time, aside from some statistical fluctuations due to the estimation process. The dynamic Allan deviation can be hence used to detect and to identify the typical anomalies of an atomic clock [21]. For this reason, it is used as a detection tool in the future European Galileo system [22, 23].

6.5 Conclusions

The deviation of the clock time with respect to a time reference is referred to as the clock noise, due to its stochastic properties. The statistical characterization of the clock noise is a fundamental problem for several reasons. First, by analyzing the clock noise we can ascertain when the clock behaves according to the specifications, and when it actually experiences an anomaly. Anomalous behaviors determine a change in the clock noise, and hence in the frequency spectrum and in the clock stability. Time-frequency analysis and the dynamic Allan variance can reveal such changes, and therefore they can be used to detect possible clock anomalies [15, 20]. Second, the statistical knowledge of the clock noise allows the development of reliable clock models. The two-state model, for example, is used in the Composite Clock algorithm to generate the time reference of the Global Positioning System [7, 8, 24]. Finally, by knowing the statistics of the clock noise we can correct it by subtracting the predictable part of its deviation. The corrected clock has superior performance than the original, uncorrected, one.

These applications prove that statistical signal processing methods represent an effective approach for the use and understanding of atomic time.

Acknowledgements This work was supported by the PRIN 2007 program.

References

1. Allan DW, Ashby N, Hodge CC (1997) The science of timekeeping. Hewlett Packard Application Note 1289
2. Parkinson BW, Spilker JJ Jr. (1996) Global positioning system: theory and applications. American Institute of Aeronautics and Astronautics, Washington
3. ESA webpage on the Galileo satellites. Available at http://www.esa.int/esaNA/galileo.html
4. Haber S, Stornetta WS (1991) How to time-stamp a digital document. J Cryptology 3(2):99–111
5. Kartaschoff P (1978) Frequency and time. Academic Press
6. Sullivan DB, Allan DW, Howe DA, Walls FL (1990) Characterization of clocks and oscillators. National Institute of Standards and Technology, Technical Note 1337
7. Galleani L, Sacerdote L, Tavella P, Zucca C (2003) A mathematical model for the atomic clock error. Metrologia 40:S257–S264
8. Galleani L (2008) A tutorial on the two-state model of the atomic clock noise, Metrologia 45:S175–S182
9. Percival DB (2006) Spectral analysis of clock noise: A primer. Metrologia 43(4):S299–S310
10. Papoulis A, Pillai SU (2002) Probability, random variables and stochastic processes. McGraw-Hill
11. Cohen L (1995) Time-frequency analysis. Prentice-Hall
12. Mark WD (1986) Power spectrum representation for nonstationary random vibration. In: Ielishakoff DH (ed) Random vibration: status and recent developments. Elsevier,
13. Wigner EP (1932) On the quantum correction for thermodynamic equilibrium. Phys Rev 40:749–759
14. Galleani L, Cohen L (2004) Direct time-frequency characterization of linear systems governed by differential equations. IEEE Signal Process Lett 11(9):721–724
15. Galleani L (2008) Detection of changes in clock noise using the time-frequency spectrum, Metrologia 45:S143–S153
16. Sesia I, Galleani L, Tavella P (submitted) Monitoring clocks in space systems with the dynamic Allan variance, IEEE Trans Aero Electron Syst
17. Allan DW (1966) Statistics of atomic frequency standards. Proc IEEE 5:221–230
18. IEEE 1139 Standard (1999) Definitions of Physical Quantities for Fundamental Frequency and Time Metrology - Random Instabilities The Institute of Electrical and Electronics Engineers, Inc.
19. Allan DW, Barnes JA (1981) A modified Allan variance with increased oscillator characterization ability. Proc 35th Ann Freq Control Symposium, Ft. Monmouth, NJ, USA, May 1981, pp. 470–474
20. Galleani L, Tavella P (2009) The dynamic Allan variance. IEEE Trans Ultra Ferro Freq Contr 56(3):450–464
21. Galleani L, Tavella P (2008) Detection and identification of atomic clock anomalies. Metrologia 45:S127–S133
22. Zanello R, Mascarello M, Tavella P. Galleani L, Detoma E, Bellotti A (2007) The Galileo precise timing facility, IEEE FCS-EFTF, Geneva, Switzerland

23. Sesia I, Galleani L, Tavella P (2007) Implementation of the dynamic Allan variance for the Galileo system test bed V2, IEEE EFTF-FCS, Geneva, Switzerland
24. Brown KR (1991) The theory of the GPS composite clock, ION GPS-91, 11–13 September Albuquerque, NM, USA, pp. 223–241

Chapter 7
Effect of Noise on Quantized Adiabatic Charge Transport in 2D Electron Systems and Nanotubes

Godfrey Gumbs

7.1 Introduction

A considerable amount of effort has been devoted to the study and successful operation of devices that employ the principles of quantized adiabatic charge transport [1–12]. The goal is to produce a device capable of delivering N electrons (or holes) in each cycle of a moving quantum dot in a controlling signal, thus giving rise to a current $I = Nef$, where f is the frequency of the signal [3–9]. However, so far, only the surface acoustic wave (SAW) on a piezoelectric substrate [3–5] and the recently developed GHZ single-electron pumps [9, 10] operate at high enough frequencies for the measured current to be suitable as a current standard. The accuracy of the quantized current on the plateaus, thus far, is one part in 10^4 for the SAW and GHz single-electron pumps [3–5, 9]. Robinson and Talyanskii [13] measured the noise produced by an approximately 3 GHz SAW single-electron pump. When the current is close to the quantized plateau corresponding to one electron being transported per SAW cycle, the noise in the current is dominated by the shot noise. Away from the plateau, i.e., either above or below a specified one under consideration, the noise is due to electron traps in the material. There are some suggestions on how to reduce the noise in single-electron pumps, as this would have important applications in setting an alternative and universal means of establishing a standard for charge. The flatness of the quantized current plateaus is one part in 10^4 and, as such, makes it a rather crude standard of measurement. Its operation needs to be further refined in order for it to be accepted as a standard that seeks an accuracy of one part in 10^8. Despite this, there are proposals to use the SAW and high frequency GHz single-electron pumps in single photon sources, single photon detectors, quantum cryptography and quantum computation [13–16].

G. Gumbs (✉)
Physics Department, Hunter College of the City University of New York, 695 Park Avenue, New York, NY 10065, USA
e-mail: ggumbs@hunter.cuny.edu

L. Cohen et al. (eds.), *Classical, Semi-classical and Quantum Noise*,
DOI 10.1007/978-1-4419-6624-7_7,

The operation of the single-electron SAW pump is as follows. A negative voltage is applied to a split gate on the surface of a piezoelectric heterostructure such as GaAs/AlGaAs. This forms a narrow depleted channel between two regions of 2DEG. The SAW is launched toward this channel from an interdigital transducer. Since the substrate is piezoelectric, a wave of electrostatic potential accompanies the SAW. The screening of the SAW potential by the metal gate is negligible and may be neglected in a model calculation [7, 8]. Electrons can be captured by the SAW from the source 2DEG and transported to the drain 2DEG. A perfectly flat plateau is obtained if the number of electrons transported in each cycle of the signal is held constant. This number is determined by the electron–electron interaction and the smallest size achieved by the moving quantum well whose energy levels are calculated from the instantaneous Hamiltonian. There are ranges of gate voltage where this minimum size is not changed appreciably, leading to quantized plateaus for the current at $I = Nef$. In the acoustoelectric current measurements reported so far, the plateaus have a slope of one part in 10^4. The many-particle interactions, the variations of the size of the quantum well minimum on the plateau and the shot noise, i.e., the statistical fluctuations in the occupation of the well, all make this problem challenging to deal with from a theoretical point of view. This article addresses the many-body physics in a simple model and makes some suggestions for reducing the shot noise in the measured current.

In addition to the turnstile and single-electron pumps, Thouless [11] suggested the quantized adiabatic charge transport that involves the employment of a one-dimensional (1D) electron system under the influence of a periodic potential. It is assumed that the periodic potential varies slowly in time and, with the help of a gate, the Fermi level can be varied so that when it lies within a minigap of the instantaneous Bloch Hamiltonian, there will be an integral multiplicity of charge Ne transported across the system during a single period [12]. If realized experimentally, such a device could also provide an important application as a current standard. We study the mechanisms of quantized adiabatic charge transport in carbon nanotubes using SAWs, as suggested by Talyanskii et al. [12]. We also consider this principle when the electron density in a two-dimensional (2D) graphene sheet is modulated by a SAW in the presence of a parallel magnetic field through the cross-section of the nanotube. The occurrence and reduction of noise in the quantized currents will also be addressed.

7.2 Theory for Current Quantization

Let us assume that a narrow channel is formed in a 2D electron gas (EG) lying in the x, y-plane. Neglect the finite thickness of the quantum well at the heterointerface in the z-direction and consider the electron motion as strictly 2D.

We will employ one of the simplest models for the gate-induced or etched [17] confining electrostatic potential that forms the channel in the 2DEG and take it as

$$V_g(x,y) = \frac{V_0}{\cosh^2(x/a)} + \frac{1}{2} m^* \Omega^2 y^2 , \tag{7.1}$$

where the x-axis is chosen along the channel, and the y-axis is across the channel. The parameter V_0 in (7.1) determines the effective height of the 1D potential barrier, that arises in the quasi-1D channel due to the applied gate voltage. The more negative the gate voltage becomes, the value of V_0 will be larger. Beyond the pinch-off voltage, the energy V_0 is greater than the electron Fermi energy in the wide 2D region. Parameter a determines the effective length of the channel, $l_{\text{eff}} = 2a$, in the x-direction. Experimentally, the SAW wavelength λ is comparable to the length of the channel. Therefore, in our numerical calculations, we set $a = \lambda/2$. In the y-direction, the electrostatic potential assumes a parabolic profile. This simple approximation is reasonable near the bottom of the well. The parameter $l_0 = (\hbar/m^*\Omega)^{1/2}$ may be interpreted as an effective width of the channel. In our numerical calculations, we chose $l_0/a = 4\times10^{-2}$, $\lambda = 1.0\,\mu\text{m}$ and $m^* = 0.067\,\text{m}_e$, appropriate for GaAs.

The SAW launched on the surface of the GaAs heterostructure along the x-axis ([011] crystallographic direction) is accompanied by the piezoelectric potential sliding along the surface with the speed of sound v_s. The piezoelectric potential accompanying the SAW was calculated previously [7, 8]. If the 2D layer is located at a distance d below the GaAs surface, then the additional SAW induced potential affecting electron motion has the form

$$V_{\text{SAW}}(x,t) = \frac{8\pi e_{14} C}{\varepsilon_b} \left[A_1 \mathrm{e}^{-\kappa q d} + A_1^* \mathrm{e}^{-\kappa q^* d} + A_2 \mathrm{e}^{-\kappa d} \right] \mathrm{e}^{\mathrm{i}\kappa(x - v_s t)}, \tag{7.2}$$

where C is the amplitude of the SAW and κ is its wave vector, e_{14} and ε_b are the piezoelectric modulus and dielectric constant of GaAlAs, respectively. The dimensionless coefficients $A_{1,2}$ and q are determined by the elastic constants of the host material [7, 8]. Our numerical calculations were done for $\text{Al}_x\text{Ga}_{1-x}\text{As}$ $(x = 0.3)$ for which nonzero elastic constants $c_{11}(x) = 11.88 + 0.14\ x$, $c_{12}(x) = 5.38 + 0.32\ x$ and $c_{44}(x) = 5.94 - 0.05\ x$ in units of 10^{10}N/m^2. For this density, $m(x) = (5.36 - 1.6\ x) \times 10^3\ \text{kg/m}^3$, this gives $v_s = 2981$ m/s, $A_1 = 1.446 + 0.512i$, $A_2 = -1.524 - 2.654i$, $q = 0.498 + 0.482i$ In (7.2), we neglected the screening effect of the gate electrodes on the piezoelectric potential. As shown in [7], the screening reduces the amplitude of the piezoelectric potential but has little effect on its shape. The SAW-induced potential in (7.2), superimposed on the gate potential in (7.1), leads to the formation of local quantum wells in the x-direction of the channel. Taking into account the confinement in the y-direction, one can conclude that the local quantum dots are formed in the channel. Statistical fluctuations in the occupancy of successive quantum dots during the course of one cycle would lead to shot noise. We now present a formalism for calculating of the tunneling probability from a moving quantum dot.

Let us consider two electrons in the quantum dot formed by the SAW-induced potential in a narrow electron channel. The Hamiltonian for two interacting electrons takes the form

$$\begin{aligned} H &= -\frac{\hbar^2}{2m}\left(\nabla_1^2 + \nabla_2^2\right) \\ &\quad + V_1(x_1) + V_1(x_2) + V_2(y_1) + V_2(y_2) + V_3(z_1) + V_3(z_2) \\ &\quad + \frac{e^2}{4\pi\varepsilon_0\epsilon_b\left[(x_1-x_2)^2 + (y_1-y_2)^2 + (z_1-z_2)^2\right]^{1/2}} \\ &\equiv H_0 + H_1, \end{aligned} \tag{7.3}$$

where H_0 describes two electrons, each of effective mass m, in the quantum dot without interaction and H_1 is the interaction term. Here, V_i ($i = 1,2,3$) are potentials in the x, y and z directions, respectively, described above. The Hamiltonian in (7.3) does not include any spin-dependent effects and is therefore diagonal in the spin variables of both electrons. In the absence of electron-electron interactions, a single-electron problem for the Hamiltonian H_0 can be easily solved. The single-electron wave functions diagonalizing the Hamiltonian H_0 takes the form

$$\langle \mathbf{r}|i,\alpha\rangle = \Phi_{i,\alpha}(x,y,z,\sigma) = \varphi_i(x)\psi(y)Z(z)\chi_\alpha(\sigma). \tag{7.4}$$

In this notation, $\varphi_i(x)$ is the localized wave function in the x direction and the subscript $i = 1,2,3$ labels the quantized energy levels in the SAW-induced quantum well. Also, $\psi(y)$ is a localized wave function in the y direction and we assume that the electrons will be in the lowest subband in the transverse direction, i.e., $\psi(y)$ is the ground state wave function of a 1D harmonic oscillator and $|Z(z)|^2 = \delta(z)$. Here, $\chi(\sigma)$ is the spin component of the wave function with σ representing the spin and $\alpha = \uparrow$ or $\downarrow$ the spin eigenvalue.

The wave function in (7.4) can be used to construct two-electron basis functions in the form of a Slater determinant, namely

$$\langle \mathbf{r}_1\mathbf{r}_2|i_1,\alpha_1;i_2,\alpha_2\rangle = \frac{1}{\sqrt{2}}[\Phi_{i_1,\alpha_1}(\mathbf{r}_1,\sigma_1)\Phi_{i_2,\alpha_2}(\mathbf{r}_2,\sigma_2) \tag{7.5}$$

$$-\Phi_{i_1,\alpha_1}(\mathbf{r}_2,\sigma_2)\Phi_{i_2,\alpha_2}(\mathbf{r}_1,\sigma_1)]. \tag{7.6}$$

The wave functions in (7.6) are normalized and orthogonal to each other. These functions form a complete set if the indices i_1, i_2 take all possible values, both discrete and continuous. From (7.4), it follows that the quantum number i labels the discrete eigenstates in the 1D quantum well along the x axis. This index may also change continuously if the energy level lies above the top of the confining potential. In general, when the Hamiltonian H in (7.3) is written in terms of the representation of functions in (7.6), we have an infinite matrix. However, we can simplify the problem if we exclude the continuous spectrum from our consideration and assume that the number of discrete energy levels within the quantum well is finite. In general, if $i = 1, \cdots, n$, then the number of independent basis functions in (7.6) is $n(2n-1)$. The Hamiltonian in (7.3) written in the basis representation of the functions in (7.6) is an $n \times n$ matrix that must then be diagonalized to get the energy eigenvalues of the two-electron problem. Calculation of the matrix elements of H with wave functions in (7.6) is very straightforward and can be very easily done if we take into account that the Hamiltonian is diagonal in the spin variables and the single-electron functions in (7.4) are eigenfunctions of the unperturbed Hamiltonian H_0.

If there are just two discrete single electron levels in the well, i.e., $i = 1, 2$, then the following six functions in (7.6) form the basis set

$$|1\uparrow;2\uparrow\rangle,\ |1\downarrow;2\downarrow\rangle, |1\uparrow;1\downarrow\rangle,\ |1\uparrow;2\downarrow\rangle, |2\uparrow;1\downarrow\rangle,\ |2\uparrow;2\downarrow\rangle. \quad (7.7)$$

In this case, the Hamiltonian is a 6×6 matrix having the form [18]

$$\hat{H} = \begin{pmatrix} E_1 & 0 & 0 & 0 & a_{21} & \cdots & \cdots & \cdots \\ 0 & 0 & \cdots & \cdots & \cdots & \cdots & & \\ 0 & 0 & \cdots & \cdots & \cdots & a_{44} & & \end{pmatrix}, \quad (7.8)$$

where $E_1 \equiv \epsilon_1 + \epsilon_2 + H^{(1)}_{12;12} - H^{(1)}_{12;21}$ with $H^{(1)}_{mn;kl} =< m| < n|H_1|k > |l >$ and a_{ij} are the elements of a 4×4 matrix $\mathbf{A}_{4\times4}$ given by

$$\mathbf{A}_{4\times4} = \begin{pmatrix} 2\epsilon_1 + H^{(1)}_{11;11} & H^{(1)}_{11;12} & H^{(1)}_{11;21} & H^{(1)}_{11;22} \\ H^{(1)}_{12;11} & \epsilon_1 + \epsilon_2 + H^{(1)}_{12;12} & H^{(1)}_{12;21} & H^{(1)}_{12;22} \\ H^{(1)}_{21;11} & H^{(1)}_{21;12} & \epsilon_1 + \epsilon_2 + H^{(1)}_{21;21} & H^{(1)}_{21;22} \\ H^{(1)}_{22;11} & H^{(1)}_{22;12} & H^{(1)}_{22;21} & 2\epsilon_2 + H^{(1)}_{22;22} \end{pmatrix}, \quad (7.9)$$

In (7.8) and (7.9), ϵ_1 and ϵ_2 are the energy eigenvalues of the ground and first excited states for a single electron in the SAW-induced quantum well and $|i>$ is defined in (7.2).

It follows from (7.8) and (7.9) that the Hamiltonian H is a block matrix. The two elements along the diagonal in the first and second rows correspond to the quantum states where both electrons have spin up or spin down. This means that the total spin is $S = 1$ and $S_z = \pm 1$. The energies of these are equal because the total Hamiltonian does not depend on the total spin orientation. The matrix $\mathbf{A}_{4\times4}$ describes the states where two electrons have opposite orientations, i.e., $S = 0$, $S_z = 0$ or $S = 1$, $S_z = 0$. The energy eigenvalues will correspond to spin singlet or spin triplet states.

7.3 Tunneling Probability and Current Quantization for Interacting Two-Electron System

Suppose there are two interacting electrons localized in the moving quantum well. We now calculate the probability for one of these electrons escaping while the other one remains in the well. This problem can be solved if we can determine the effective single electron potential for each electron. Then the procedure for single electron tunneling can be used. This can be accomplished by using a simplified version of the density functional method if the two interacting electrons in the ground state have opposite spin.

7.3.1 *Spin Unpolarized Case*

Let us denote the wave function describing two interacting electrons in the well with energy E and coordinates x_1 and x_2 in the ground state by $\Psi(x_1, x_2)$. The electron density in the well is given by

$$N(x) = 2\int_{-\infty}^{\infty} \mathrm{d}x_1 \; |\Psi(x, x_1)|^2 \,. \tag{7.10}$$

Making use of the single electron description for noninteracting electrons, each electron has the same energy $\epsilon = E/2$ and wave function $\varphi(x)$ since they have opposite spin. Since the wave function $\varphi(x)$ gives the same electron density distribution as the two electron wave function $\Psi(x_1, x_2)$, we have $|\varphi(x)|^2 = N(x)/2$ where $\varphi(x)$ is a solution of the one electron Schrödinger equation with effective electric potential

$$V_{\mathrm{eff}} = \epsilon + \frac{\hbar^2}{2m}\frac{\mathrm{d}^2\varphi(x)/\mathrm{d}x^2}{\varphi(x)}. \tag{7.11}$$

We can now calculate the probability of tunneling as we have done previously [7,8]. When two electrons are captured in the local SAW induced quantum well, they can escape from the well in two possible ways. First, one of the electrons can tunnel back to the source and the probability of the tunneling $P_2(\tau, \beta)$ can be calculated for arbitrary τ and β using

$$P_2(\tau, \beta) = \exp\left(-\frac{1}{\hbar}\int_{X_1}^{X_2} \mathrm{d}x \; \sqrt{2m(V_{\mathrm{eff}}(x) - \varepsilon)}\right), \tag{7.12}$$

where $\varepsilon \equiv E_0/2$ with E_0 equal to the ground state energy of the two-electron system. When this electron (electron 2, say) escapes, the potential profile returns to its original shape $v_c(x)$ and the remaining electron (electron 1, say) is in its ground state with energy ϵ_0. The probability for this electron to tunnel out of the well to the source is $P_1(\tau, \beta)P_2(\tau, \beta)$ where

$$P_1(\tau, \beta) = \exp\left(-\frac{1}{\hbar}\int_{X_1}^{X_2} \mathrm{d}x \sqrt{2m(V_c(x) - \epsilon_0)}\right). \tag{7.13}$$

This calculation is based on the assumption that two electrons never tunnel out of the well simultaneously, but always one after another. The total probability for electron 2 to escape from the well in one SAW cycle is

$$P_{2T}(\beta) = \int_0^{2\pi} \mathrm{d}\tau \; P_2(\tau, \beta), \tag{7.14}$$

whereas the corresponding probability for electron 1 is

$$P_{1T}(\beta) = \int_0^{2\pi} \mathrm{d}\tau \; P_1(\tau, \beta)P_2(\tau, \beta). \tag{7.15}$$

As a function of β, the acoustoelectric current is given by

$$I(\beta) = (1 - P_{1T}(\beta))\, ef + (1 - P_{2T}(\beta))\, ef. \tag{7.16}$$

We solved the single particle problem numerically and obtained the wave functions $\varphi_i(x)$ for each value of τ in the interval $0 \leq \tau \leq 2\pi$ and different values of the dimensionless parameter $\beta = 8\pi e_{14}C/\varepsilon_s V_0$ that is the ratio of the SAW potential amplitude to the height of the gate induced potential barrier in the channel. The probability distributions for the two lowest eigenstates, along with the ground state energy ε_0 for one electron captured in the dot, are shown in Fig. 7.1a. Using the procedure described above, the ground state energies E_0 and the x components of the ground state eigenfunctions $\Psi(x_1, x_2)$ were found numerically for each value of τ for $0 \leq \tau \leq 2\pi$ and for different values of β in the range $0 < \beta < 2.0$. An example of the electron density distribution $\rho(x) = 2\int_{-\infty}^{\infty} \mathrm{d}x_1 \, |\Psi(x, x_1)|^2$ within the local well in the x-direction for fixed β and τ is shown in Fig. 7.1b.

7.4 Adiabatic Charge Transport in Carbon Nanotubes

An intrinsic carbon nanotube has a linear band structure for the valence and conduction bands which intersect at the Fermi level ($E_F = 0$). When a magnetic field is applied, minigaps open close to the Fermi level. This is simulated through a parameter Δ. One can align the chemical potential with one of the minigaps with the use of a gate or through doping. Talyanskii et al. [12] employed a magnetic field-dependent Hamiltonian to study the effects of electrostatic modulation due to a SAW

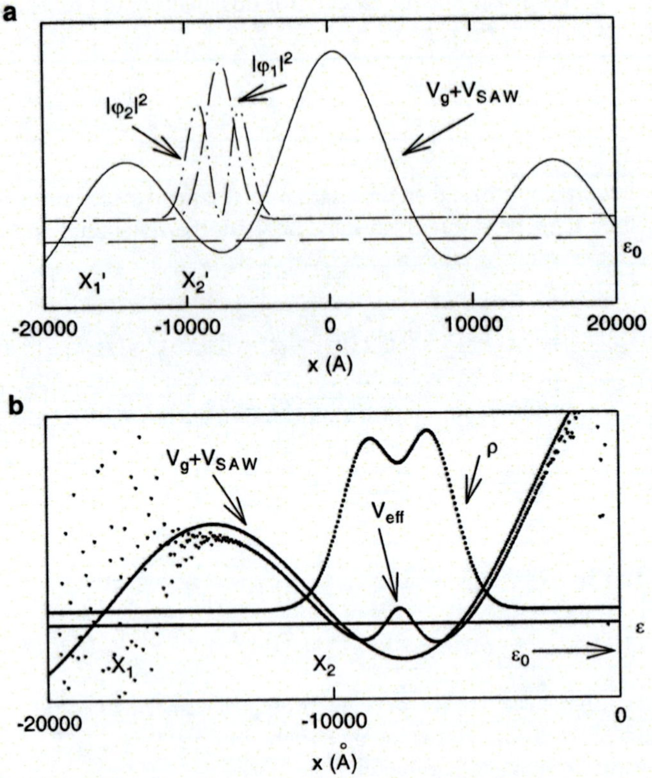

Fig. 7.1 (**a**) Schematic plots for fixed β and τ of the probability distributions for the ground (φ_1) and first excited (φ_2) states for one electron captured in the local quantum dot and the potential $V_c = V_g + V_{\mathrm{SAW}}$ in the channel. The ground state energy ε_0 is shown as a horizontal line. The boundaries of the classically forbidden region under the barrier at energy ε_0 are denoted by X_1' and X_2'. (**b**) The effective potential $V_{\mathrm{eff}}(x)$ and the density distribution ρ for a pair of interacting electrons are shown. For comparison, $V_g + V_{\mathrm{SAW}}$ is plotted. The horizontal line is $\varepsilon = E_0/2$ where E_0 is the ground state energy for the two-electron system. The boundaries of the classically forbidden region under the barrier at energy ε are denoted by X_1 and X_2; the position of ϵ_0 is also indicated

on the band structure of a carbon nanotube. The SAW velocity $u \approx 3 \times 10^3$ m/s is much less than $v_F \approx 10^5$ m/s. Thus, the 1D energy spectrum can be treated in a stationary/adiabatic approximation. In this approximation, the low-lying eigenstates are solutions of a Dirac equation for massless particles [12]

$$\hat{H} = -i\hbar v_F \frac{\mathrm{d}}{\mathrm{d}x}\sigma_3 + \sigma_1 \Delta + A \sin kx, \tag{7.17}$$

where σ_1 and σ_3 are Pauli matrices. The amplitude of the SAW is denoted by A and $2\pi/k$ is its wavelength. The Fermi velocity is denoted by v_F. Since

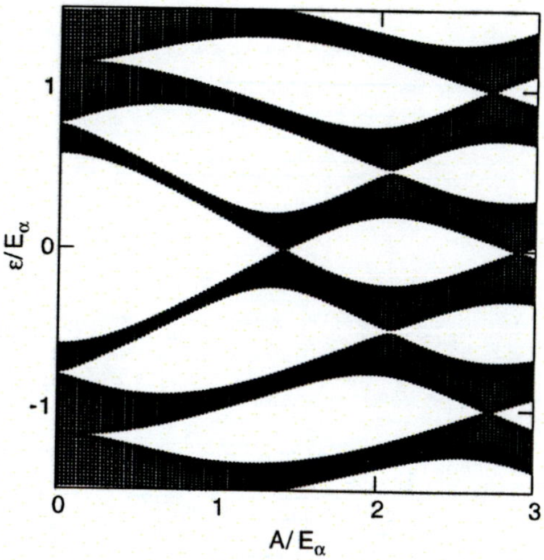

Fig. 7.2 Electron energy bands obtained by solving (7.19) as a function of the SAW field A/E_α, in units of $E_\alpha = \hbar k v_F$ with $2\pi/k$ the SAW wavelength and v_F the Fermi velocity. We chose $\Delta/E_0 = 0.6$ for the parameter representing the effect of a parallel magnetic field on the electronic band structure of the nanotube

the wavelength of the SAW $2\pi/k$ ($\sim$1 μm) is much larger than the diameter ($\sim$1 nm) of the nanotube, we are justified in ignoring the variation of the SAW potential over the cross-section of the nanotube. The role played by a parallel magnetic field that produces a gap near the Fermi level in the energy bands of the nanotube and is denoted by the parameter Δ couples the two components of the spinor wave function. Writing the two-component wave function as

$$\begin{pmatrix} \psi_1(x) \\ \psi_2(x) \end{pmatrix} = \exp\left(\frac{i}{2}\sigma_3\,\lambda \cos\,kx\right)\begin{pmatrix} \Psi_1(x) \\ \Psi_2(x) \end{pmatrix}, \tag{7.18}$$

where $\lambda = \dfrac{2A}{\hbar k v_F}$, we obtain the following pair of coupled ordinary differential equations when the Hamiltonian in (7.17) operates on the column vector in (7.18)

$$\begin{aligned} -\,\mathrm{i}\hbar v_F\,\frac{\mathrm{d}}{\mathrm{d}x}\Psi_1(x) + \Delta \mathrm{e}^{-\mathrm{i}\lambda\cos kx}\Psi_2(x) &= \epsilon\Psi_1(x) \\ \mathrm{i}\hbar v_F\,\frac{\mathrm{d}}{\mathrm{d}x}\Psi_2(x) + \Delta \mathrm{e}^{\mathrm{i}\lambda\cos kx}\Psi_1(x) &= \epsilon\Psi_2(x). \end{aligned} \tag{7.19}$$

This (7.19) was solved numerically and the energy bands have minigaps between which the Fermi energy can lie to produce current quantization.

In Fig. 7.2,we present our results obtained by solving (7.19) for ϵ/E_α as a function of A/E_α for fixed $\Delta/E_\alpha = 0.6$. Here, energies are measured in units of $E_\alpha = \hbar k v_F$. In Fig. 7.3, these calculations were repeated for fixed $\Delta/E_\alpha = 1.2$. Figure 7.4 is a plot of the energy spectrum as a function of Δ/E_α for fixed $A/E_\alpha = 0.8$.

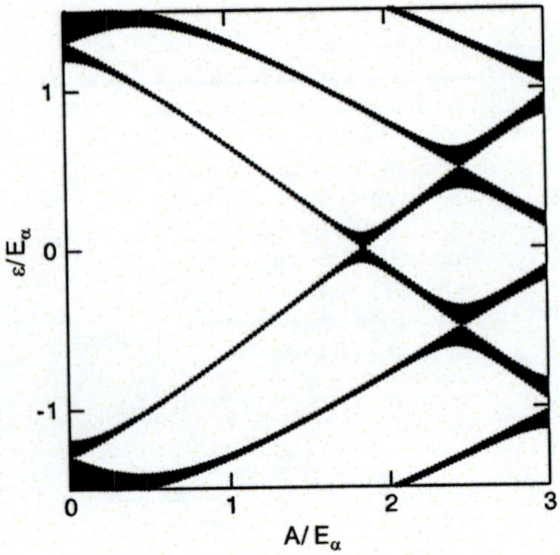

Fig. 7.3 The same as Fig. 7.2, except that we chose $\Delta/E_\alpha = 1.2$

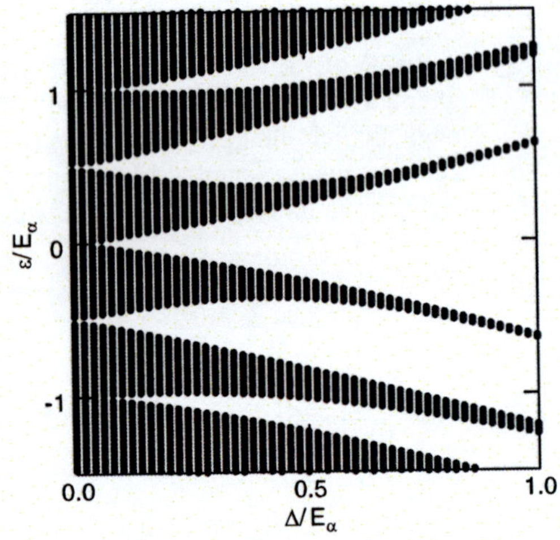

Fig. 7.4 Calculated electron energy spectrum obtained as solutions of (7.19) as a function of the magnetic field parameter Δ in units of the energy parameter $E_\alpha = \hbar k v_F$. The SAW field is chosen as $A/E_\alpha = 0.8$

It is shown in Figs. 7.2–7.4 that minigaps are formed in the energy spectrum. These minigaps oscillate as a function of the SAW amplitude A and vanish at specific values that are determined by the choice of the magnetic field parameter Δ. The widths of the minibands are reduced as Δ is increased. In addition, Fig. 7.4 demonstrates that there are no minigaps when $\Delta = 0$. Since we have quantized charge transport when the chemical potential falls within one of the minigaps, the widths of the plateaus depend on the SAW power and the magnetic field. The value of the quantized current will remain the same within a whole range of values of Δ and A that stay within a gap. The energy bands in Figs. 7.2–7.4 $\varepsilon \rightarrow -\varepsilon$ electron-hole symmetry. This is due to the symmetry of the Hamiltonian for left and right-travelling carriers when $\Delta = 0$.

7.5 Summary and Concluding Remarks

As stated above, the accuracy of the SAW pumps has to be improved substantially if they are to find applications in electrical metrology, for example. A step toward achieving this improvement will require reduction of the *shot* noise in the current produced by the SAW or high-frequency single-electron pumps. This means improving the probability of transporting an electron in a cycle.

In a recent paper, Huang et al. [15] considered the overall efficiency for using SAWs in a photon detector. The process would involve photon absorption, the degeneration of a photon into an electron-hole pair and the capture and transport of the electron and hole by the SAW. In our dual-charged-fluid model for the steady-state transport of SAWs, dragged photocurrents of 1D confined-state carriers were considered. This model incorporated the effects of the quantum confinement and the tunneling escape of SAW-dragged 1D carriers, as well as the effects of the inelastic capture of 2D continuous-state carriers and the self-consistent space-charge field. Our results show that as a function of SAW power and frequency, as well as temperature, the heavier hole was more likely to be transported during a SAW cycle. Based on this model, the shot noise from the SAW-dragged hole currents should be diminished from their levels measured experimentally involving electrons.

As we demonstrated by the minigaps in a 1D conductor, quantized charge transport may be observed in a noninteracting model of electrons. However, for a strongly interacting Luttinger liquid in which the second-order perturbations are treated as bosons, the current quantization may not be so sharp. Unlike the Fermi liquid for which the quasiparticles carry both spin and charge, the elementary excitations of the Luttinger liquid are charge and spin waves for which impurities and other types of mechanisms producing *backscattering* are important. Even at low temperature, the distribution function for the particle momentum does not have a sharp jump. This is in contrast with Fermi liquid where this jump indicates the Fermi surface. In the momentum-dependent spectral function, there is no "quasiparticle peak" whose width becomes much narrower than the excitation energy above the Fermi level, as occurs for the Fermi liquid. Instead, there is a power-law singularity, with a "non-universal" exponent which depends on the strength of the interaction.

Acknowledgments This work is supported by contract FA9453-07-C-0207 of AFRL. I would also like to thank Dr. Tibab McNeish for a careful reading of the manuscript and helpful comments.

References

1. Kouwenhoven LP, Johnson AT, van der Vaart NC, Harmans CJPM, Foxon CT (1991) Phys Rev Lett 67:1626
2. Mark W. Keller, Ali L. Eichenberger, John M. Martinis, Neil M. Zimmerman (1999) Science 285:1706
3. Shilton JM, Talyanskii VT, Pepper M, et al. (1996) J Phys Condens Matter 8:L531

4. Talyanskii VI, Shilton JM, Pepper M, et al. (1997) Phys Rev B 56:15180
5. Cunningham J, Talyanskii VI, Shilton JM, Pepper M, Simmons MY, Ritchie DA (1999) Phys Rev B 60:4850
6. Aĭzin GR, Godfrey Gumbs, Pepper M (1998) Phys Rev B 58:10589
7. Godfrey Gumbs, Aĭzin GR, Pepper M (1999) Phys Rev B 60:R13954
8. Godfrey Gumbs, Aĭzin GR, Pepper M (1998) Phys Rev B 57:1654
9. Blumenthal MD, Kaestner B, Li L, Giblin S, Janssen TJBM, Pepper M, Anderson D, Jones G, Ritchie DA (2007) Nat Phys 3:343
10. Kaestner B, Kashcheyevs V, Amakawa S, Blumenthal MD, Li L, Janssen TJBM, Hein G, Pierz K, Weimann T, Siegner U, Schumacher HW (2008) Phys Rev B 77:153301
11. Thouless DJ (1983) Phys Rev B 27:6083
12. Talyanskii VI, Novikov DS, Simons BD, LS Levitov (2001) Phys Rev Lett 87:276802
13. Robinson AM, Talyanskii VI (2005) Phys Rev Lett 95:247202
14. Godfrey Gumbs (2008) AIP Conf Proc 991:57
15. Huang D, Godfrey Gumbs, Pepper M (2008) J Appl Phys 103:083714
16. Barnes CHW, Shilton JM, Robinson AM (2000) Phys Rev B 62:8410
17. Cunningham J, Talyanskii VI, Shilton JM, Pepper M (2000) Phys Rev B 62:1564
18. Gumbs G (2003) Solid State Commun 128:443

Chapter 8
The Ubiquitous Matched Filter: A Tutorial and Application to Radar Detection

Steven Kay and Muralidhar Rangaswamy

8.1 History

In 1946, David Middleton, along with his PhD advisor J.H. Van Vleck (later to be awarded the Nobel prize), published the paper "A Theoretical Comparison of the Visual, Aural, and Meter Reception of Pulsed Signals in the Presence of Noise" in the Journal of Applied Physics [23]. This paper evolved from earlier work of Middleton in 1943. In it, a new type of device was derived and analyzed. It was to be called the *matched filter*. It allowed filter designers a straightforward method for detecting the presence of a signal that was obscured by additive noise. Coming at a time when radar was being developed and its important role in World War II recognized, the matched filter was to have a lasting impact on humanity. Although principally motivated by the radar application, it has found widespread use in commercial and military applications. From locating blood vessels from retinal images [11], to monitoring water pollution [1], to the design of cell phones [7], this once "military " concept has proven to be a fundamental tool in modern signal processing. A recent Google search produced more than 568,000 hits in response to the input "matched filter", with application fields that are widely diverse.

As great contributions to society are usually the cumulative work of many researchers so it is with the matched filter. In addition to Middleton and Van Vleck, D.O. North of RCA laboratories is also credited with the invention. His independent

S. Kay (✉)
Department of Electrical, Computer, and Biomedical Engineering, University of Rhode Island, Kingston, RI 02881, USA
e-mail: kay@ele.uri.edu

M. Rangaswamy
Principal Electronics Engineer, Air Force Research Laboratory Sensors Directorate, AFRL/RYHE 80 Scott Drive, Hanscom AFB, MA 01731-2909, USA
e-mail: Muralidhar.Rangaswamy@hanscom.af.mil

L. Cohen et al. (eds.), *Classical, Semi-classical and Quantum Noise*,
DOI 10.1007/978-1-4419-6624-7_8,

work, as reported in an RCA report, "An Analysis of Factors which Determine Signal/Noise Discrimination in Pulsed Carrier Systems" [17], was published at about the same time, and although not generally available, contained the essence of the matched filter. That such an important discovery was brought forth by several prominent researchers at the same time, was probably a response to the immediate need to improve radar performance during World War II.

In the remainder of this chapter, we explain the essence of the matched filter, including some extensions, and finally its important application to radar. It is fitting and perhaps telling that the area of interest that spawned the matched filter still relies heavily on its use today.

8.2 Introduction

The central problem in radar is to determine when a signal is present. This signal is the result of an echo from a target. Since the target may be many miles away, the received echo exhibits very little power, and hence is easily obscured by the many noises present in a radar receiver. It is intuitively clear that one should always filter the received waveform to attenuate any frequency bands for which there is known to be only noise. However, what is one to do when a frequency band contains possibly the echo as well as the ever present noise? The usual assumption concerning the noise is that it has equal power at all frequencies, i.e., it is *white* noise. The signal, on the contrary, will have an unequal distribution of power with frequency. For example, if the transmitted signal is a pulse given by

$$s(t) = \begin{cases} 1 & 0 \leq t \leq T \text{ sec} \\ 0 & \text{otherwise} \end{cases}$$

and if the received signal is identical to the transmit signal (a simplification, of course), then the received signal will have spectral content given by the Fourier transform of $s(t)$. This is easily shown to be

$$S(F) = T \frac{\sin(\pi F T)}{\pi F T} \exp\left[-j2\pi F(T/2)\right],$$

where F is the frequency in Hertz. The Fourier transform is a "sinc" pulse and occupies the entire frequency spectrum. Practically speaking, however, its bandwidth is about k/T Hz, where k is an integer that is usually chosen as 2 or 3.

This is shown in Fig. 8.1 for $T = 1$, where it should be noted that the power in the signal is proportional to $|S(F)|^2$ and is not constant over the signal bandwidth. The white noise, on the other hand, does exhibit the same power versus frequency. It therefore seems reasonable to design a filter that emphasizes the frequencies near zero and attenuates those away from zero frequency. In fact, a filter frequency

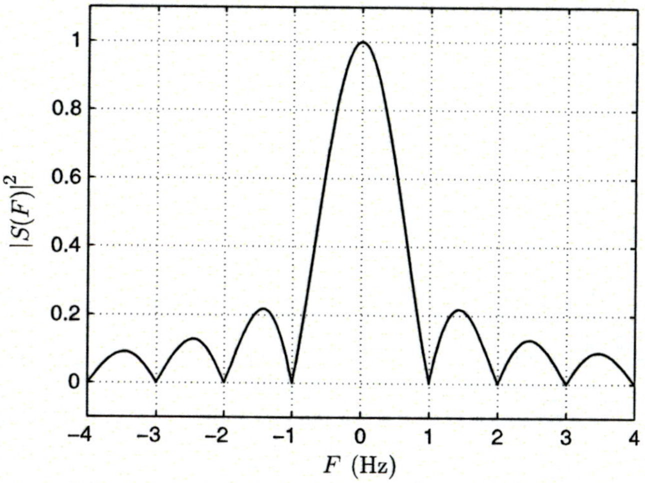

Fig. 8.1 Fourier transform magnitude-squared of a pulse with width $T = 1$

magnitude response of $|H(F)| = |S(F)|$ might just do the trick! Here, the filter response is *matched* to the signal. But is this the best choice? Middleton and others have provided the answer.

8.3 How to Design the Best Filter?

Before we can design the *best* filter, we need to know what best means. It is generally accepted that we should design a filter that yields the maximum *signal-to-noise ratio* (SNR) at its *output*, which we denote by $\mathrm{SNR_O}$. For many applications, this is indeed the design criterion we should use. As an aside, if the noise is *white Gaussian noise*, then maximizing SNR is equivalent to maximizing the probability of detection for a fixed probability of false alarm [13]. To proceed, we next define the $\mathrm{SNR_O}$.

As shown in Fig. 8.2, we envision a filter that processes the received waveform $x(t)$, consisting of the signal $s(t)$ and the white noise $w(t)$. Since the signal has a duration of T sec, it seems reasonable to examine the output at $t = T$. The output of the filter is

$$y(t) = \int_0^t h(t-\tau)x(\tau)\mathrm{d}\tau, \tag{8.1}$$

where $h(\tau)$ is the impulse response of a causal filter. The upper limit is t since for the filter to be causal, and hence physically realizable, the impulse response must satisfy $h(\tau) = 0$ for $\tau < 0$, and the lower limit is 0 since the received signal $s(t)$ is zero prior to $t = 0$. Note that if we had decided to examine the output at $t > T$,

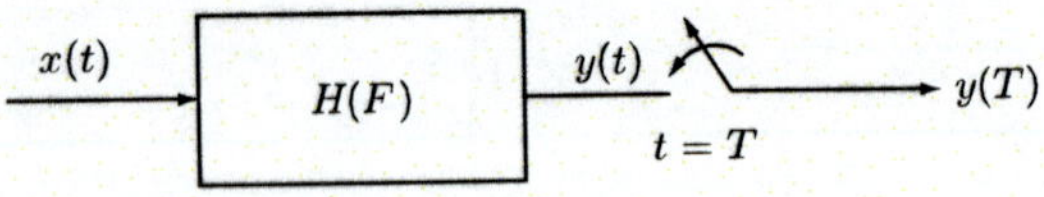

Fig. 8.2 Setup for derivation of matched filter

then since $s(t) = 0$ for $t > T$, we would be processing a portion of the waveform that contained only noise. This certainly would degrade the output SNR. The signal output of the filter is therefore given by

$$s_o(t) = \int_0^t h(t-\tau)s(\tau)\mathrm{d}\tau \tag{8.2}$$

while the noise at the output is

$$w_o(t) = \int_0^t h(t-\tau)w(\tau)\mathrm{d}\tau. \tag{8.3}$$

We next define the SNR at the output of the filter and sampler as the value of $s_o^2(T)$ divided by the average noise power at the same time, which is $E[w_o^2(T)]$, so that

$$\mathrm{SNR_o} = \frac{s_o^2(T)}{E[w_o^2(T)]}. \tag{8.4}$$

But the signal output can be written as an inverse Fourier transform

$$s_o(T) = \int_{-\infty}^{\infty} H(F)S(F)\exp(\mathrm{j}2\pi FT)\mathrm{d}F \tag{8.5}$$

and the average noise power at time $t = T$ (actually at any time, assuming we have wide sense stationary white noise) is given by the power spectral density (PSD) of the output process when integrated over all frequencies. It is well known that the output PSD of a linear time invariant filter is given by $P_{w_o}(F) = |H(F)|^2 P_w(F)$, where $P_w(F)$ is the PSD of $w(t)$ [14]. If the latter is given as $P_w(F) = N_0/2$ (note that it is flat with frequency, i.e., white), then

$$P_{w_o}(F) = |H(F)|^2 \frac{N_0}{2}. \tag{8.6}$$

As a result, we have that

$$\begin{aligned} E[w_o^2(T)] &= \int_{-\infty}^{\infty} P_{w_o}(F)\mathrm{d}F \\ &= \int_{-\infty}^{\infty} |H(F)|^2 \frac{N_0}{2}\mathrm{d}F \end{aligned} \tag{8.7}$$

and finally using (8.5) and (8.7) in (8.4) produces the output SNR in terms of the filter frequency response as

$$\mathrm{SNR_o} = \frac{\left(\int_{-\infty}^{\infty} H(F)S(F)\exp(\mathrm{j}2\pi FT)\mathrm{d}F\right)^2}{\int_{-\infty}^{\infty} |H(F)|^2 \frac{N_0}{2} \mathrm{d}F}. \tag{8.8}$$

It should be noted that the optimal filter frequency response is not unique since scaling by a constant produces the same output SNR, an intuitively obvious result. To maximize (8.8), we can use the Cauchy–Schwarz inequality, which in its integral form, appears as

$$\left| \int_{-\infty}^{\infty} G^*(F)H(F)\mathrm{d}F \right|^2 \le \int_{-\infty}^{\infty} |G(F)|^2 \mathrm{d}F \int_{-\infty}^{\infty} |H(F)|^2 \mathrm{d}F$$

with equality if and only if $H(F) = cG(F)$ for c some complex constant. Expressed as

$$\frac{\left|\int_{-\infty}^{\infty} G^*(F)H(F)\mathrm{d}F\right|^2}{\int_{-\infty}^{\infty} |H(F)|^2 \mathrm{d}F} \le \int_{-\infty}^{\infty} |G(F)|^2 \mathrm{d}F$$

and now identifying $G(F)$ with $S(F)\exp(\mathrm{j}2\pi FT)$ in (8.8) (and noting that this results in $\int_{-\infty}^{\infty} G^*(F)H(F)\mathrm{d}F$ being real), we have the inequality

$$\begin{aligned}
\mathrm{SNR_o} &= \frac{\left(\int_{-\infty}^{\infty} H(F)S(F)\exp(\mathrm{j}2\pi FT)\mathrm{d}F\right)^2}{\frac{N_0}{2}\int_{-\infty}^{\infty} |H(F)|^2 \mathrm{d}F} \\
&\le \frac{1}{N_0/2} \int_{-\infty}^{\infty} \left|S^*(F)\exp(-\mathrm{j}2\pi FT)\right|^2 \mathrm{d}F \\
&= \frac{\int_{-\infty}^{\infty} |S(F)|^2 \mathrm{d}F}{N_0/2} \\
&= \frac{\mathcal{E}}{N_0/2},
\end{aligned}$$

where $\mathcal{E}$ denotes the *energy of the signal*. To achieve this maximum, we need to only choose $H(F) = cG(F)$ or finally, the optimal filter receiver frequency response is

$$H_{\mathrm{opt}}(F) = cS^*(F)\exp(-\mathrm{j}2\pi FT). \tag{8.9}$$

This derivation and result is the great contribution of David Middleton. Why it is called the *matched filter* is described next. Note first that the maximum SNR is

$$\mathrm{SNR_o} = \frac{\mathcal{E}}{N_0/2} \tag{8.10}$$

and it is the *output SNR of a matched filter*. It is seen to actually be an *energy-to-noise ratio* (ENR). Finally, a very important result is that the maximum output SNR *only depends on the signal energy* but not the signal shape. This says that the shape can be chosen as desired as long as the energy is maintained. This has important practical ramifications [8].

8.4 So Why Is it Called a Matched Filter?

Since the constant c in (8.9) is arbitrary, we set it equal to unity. The impulse response of the optimal filter then becomes

$$h(\tau) = \mathcal{F}^{-1}\{S^*(F)\exp(-j2\pi FT)\}, \tag{8.11}$$

where $\mathcal{F}^{-1}$ denotes the inverse Fourier transform. Since $\mathcal{F}\{s(-t)\} = S^*(F)$ and the exponential factor in (8.11) represents a time delay, we have that

$$h(\tau) = s(T - \tau) \tag{8.12}$$

and it is seen that *the impulse response of the filter is matched (although flipped around) to the signal*. As an example, in Fig. 8.3 a signal and the corresponding impulse response of the matched filter is shown. Note that because $s(t)$ is nonzero only over the time interval $0 \le t \le T$, then so is the impulse response. Continuing with this line of thought, we examine the output of the filter for all times. As seen in Fig. 8.4, the nonzero output begins at time $t = 0$ and continues until it is maximum at $t = T = 1$ (recall that this is the sampling time). If we continue to examine the output, it decreases to zero at time $t = 2T = 2$. It is clear that we need, only examine the output up to the sampling time.

Since $h(\tau) = s(T - \tau)$, and therefore $h(t - \tau) = s(T - (t - \tau))$, the signal output of the matched filter is from (8.2)

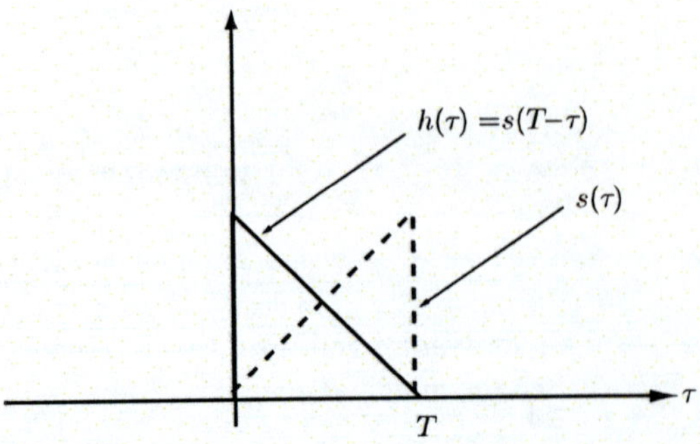

Fig. 8.3 Signal and the corresponding impulse response of the matched filter

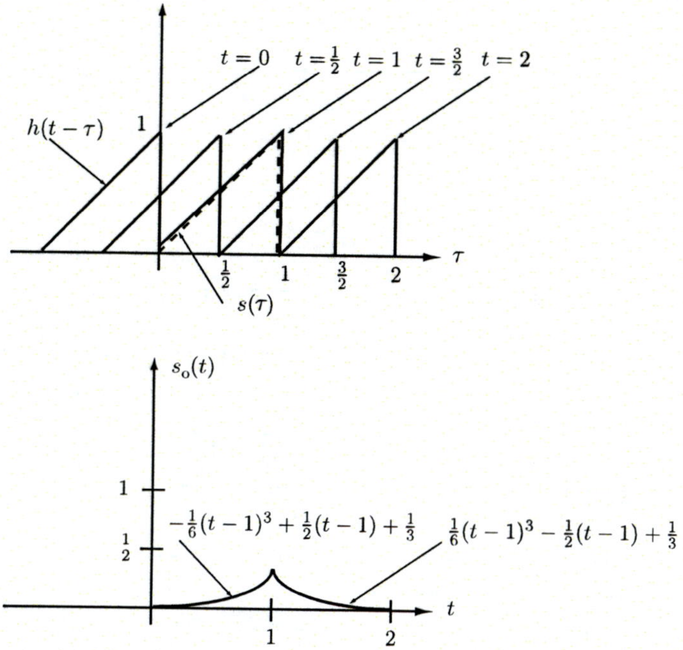

Fig. 8.4 Convolution operation of (8.13) and resulting signal output of the matched filter for the signal shown in Fig. 8.3

$$
\begin{aligned}
s_o(t) &= \int_0^t h(t-\tau)s(\tau)\mathrm{d}\tau \\
&= \int_0^t s(T-(t-\tau))s(\tau)\mathrm{d}\tau.
\end{aligned} \tag{8.13}
$$

At $t = T$, we have the matched filter output

$$
s_o(T) = \int_0^T s(\tau)s(\tau)\mathrm{d}\tau = \int_0^T s^2(\tau)\mathrm{d}\tau,
$$

which is again the signal energy, although now expressed in the time domain.

8.5 The Modern Matched Filter

Since most modern signal processing systems are digital, the processing is done within a digital computer or specialized digital hardware. As such, we operate on sampled waveforms or *discrete-time* waveforms, also known as sequences.

The analogous result in discrete-time is that the impulse response of the matched filter is

$$h[n] = s[N-1-n] \qquad n = 0, 1, \ldots, N-1 \tag{8.14}$$

for a signal sequence $\{s[0], s[1], \ldots, s[N-1]\}$. In analogy with (8.1), the matched filter output becomes

$$y[n] = \sum_{k=0}^{n} h[n-k]x[k]$$

so that using (8.14) we have

$$y[n] = \sum_{k=0}^{n} s[N-1-(n-k)]x[k].$$

Sampling as before at the end of the signal interval, which is $n = N-1$, we have for the matched filter output

$$y[N-1] = \sum_{k=0}^{N-1} s[k]x[k]. \tag{8.15}$$

Note that since we are doing this computation in a digital computer and not as the output of an analog filter, there is really no need to compute the outputs for times other than at $n = N-1$. Thus, (8.15) only needs to be computed and because of its form, it is sometimes referred to as a *replica correlator or correlation receiver* [6].

8.6 A More Realistic Example

A detection problem that arises in many fields is that of detecting a sinusoidal pulse of known duration but unknown amplitude, frequency, and phase. For example, the signal may take the form in discrete-time as $s[n] = A\cos(2\pi f_0 n + \phi)$, where A, the amplitude, f_0, the frequency, and ϕ, the phase are all unknown. How is one to design a detector? The solution is to use a pair of matched filters as follows. First expand the signal into its "sine" and "cosine" components as

$$s[n] = A\cos(\phi)\cos(2\pi f_0 n) - A\sin(\phi)\sin(2\pi f_0 n) \qquad n = 0, 1, \ldots, N-1.$$

Since both A and ϕ are unknown, we let $a = A\cos(\phi)$ and $b = -A\sin(\phi)$ so that the received signal becomes

$$s[n] = a\cos(2\pi f_0 n) + b\sin(2\pi f_0 n).$$

Noting that depending upon the value of ϕ, a or b could be small or even zero, but not at the same time, the solution then is to use *two matched filters*, one matched to the signal $\cos(2\pi f_0 n)$ and another matched to the signal $\sin(2\pi f_0 n)$. Then the outputs are squared (since the output can be positive or negative) and the results are added. Thus, we have the *quadrature matched filter*, which from (8.15) is

$$\left(\sum_{k=0}^{N-1} \cos(2\pi f_0 k)x[k]\right)^2 + \left(\sum_{k=0}^{N-1} \sin(2\pi f_0 k)x[k]\right)^2. \tag{8.16}$$

Since any scaling is arbitrary, we change this by a scale factor of $1/N$ to yield

$$\frac{1}{N}\left[\left(\sum_{k=0}^{N-1} \cos(2\pi f_0 k)x[k]\right)^2 + \left(\sum_{k=0}^{N-1} \sin(2\pi f_0 k)x[k]\right)^2\right],$$

which as we will see is a convenient normalization. Next, we use the identity that the sum of squares of two numbers can be viewed as the magnitude-squared of a complex number or $\alpha^2 + \beta^2 = |\alpha - j\beta|^2$ to yield

$$\frac{1}{N}\left|\sum_{k=0}^{N-1} x[k]\exp(-j2\pi f_0 k)\right|^2. \tag{8.17}$$

Finally, since the frequency was also assumed unknown, and hence we cannot compute (8.17), we try *all frequencies* from $f_0 = 0$ to $f_0 = 1/2$ in (8.17) and choose the one that yields the maximum output. This is similar to the lesson learned using convolution in Fig. 8.4 in which the maximum output was used. This extended matched filter is a standard one in radar and sonar and should look familiar to the astute reader in that it is actually the *discrete-time Fourier transform of the received sequence, after taking the magnitude-squared to indicate power and normalized by* $1/N$ *to yield an estimate of the power spectral density*. It is also called the *periodogram* in spectral analysis [12], and sometimes goes by the name of the *narrowband detector* and the *FFT detector*, in that the FFT is usually used to compute (8.17) for all frequencies.

As an example, let $A = 1$, $f_0 = 0.22$, $\phi = 0$, $N = 400$ for the signal and assume that the additive noise is white Gaussian noise with a total power of unity. As a result, the noise PSD is just $P_w(f) = 1$ for $0 \le f \le 1/2$. A realization of the received sequence for noise only and the sinusoid plus noise is shown in Fig. 8.5. Note that the same realization of noise is used in both figures. The sinusoidal signal is not observed in the lower plot because the SNR is very low. It is given by $10\log_{10} A^2/(2\sigma^2) = 10\log_{10}(1/2) = -3$ dB, where $\sigma^2 = 1$ is the total noise power. If, however, we plot the quadrature matched filter output as given by (8.17) versus the frequency f_0, we obtain the periodograms shown in Fig. 8.6.

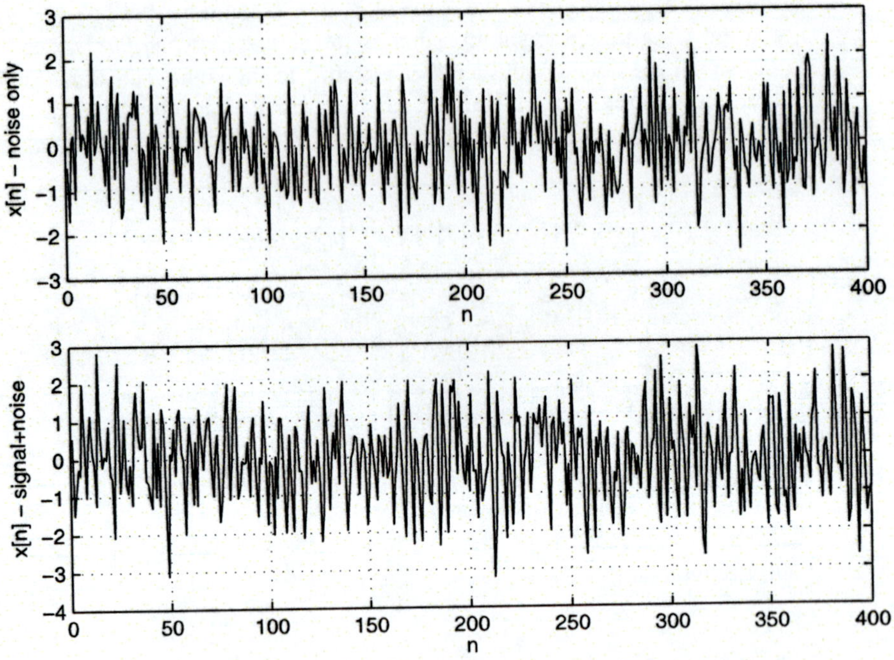

Fig. 8.5 A realization of noise only in the upper plot and a sinusoid plus noise in the lower plot

It is clear now that the sinusoid is easily detected since the peak at $f_0 = 0.22$ is easily discerned from the background of noise. In fact, the detectability is directly dependent on the *peak-to-background ratio*. The peak output at $f_0 = 0.22$ can be easily found by substituting $s[n] = A[\exp(\mathrm{j}(2\pi f_0 k + \phi)) + \exp(-\mathrm{j}(2\pi f_0 k + \phi))]/2$ into (8.17) and discarding the very small contribution of the complex sinusoid with the negative frequency. This will yield approximately $NA^2/4$ for the peak value of the periodogram. In this case, it is $400/4 = 100$ or 20 dB. Also, it can be shown that the periodogram background estimates the white noise PSD of $P_w(f) = \sigma^2 = 1$ (and this is why we chose to include the normalization factor of $1/N$ in (8.17)) so that the average background value is 0 dB as seen in Fig. 8.6. Hence, the overall peak to background ratio is given by $NA^2/(4\sigma^2) = 20$ dB and is the key factor that determines detectability of a sinusoid in white noise. Also, note that the peak-to-background ratio is a factor of $N/2$ larger than the input SNR. One may therefore view the peak to background ratio as a type of output SNR of this quadrature matched filter. Thus, the gain in SNR due to processing is $10\log_{10} N/2 = 23$ dB, which accounts for the ease with which the sinusoid can be detected from Fig. 8.6 but not in Fig. 8.5. This gain is sometimes termed the *processing gain*.

We next discuss the use of the matched filter in modern radar.

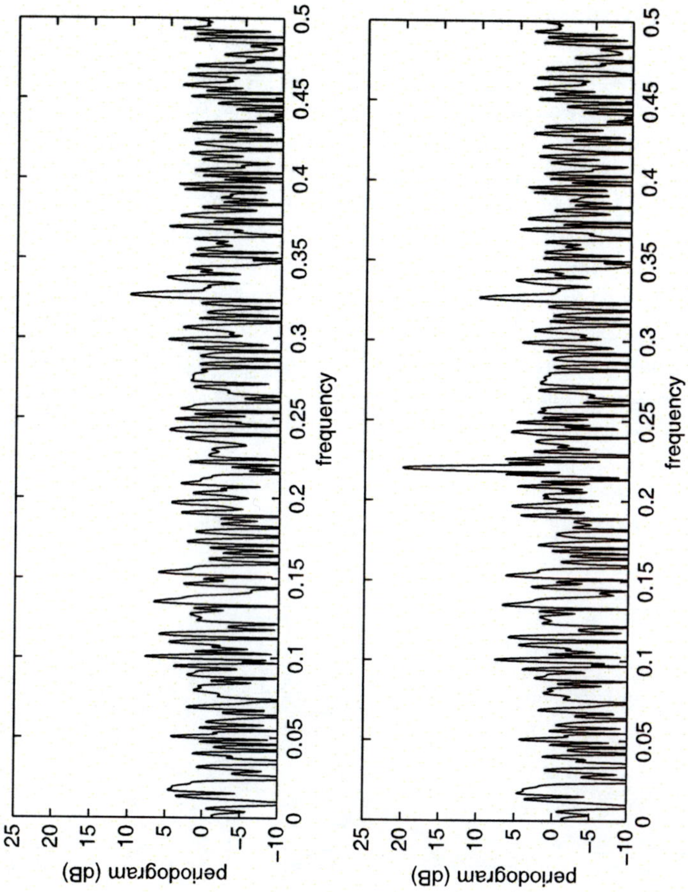

Fig. 8.6 The corresponding periodograms (shown in dB) for the sequences in Fig. 8.5

8.7 The Radar Problem

A radar is a sensor, in our case an antenna array on an airborne platform, which transmits and receives electromagnetic radiation. The transmitted electromagnetic signal impinges on various objects such as buildings, land, water, vegetation, and one or more targets of interest. The illuminated objects reflect the incident wave, which is received and processed by the radar receiver. The reflected signal includes not only desired signals (targets) but also undesired returns from extraneous objects, designated as clutter. Additionally, there could be one or more jammers, high-powered noise-like signals transmitted as electronic countermeasure (ECM), masking the desired target signals. Finally, the received returns include the ubiquitous background white noise caused by the radar receiver circuitry as well as by man-made sources and machinery. Typically, if it exists, the power of the desired signal return is a very small fraction of the overall interference power (due to clutter, jamming, and noise). The problem at hand is to detect the target, if it exists, within the background of clutter and jammer returns. The key to solving this problem is the availability of suitable models for targets, clutter, and jammers [26]. These models account for the angular position of the target in relation to the receiving array. If moving, the target signature includes the effect of the resulting Doppler frequency.

More precisely, the radar receiver front end consists of an array of J antenna elements, which receives signals from targets, clutter, and jammers. These radiations induce a voltage at each element of the antenna array, which constitutes the measured array data at a given time instant. Snapshots of the measured data collected at N successive time epochs give rise to the spatio-temporal nature of the received radar data. The spatio-temporal product $JN = M$ is defined to be the system dimensionality. Figure 8.7 uses the angle-Doppler space to illustrate the need for

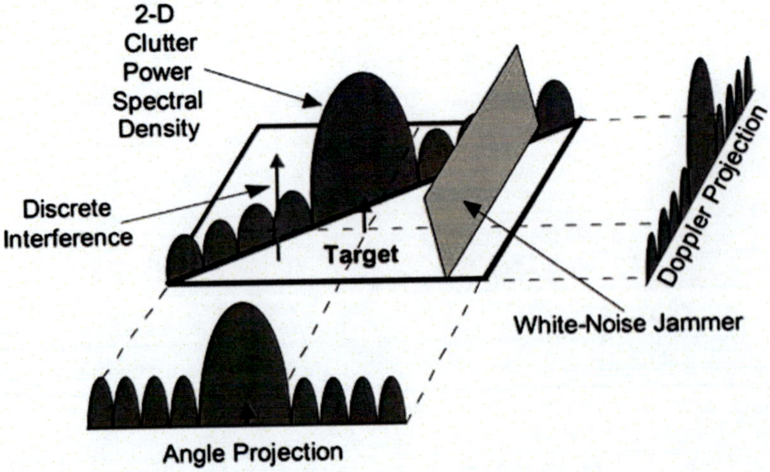

Fig. 8.7 The target and interference scenario in an airborne radar

space-time (joint domain) processing. A target at a specific angle and travelling at a specific velocity (corresponding to a Doppler frequency) occupies a single point in this space. A jammer originates from a particular angle, but is temporally white (noise like). The clutter, due to the motion of the platform, occupies a *ridge* in this 2D-space [26] – a clutter patch in front of the moving aircraft has the highest Doppler frequency while one at broadside has zero Doppler (no relative velocity). The clutter spectrum reflects the two-way beampattern of the transmitted signal.

The figure also illustrates the effect of strictly temporal (Doppler) or spatial processing (in angle). The former is equivalent to a projection of the two-dimensional target plus interference spectrum onto the Doppler plane – however, the target signal is masked by the temporally white jamming. The latter is equivalent to a projection of the said spectrum onto an angular plane, but since the clutter power is strongest at the look angle, the target cannot be distinguished from clutter. However, joint domain processing identifies clear regions in the two-dimensional plane, which affords recovery of the target from the interference background.

The detection problem can be formally cast in the framework of a statistical hypothesis test of the form:

$$H_0: \quad \mathbf{x} = \mathbf{d} = \mathbf{c} + \mathbf{j} + \mathbf{n} \tag{8.18}$$

$$H_1: \quad \mathbf{x} = \alpha\mathbf{e}(\theta_t, f_t) + \mathbf{d} = \alpha\mathbf{e}(\theta_t, f_t) + \mathbf{c} + \mathbf{j} + \mathbf{n}, \tag{8.19}$$

where $\mathbf{x} \in \mathcal{C}^{JN\times 1}$ denotes the received data under either hypothesis, $\mathbf{d}$ represents the overall interference being the sum of $\mathbf{c}$, the clutter vector, $\mathbf{j}$, the jammer vector, and $\mathbf{n}$, the background white noise. Finally, $\mathbf{e}$ is a known spatio-temporal steering vector that *models* the target return for a specific angle-Doppler, and α is the unknown target complex amplitude. For the popular case of a linear array of equispaced elements,

$$\mathbf{e} = \mathbf{e}_t \otimes \mathbf{e}_s \tag{8.20}$$

$$\mathbf{e}_t = \begin{bmatrix} 1 & z_t & z_t^2 & \dots & z_t^{(N-1)} \end{bmatrix}^{\mathrm{T}}, \tag{8.21}$$

$$\mathbf{e}_s = \begin{bmatrix} 1 & z_s & z_s^2 & \dots & z_s^{(J-1)} \end{bmatrix}^{\mathrm{T}}, \tag{8.22}$$

$$z_s = \mathrm{e}^{\mathrm{j}2\pi f_s} = \mathrm{e}^{(\mathrm{j}2\pi \frac{\mathrm{d}}{\lambda} \sin\phi_t)}, \qquad z_t = \mathrm{e}^{\mathrm{j}2\pi f_t/f_R}, \tag{8.23}$$

where ϕ_t and f_t represent the look angle, measured from broadside, and Doppler frequency respectively, $\otimes$ represents the Kronecker product of two vectors, f_R is the pulse repetition frequency (PRF), and λ the wavelength of operation. The vectors $\mathbf{e}_t$ and $\mathbf{e}_s$ represent the temporal and spatial steering vectors, respectively. Note that from one pulse to the next and from one element to the next the steering vectors represent a *constant phase shift*.

Adaptive algorithms generally determine a weight vector $\mathbf{w}$ to obtain a test statistic, Λ, i.e.,

$$\Lambda = \left|\mathbf{w}^H \mathbf{x}\right|^2 \underset{H_0}{\overset{H_1}{\gtrless}} \gamma, \tag{8.24}$$

where the superscript ^{H} represents the Hermitian transpose of a vector/matrix and γ represents a *threshold* above which a target is declared present. This threshold determines the probability of false alarm, the rate at which a target is *detected* "by mistake". For Gaussian interference statistics, the optimum processing method, corresponding to the case of a known interference covariance matrix $\mathbf{R}_d$, is the whiten-and-match filter for detecting a rank-1 signal given by [22]

$$\mathbf{w} = \frac{\mathbf{R}_d^{-1}\mathbf{e}}{\sqrt{\mathbf{e}^H \mathbf{R}_d^{-1}\mathbf{e}}} \Rightarrow \Lambda_{\mathrm{MF}} = \frac{|\mathbf{e}^H \mathbf{R}_d^{-1}\mathbf{x}|^2}{\mathbf{e}^H \mathbf{R}_d^{-1}\mathbf{e}} \underset{H_0}{\overset{H_1}{\gtrless}} \gamma_{\mathrm{MF}}, \tag{8.25}$$

which represents the matched filtering of the whitened data $\check{\mathbf{x}} = \mathbf{R}_d^{-1/2}\mathbf{x}$ and whitened steering vector $\check{\mathbf{e}} = \mathbf{R}_d^{-1/2}\mathbf{e}$. It can be readily shown that Λ_{MF} is simply the output signal-to-interference-plus-noise ratio (SINR) of the minimum variance distortionless response (MVDR) beamformer, the maximum likelihood *estimate* of the target complex amplitude. The relationship between Λ_{MF} and the MVDR beamformer output SNR thus provides a unified perspective of detection and estimation in the context of space-time adaptive processing (STAP).

In practice, the covariance matrix, $\mathbf{R}_d$, is unknown and must be estimated. Early work on antenna arrays by Widrow (least squares method) [27] and Applebaum (maximum SNR criterion) [2] suggest use of feedback loops to ensure convergence of iterative methods for calculating the weight vector. However, these methods were slow to converge to the steady-state solution. Fundamental work by Reed, Mallet, and Brennan (RMB beamformer) [21] showed that the sample matrix inverse (SMI) method offered considerably better convergence. In the SMI approach, the basis for most modern STAP algorithms, the interference covariance matrix is *estimated* using K data ranges for training,

$$\hat{\mathbf{R}}_d = \frac{1}{K}\sum_{k=1}^{K} \mathbf{x}_k \mathbf{x}_k^H = \frac{1}{K}\mathbf{X}\mathbf{X}^H, \tag{8.26}$$

where $\mathbf{X} = [\mathbf{x}_1, \mathbf{x}_2, \ldots, \mathbf{x}_K]$ and the adaptive weights are obtained as $\mathbf{w} = \hat{\mathbf{R}}_d^{-1}\mathbf{e}$. A drawback of the RMB approach is the lack of a constant false alarm rate (CFAR), i.e., the false alarm rate varies with the interference level, an important consideration in practical systems. Variants of the RMB beamformer to obtain CFAR, such as the Kelly GLRT [15], the adaptive matched filter [22] and the adaptive coherence estimator (ACE) [16], were the focus of a number of efforts in the 1980s and early 1990s. Interestingly, the whiten-and-match filter of (8.25), with the true covariance matrix, $\mathbf{R}_d$, replaced with the estimated covariance matrix, $\hat{\mathbf{R}}_d$, has CFAR. There are, however, three fundamental problems with this approach when applied in the real world: the associated computation load, the need for an adequate number of training samples and finally, most importantly, the heterogeneity of the available data.

The SMI algorithm requires the solution to a system of linear equations involving a $JN \times JN$ matrix in real time, an $\mathcal{O}(J^3N^3)$ operation. The fact that the algorithm must be executed for each range and Doppler bin of interest exacerbates the problem. Furthermore, to obtain performance within 3 dB of optimum, one requires approximately $K \geq 2JN$ training samples to estimate the $JN \times JN$ matrix $\mathbf{R}_d$. Such a large number of samples are generally not available.

Finally and most importantly, the training data must be *homogeneous*, i.e., statistically representative of the interference within the range cell of interest. This is generally impossible to obtain in practice due to limitations imposed by the spatio-temporal nonstationarity of the interference as well as by system considerations such as bandwidth and fast scanning arrays. For example with $J = 11$ and $N = 32$, the parameters for the KASSPER dataset [3], the training data support for 3 dB performance is 704. Assuming an instantaneous RF bandwidth of 500 KHz, this calls for the wide-sense stationarity (homogeneity) over a 400 km range! The scarcity of training data is exacerbated by system errors such as aircraft crabbing and internal clutter motion [26] and environmental considerations such as strong clutter discretes [19], range varying interference spectra and power levels [18], and outlier contamination of training data by target-like signals [20] occurring in dense target scenarios caused by flight formations.

These three issues are interlinked – the computation load is a function of the JN degrees of freedom (DOF) in the adaptive process and the number of training samples are approximately twice the DOF, i.e., reducing the computation load also reduces the required training. Clearly, reducing the required training also addresses the heterogeneity problem, making it easier to acquire an adequate number of training samples.

As is clear from the above discussion, adequately and effectively training the adaptive filter is essential. The central theme of the following discussion is the use of pre-existing and the development of real-time knowledge bases to help in the training process. This knowledge base comprises many aspects – using a priori knowledge in choosing the secondary data, using real-time processing to identify homogeneous data samples and choosing the most effective algorithm based on the available information. The use of knowledge-based processing has resulted in the development of the Knowledge Aided Sensor Signal Processing Expert Reasoning (KASSPER) program [3]. Using simulated and measured data, the preliminary results, now appearing in the literature, show both the importance and improvements from using knowledge aided processing [4, 5, 20].

8.8 Issues of Data Support

As already discussed in this chapter, an extremely important issue in STAP is the formation and inversion of the covariance matrix underlying the disturbance. In practice, the unknown interference covariance matrix is estimated from a set of independent identically distributed (iid) target-free training data, which is assumed to be representative of the interference statistics in a cell under test. Frequently, the training data is subject to contamination by discrete scatterers or interfering targets. In either event, the training data becomes nonhomogeneous. As a result, it is not representative of the interference in the test cell. Hence, standard estimates of the covariance matrix from nonhomogeneous training data result in severely under-nulled clutter. Consequently, CFAR and detection performance suffer. Significant performance improvement can be achieved by employing pre-processing to select representative training data.

Figure 8.8, borrowed from [19], illustrates the importance of homogeneous data. The figure plots the probability of detection (P_d) versus SINR for a false alarm probability of $P_{\text{fa}} = 0.01$ and a clutter-to-noise ratio of 40 dB. The system uses $M = JN = 64$ and $K = 128$ range cells to estimate the interference covariance matrix. The matched filter (MF) curve is theoretical, corresponding to optimal performance in Gaussian clutter. The curve for the adaptive matched filter [see curve AMF (analytical)], operating with homogeneous interference has performance within 3 dB of the MF. The curve corresponding to nonhomogeneous data is obtained using Monte Carlo simulations(AMF discretes, no NHD). The training data is corrupted using 30 high-amplitude discrete targets. As is clear, the nonhomogeneity of the data arising from the discrete contamination of the training data significantly

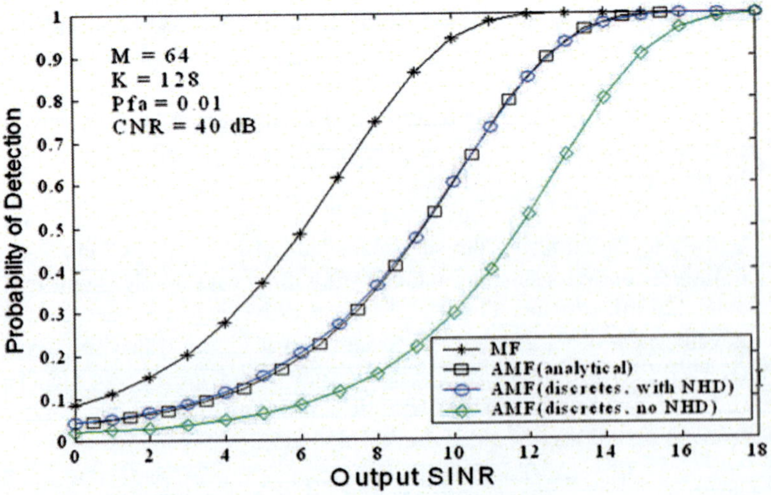

Fig. 8.8 Impact of non-homogeneous data on detection performance

worsens detection performance (a 3–5 dB detection performance loss). Using a non-homogeneity detector (NHD) [19], the performance of the AMF algorithm is restored (see curve for AMF discretes, with NHD).

In general, nonhomogeneity of training data is caused by environmental factors, such as the presence of strong discrete scatterers, dense target environments, nonstationary reflectivity properties of the scanned area, and radar system configurations such as conformal arrays, and bistatic geometries. A variety of robust adaptive signal processing methods to combat specific types of nonhomogeneities have been developed in [9, 10, 20, 24, 25].

8.9 Conclusions

This review has attempted to provide the reader an intuitive and theoretical basis of space-time adaptive processing. First, the pioneering matched filter invention plays a pivotal role in almost every radar system. In particular, the matched filter represents the upper bound on detection performance in a Gaussian interference environment. In practice, knowledge of parameters underlying the interference scenario is not available and hence must be inferred from representative training data. Attempts to attain the matched filter performance in practice are governed by the quantity and quality of training data available to the system designer. The quality and quantity of training data in a given scenario are dictated by the scale of spatio-temporal variation of the scenario and systems considerations such as bandwidth and fast scanning arrays used to sample the interference process.

Acknowledgments Dr. Rangaswamy's work on this chapter was supported by the Air Force Office of Scientific Research under project 2311. The material in Sections 7 and 8 is reproduced from the following publication with permission from the IEEE.

M.C. Wicks, M. Rangaswamy, R.S. Adve, and T.B. Hale, "Space-Time Adaptive Processing: A Knowledge-Based Perspective for Airborne Radar," IEEE Signal Processing Magazine, Vol. 23, no. 1, January 2006, pp. 51–65.

References

1. Almeida S, Eu J (1976) Water pollution monitoring using matched spatial filters. Appl Optics
2. Applebaum S (1966) Adaptive arrays. Tech. rep., Syracuse University Reasearch Corporation
3. Bergin J, Techau P (2002) High-fidelity site-specific radar simulation: Kassper'02 workshop datacube. Tech. Rep. ISL-SCRD-TR-02-105, Defense Advanced Research Projects Agency
4. Bergin J, Teixeira C, Techau P, Guerci J (2004) STAP with knowledge aided data pre-whitening. In: Proceedings of the IEEE Radar Conference, Philadelphia, PA
5. Blunt S, Gerlach K, Rangaswamy M (2004) The enhanced FRACTA algorithm with knowledge aided covariance estimation. In: Proceedings of the Third IEEE Workshop on Sensor Array and Multichannel Processing, Barcelona, Spain
6. Burdic W (1984) Underwater acoustic system analysis. Prentice-Hall, Englewood Cliffs, NJ

7. Clark M, Greenstein L, Kennedy W, Shafi M (1992) Matched filter performance bounds for diversity combining receivers in digital mobile radio. IEEE Trans Vehicular Tech
8. Cook C, Bernfeld M (1993) Radar signals. Artech House, Boston, MA
9. Gerlach KR (2002) Outlier resistant adaptive matched filtering. IEEE Trans Aerospace Electron Syst 38(3):885–901
10. Guerci J, Bergin J (2002) Principal components, covariance matrix tapers, and the subspace leakage problem. IEEE Trans Aerospace Electron Syst 38(1):152–162
11. Hoover A, Kouznetsova V, Goldbaum M (2000) Locating blood vessels in retinal images by piecewise threshold probing of a matched filter response. IEEE Trans Medical Imaging
12. Kay S (1988) Modern spectral estimation: theory and application. Prentice-Hall, Englewood Cliffs, NJ
13. Kay S (1998) Fundamentals of statistical signal processing: detection theory. Prentice-Hall, Upper Saddle River, NJ
14. Kay S (2006) Intuitive probability and random processes using MATLAB. Springer, New York
15. Kelly E (1986) An adaptive detection algorithm. IEEE Trans Aerospace Electron Syst AES-22:115–127
16. Kraut S, Scharf LL, McWhorter L (2001) Adaptive subspace detectors. IEEE Trans Signal Process 49:1–16
17. North D (1943) An Analysis of Factors which Determine Signal/Noise Discrimination in Pulsed Carrier Systems. Tech. rep., RCA Laboratory, Reprinted in Proceedings of the IEEE, pp. 1016–1027, July 1963
18. Rangaswamy M (2005) Statistical analysis of the non-homogeneity detector for non-Gaussian interference backgrounds. IEEE Trans Signal Process SP-53
19. Rangaswamy M, Michels JH, Himed B (2004a) Statistical analysis of the nonhomogeneity detector for STAP applications. Digital Signal Process 14(3):253–267
20. Rangaswamy M, Lin F, Gerlach K (2004b) Robust adaptive signal processing methods for heterogeneous radar clutter scenarios. Signal Process 84:1653–1665
21. Reed I, Mallett J, Brennan L (1974) Rapid convergence rate in adaptive arrays. IEEE Trans Aerospace Electron Syst AES-10:853–863
22. Robey F, Fuhrmann D, Kelly E, Nitzberg R (1992) A CFAR adaptive matched filter detector. IEEE Trans Aerospace Electron Syst AES-28:208–216
23. Vleck JV, Middleton D (1946) A theoretical comparison of the visual, aural, and meter reception of pulsed signals in the presence of noise. J Appl Phys
24. Vorobyov S, Gershman A, Luo Z (2003) Robust adaptive beamforming using worst-case performance optimization: a solution to the signal mismatch problem. IEEE Trans Signal Process 51:313–324
25. Wang YL, Chen JW, Bao Z, Peng YN (2003) Robust space-time adaptive processing for airborne radar in nonhomogeneous clutter environments. IEEE Trans Aerospace Electron Syst 39(1):70–81
26. Ward J (1994) Space-time adaptive processing for airborne radar. Tech. Rep. Technical Report 1015, MIT Lincoln Laboratory
27. Widrow B, Mantey PE, Griffiths LJ, Goode BB (1967) Adaptive antenna systems. Proc IEEE 55(12):2143–2161

Chapter 9
Noise-Driven Informatics: Secure Classical Communications via Wire and Noise-Based Computing[1]

Laszlo B. Kish

9.1 Introduction

Very recently, it has been shown that thermal noise and its artificial versions (Johnson-like noises) can be utilized as an information carrier [1] with peculiar properties; therefore, it may be proper to call this topic *Thermal Noise Informatics* [2]. *Thermal Noise-Driven Computing*, *Zero Power Communication*, and *Totally Secure Classical Communication* are relevant examples.

9.2 Zero Power and Zero-Quantum Communications, Stealth Communications

Recently, it has been shown [1] that the equilibrium thermal noise in information channels can be utilized to carry information. In this case, the transmitter does not emit any signal energy into the channel; however, it only modulates the existing noise there. This issue is completely different from the earlier Porod–Landauer debate [3] about the question if communication without net energy cost is possible by gaining back the energy spent in the communicator devices. (In our opinion, Porod is right, energy-free communication is impossible just like energy-free computing; however, those debates are irrelevant here). In our system, the noise is

[1]A short review of earlier results re-edited and expanded, see the references. The noise-based logic results are new and still unpublished at the time of submitting this chapter.

L.B. Kish (✉)
Department of Electrical and Computer Engineering, Texas A&M University, College Station, TX 77843-3128, USA
e-mail: Laszlo.Kish@ece.tamu.edu

L. Cohen et al. (eds.), *Classical, Semi-classical and Quantum Noise*,
DOI 10.1007/978-1-4419-6624-7_9,

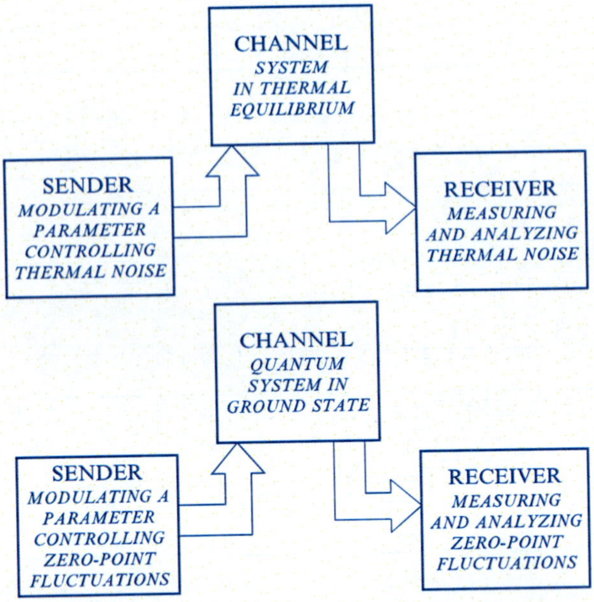

Fig. 9.1 Stealth communications. Zero (signal) power classical communication (*left*) and zero-quantum quantum communication (*right*) [1]

used as information carrier and no effort is made to restore the energy dissipated in the communicator devices. Therefore, this communicator is *not energy-free communication* but *it is free of emitted signal energy*, except perhaps a negligible energy in the order of kT/bit, a presently open problem.

Zero (signal) power classical communication can utilize the modulation of background thermal noise in the information channel and zero-quantum quantum communication can utilize the modulation of the zero-point fluctuations in the quantum channel, see Fig. 9.1 [1].

In [1], some possible realizations were shown. Figure 9.2 shows one of the examples where the Johnson noise of resistors and bandwidth modulation are used in the classical and the quantum limits [1]. In the classical limit, $kT \gg h/RC$, the Johnson noise voltage spectrum is:

$$S_{u,\mathrm{class}}(f) = 4kT\mathrm{Re}[Z(f)], \tag{9.1}$$

and in the quantum limit, $kT \ll h/RC$, the zero point voltage noise spectrum is:

$$S_{u,\mathrm{quan}}(f) = 2hf\mathrm{Re}[Z(f)]. \tag{9.2}$$

The contacted impedance $Z(f)$ is different in the two positions of the switch because of the different capacitance values. The receiver can be a simple noise spectrum analyzer or just a simple AC voltmeter.

In conclusion, it is possible to execute electronic data communication without injecting signal energy in the information channel. Because this is the most invisible way of communication with basic background noise in the channel, it is proper to call these communications *stealth communication*.

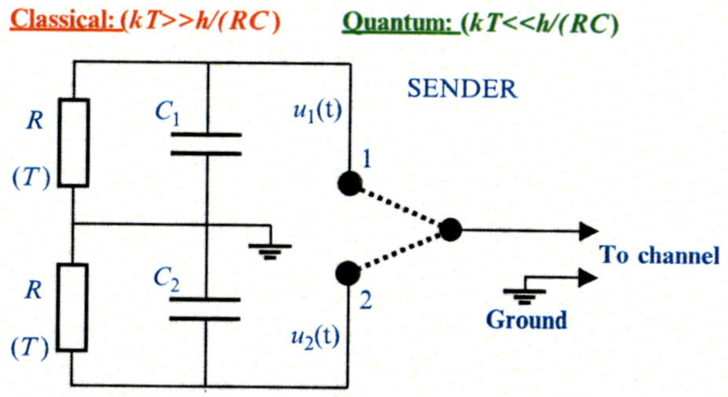

Fig. 9.2 A possible realization of stealth communication with zero power classical communication or zero-quantum quantum communication utilizing classical thermal noise and zero-point quantum fluctuations with frequency bandwidth modulation. Classical or quantum, it depends on the upper cutoff frequency f_c and the temperature. Classical: $hf_c \ll kT$; quantum: $hf_c \gg kT$. The receiver can be a simple noise spectrum analyzer or just a simple AC voltmeter

9.3 Noise-Driven Computing

Because noise turned out to be a special information carrier with low-energy density in the information channel, it is natural to pose the question: can we use it for information processing and computing?

9.3.1 Thermal Noise-Driven Computing?

The first question we addressed is a computer with extraordinarily large error rate, see Fig. 9.3 which we called "thermal noise-driven computer". At thermal noise-driven computing [4], an idea further inspired by the fact that the neural signals in the brain are noise, certain statistical parameters of an electrical noise, including thermal noise, is the information carrier.

The information may be carried by the bandwidth as at the zero power communication examples above or in other way. To study the minimal energy requirement of requirement/dissipation, another specific realization was used, a simple digital system with zero threshold voltage, when the thermal noise is equal to or greater than the digital signal [4]. Under these conditions, when the digital signal amplitude is less than the variance of the noise, classical digital information is usually considered to be zero or useless, see Fig. 9.3.

We would like to note that in a different class of efforts many notable scientists, including John von Neumann [5]; Forshaw and coworkers [6, 7] and references therein; and Palem and coworkers ([8] and references therein, [9]), have been proposing efficient ways of working with probabilistic switches, which are noisy digital logic units with relatively high error probability. For example, Palem and

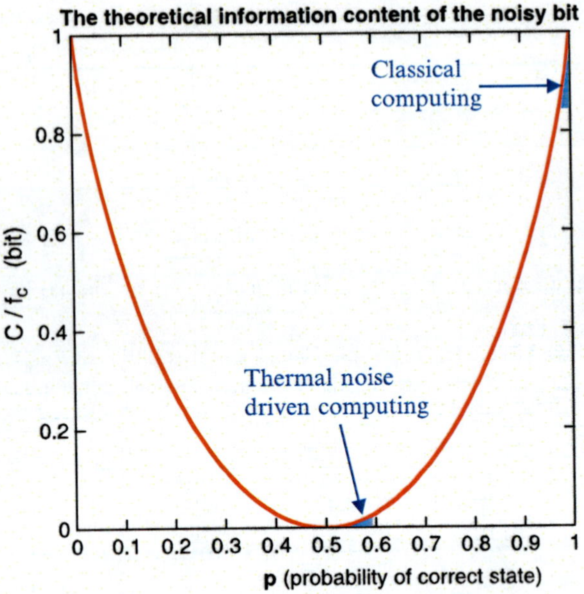

Fig. 9.3 Illustration of the range of error probabilities of classical computing and that of the thermal noise-driven computing. Y-axis: bit content/single clock cycle; plot of Shannon's digital channel coding theorem. A thermal noise-driven computer will supposedly work in that range where the information content/single switching is low, the signal looks like noise and no usual classical computer could work

coworkers have pointed out that this may be a way to reduce power dissipation ([8] and references therein, [9]) if we give up our requirement of data accuracy. However, though these approaches can be relevant to future developments of thermal noise-driven computing, they are very different from our present approach. In a *thermal noise-driven computer*, the voltage in the channel is dominantly noise and the modulation of the statistical properties of this noise carry the information. Therefore, the way of extracting the information is not by "*error correcting the information*" like the efforts mentioned above do but by "*decoding the information in the noise*". The realization example we analyzed is working in the regime of huge error probability, $p \approx 0.5$, with zero logic threshold voltage and in the subnoise signal amplitude limit; and all these parameters are very unusual.

Even though that at this stage there are more open questions than answers and we were are at the moment unable to show a functioning thermal noise-driven logical circuitry, certain questions can already be answered about it. Estimations were given about the information channel capacity and the minimal energy dissipation E_1 given in *Energy/bit* unit which is given as:

$$\eta = \frac{P_s}{C_{dig}\big|_{p\approx 0.5}} = \frac{\pi \ln 2}{2} kT/\text{bit} \approx 1.1\text{bit}/kT, \tag{9.3}$$

where P_s is the power needed to run the subnoise level digital signal, $C_{dig}\big|_{p\approx 0.5}$ is the information channel capacity of this signal.

The main advantage of such a hypothetical thermal noise-driven computer would be a potentially improved energy efficiency and an obvious lack of leakage current, cross-talk and ground EMI problems due to the very low DC voltages. An apparent disadvantage is the large number of extra (redundancy) elements required for error reduction [5–7] if we want a Turing machine.

9.3.2 *Noise-Based Logic: Deterministic, with Low Error Rate*

The thermal noise-driven computer with the extraordinary error rate shown above is interesting; however, it is unclear what type of computer it would be. Certainly not a Turing machine because Turing machines must operate error-free, and error correction is hopeless here. In this section, we briefly mention a different solution, which excellently fits Turing machines because its error rate can be set arbitrarily low.

In noise-based logic [10], the distinct values are represented by independent stochastic processes: independent voltage (or current) noises. The orthogonality of these processes provides a natural way to construct binary or multi-valued logic circuitry with arbitrary number N of logic values by using analog circuitry. Moreover, the logic values on a single wire can be made a (weighted) superposition of the N distinct logic values. Fuzzy logic is also naturally represented by a two-component superposition within the binary case ($N = 2$). Error propagation and accumulation are suppressed. Other relevant advantages are reduced energy dissipation and leakage current problems, and robustness against circuit noise and background noises such as $1/f$, Johnson, shot and crosstalk noise. Variability problems are also nonexistent because the logic value is an *AC* signal. A similar logic system can be built with orthogonal sinusoidal signals (different frequency or orthogonal phase), however that has an extra $1/N$ type slowdown compared to the noise-based logic system with increasing number of N; furthermore, it is less robust against time delay effects than the noise-based counterpart.

The noise-based logic we have introduced in [10], in its binary working mode, has the potential to outperform normal digital circuitry even though it is based on analog circuit elements. Still the most far reaching potential of it maybe the high dimensional multi-value logic type working modes provided their mathematical theory is available.

9.4 Secure Classical Communication via Wire

9.4.1 *The Kirchhoff–Loop–Johnson(-Like)-Noise Cipher*

Recently, a totally secure classical communication scheme, a statistical–physical competitor of quantum communicators, was introduced, see Fig. 9.4 [11, 12], the Kirchhoff–Loop–Johnson(-like)-Noise (KLJN) communicator, with two identical

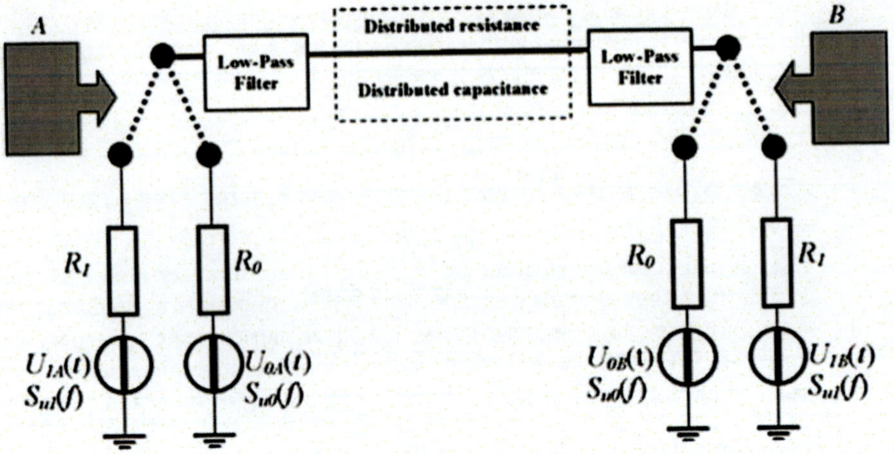

Fig. 9.4 The outline of the KLJN communicator. The practical line can generally be represented by distributed resistors and capacitors; however, with the proper choice of the cable, the driving elements (resistors and capacitance killer, see below), and the frequency range, it behaves as a simple resistor. The instantaneous current and voltage data are measured and compared at the two ends and, in the case of deviance; the eavesdropping alarm goes on and the currently exchanged bit is not used. The low-pass line filters are necessary to protect against out-of-alarm-frequency-band breaking attempts and false alarms due to parasite transients. False alarms would occur due to any wave effect (transient or propagation effects), illegal frequency components or external disturbance of the current-voltage-balance in the wire

pairs of resistors and corresponding Johnson(-like) noise voltage generators. The KLJN communicators are utilizing the physical properties of an idealized Kirchhoff-loop and the statistical physical properties thermal noise (Johnson noise). The resistors (low bit = small resistor, high bit = large resistor) and their thermal-noise-like voltage generators (thermal noise voltage enhanced by a pre-agreed factor) are randomly chosen and connected at each clock period at the two sides of the wire channel. A secure bit exchange takes place when the bit states at the two line ends are different, which is indicated by an intermediate level of the *rms* noise voltage on the line, or that of the *rms* current noise in the wire. The most attractive properties of the KLJN cipher are related to its security [11, 12] and to the extraordinary robustness of classical information when compared that to the fragility of quantum information. To provide security against arbitrary types of attacks, the instantaneous currents and voltages are measured at both ends by Alice and Bob and they are published and compared. In the idealized scheme of the KLJN cipher, the passively observing eavesdropper can extract zero bit (zero-bit security) of information and the actively eavesdropping observer can extract at most one bit before getting discovered (one-bit security) [11]. The system has a natural zero-bit security against the *man-in-the-middle attack,* which is a unique property among secure communicators [13], see Fig. 9.5. The KLJN system has recently became network-ready [14]. This new property opens a large scale of practical applications

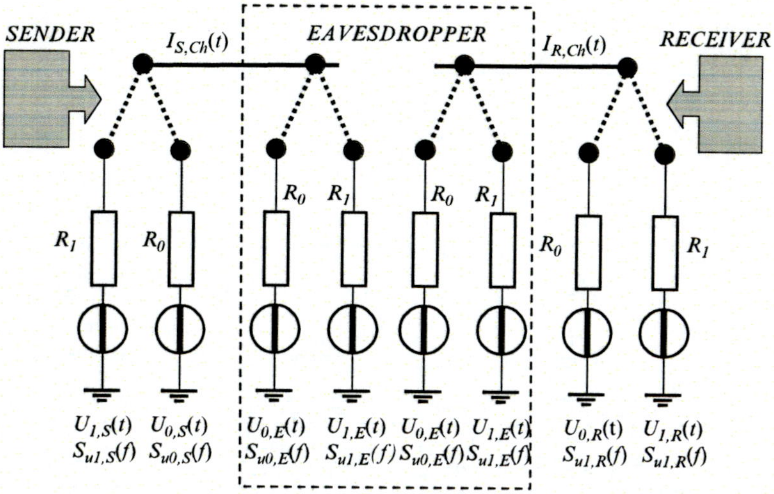

Fig. 9.5 An example for the man-at-the-middle-attack by using resistors with the same values and noise voltage generators with the same parameters as those of the sender and the receiver [11]. In the two separate loops, the current noises are totally independent, and thus the attack is immediately discovered within a fraction of the clock duration. Zero bit can be extracted and the alarm goes on. The man-in-the-middle-attack is one of the weakest type of attacks against the KLJN cipher. This property is unique

because the KLJN cipher can be installed as a computer card [14], similar to ethernet network cards. Other practical advantages compared to quantum communicators are the high speed, resistance against dust, vibrations, temperature gradients, and the low price [11]. It has been shown [15] that the KJLN communicator may use currently used wire lines, such as power lines, phone lines, internet wire lines by utilizing proper filtering methods.

Concerning the security of the KLJN system, the basic rule of any physical secure communication holds here, too: *If you compromise it you lose it.* Although there have been several attempts to break in the KJLN line, so far, no proposed method has been able to challenge the total security of the idealized KLJN system. All these breaking attempts have used certain assumptions which are *directly violating the basic model* of the idealized KLJN method. These assumptions are as follows:

1. Allowing nonnegligible wire resistance (Bergou [12], Scheuer and Yariv [16]). For the evaluation of this claim at practical conditions, see the response in [17].
2. Allowing high-enough bandwidth for wave (transient propagation) effects (Scheuer and Yariv [16]). This possibility has been explicitly excluded since the original article [11]. For further refusal of this claim, see the response in [17]. Moreover, *and this is the most direct practical argument*, in this case the standard current and voltage protection of the KLJN system would *immediately alarm and*

Table 9.1 Summary of the security analysis of the realized KLJN communicator pair and a quick comparison with quantum security

Type of breaking	Measured number, or ratio, of eavesdroppable bits without setting on the current-voltage alarm (tested through 74497 bits)	Remarks
BSchY (*i*) [12, 16] attack in the present KLJN system	0.19%	0.00000019% at 10 times thicker wire (theoretical extrapolation). Arbitrarily can be enhanced by privacy amplification [20, 21]; the price is slowing down
Hao (*iii*) [16] attack in the present KLJN system	Zero bit	Below the statistical inaccuracy. Considering the 12 bit effective resolution of noise generation accuracy, it is theoretically: $<10^{-10}$
Kish (*iv*) [17] attack utilizing resistor inaccuracies in the present KLJN system	Zero bit	Below statistical inaccuracy. Theoretically, when pessimistically supposing 1% resistance inaccuracy: <0.01%
Current pulse injection (Kish) [9] in the present KLJN system	Zero bit	One bit can be extracted while the alarm goes on thus the bit cannot be used

shut down the communication because the existence of transient and wave effects *per-definition* yield different instantaneous current and/or voltage values at the two ends.

3. Assuming inaccuracy of (noise) temperatures (Hao [18]). For the theoretical refusal of this claim, see [19].
4. Utilizing practical inaccuracy of resistors (Kish [19]). For the mathematical evaluation of this claim at practical conditions, see there [19].

Let us here consider an analogy in the field of quantum communication. Although all the practical quantum communicators suffer from the inability of producing *totally single-photon* output and this fact is unavoidably compromising the security of practical quantum communicators, we do not say that the idealized/mathematical quantum communicator schemes are insecure. In discussing the unconditional security of idealized quantum communication schemes, we suppose that single-photon sources do exist, which is a similar though more difficult claim than to assume the existence of zero wire resistance (c.f. superconductors), and then we conclude that the quantum system is therefore unconditionally secure, at least conceptually. Therefore, the practical claim that idealized single-photon source does not exist cannot compromise the claim about the security of an idealized, conceptual quantum communicator. On the other hand, as soon as the security of a *practical communicator system* is discussed, the aspects nonideal single-photon sources, optical fiber absorption, etc., for quantum communicators and similarly the aspects (*i-iv*) listed above for the KLJN systems cannot be neglected any more and the practical design must keep these effects under control (Table 9.1).

The realized KLJN communicator shows excellent performance through the model-line through communication ranges orders of magnitude beyond the ranges of known direct quantum communication channels [21]. While security claims of quantum communicators are usually theoretical because of the expensive nature of relevant breaking tests, in our practical system we have been able to carry out extensive security tests with various ways of breaking attempts. The study indicates that the most important type of security leak, which will need most of resources to control, is the Bergou–Scheuer–Yariv type wire resistance effect [12, 16, 17]; and the rest of the practical security compromises can be neglected, such as the transient/wave arguments of Scheuer–Yariv [16, 17], the noise strength argument of Hao [18] and the resistor inaccuracy argument of Kish [19]. The results indicate unrivalled fidelity and security levels among existing physical secure communicators and there are straightforward ways to further improve security, fidelity and range, if it is necessary and resources are available, such as thicker cable.

Acknowledgments Comments of Gabor Schmera about the draft are appreciated. Zoltan Gingl is grateful for the Bolyai Fellowship of Hungarian Academy of Sciences. The travel of LBK to the University of Szeged for the startup phase of the experiments was covered by the Swedish STINT foundation and the cost of staying (10–15 December 2006) was partially covered by the European Union's SANES grant. The costs of the KLJN system design were partially covered by the TAMU Information Technology Task Force (TITF, grant 2002). Part of these results was submitted to Physics Letters A [10].

References

1. Kish LB (2005) Stealth communication: zero-power classical communication, zero-quantum quantum communication and environmental-noise communication. Appl Phys Lett 87:234109
2. Kish LB, Mingesz R, Gingl Z (2007) Thermal noise informatics: totally secure communication via a wire; Zero-power communication; and thermal noise driven computing. Plenary talk, Proceedings of the 4th SPIE Symposium on Fluctuation and Noise (SPIE)
3. Porod W (1988) Appl Phys Lett 52:2191
4. Kish LB (2006) Thermal noise driven computing. Appl Phys Lett 89:144104
5. von Neumann J (1956) In: Shannon CE, McCarthy J (eds) Automata studies. Princeton University Press, Princeton, pp 329–378
6. Nikolic K, Sadek A, Forshaw M (2002) Nanotechnology 13:357–362
7. Sadek AS, Nikolic K, Forshaw M (2004) Nanotechnology 15:192–21
8. Palem KV (2005) IEEE Trans Comput 54:1123
9. Korkmaz P, Akgul BES, Palem KV, Chakrapani LN (2006) Japan J Appl Phys 45:3307–3316
10. Kish LB. Noise-based logic: binary, multi-valued, or fuzzy, with optional superposition of logic states. Submitted, preprint: http://arxiv.org/abs/0808.3162
11. Kish LB (2006) Totally secure classical communication utilizing Johnson (-like) noise and Kirchhoff's law. Phys Lett A 352:178–182
12. Cho A (2005) Simple noise may stymie spies without quantum weirdness. Science 309:2148
13. Kish LB (2006) Protection against the man-in-the-middle-attack for the Kirchhoff-loop-Johnson(-like)-noise cipher and expansion by voltage-based security. Fluctuation Noise Lett 6:L57–L63

14. Kish LB, Mingesz R (2006) Totally secure classical networks with multipoint telecloning (teleportation) of classical bits through loops with Johnson-like noise. Fluctuation Noise Lett 6:C9–C21
15. Kish LB (2006) Methods of using existing and currently used wire lines (power lines, phone lines, internet lines) for totally secure classical communication utilizing Kirchoff's Law and Johnson-like Noise, draft. http://arxiv.org/abs/physics/0610014
16. Scheuer J, Yariv A (2006) A classical key-distribution system based on Johnson (like) noise—How secure? Phys Lett A 359:737
17. Kish LB (2006) Response to Scheuer-Yariv: "a classical key-distribution system based on johnson (like) noise - how secure?" Phys Lett A 359:741–744
18. Hao F (2006) Kish's key exchange scheme is insecure. IEE Proc Inform Sec 153:141–142
19. Kish LB (2006) Response to Feng Hao's paper "Kish's key exchange scheme is insecure. Fluctuation Noise Lett 6:C37–C41
20. Mingesz R, Gingl Z, Kish LB (2008) Johnson(-like)–Noise–Kirchhoff-loop based secure classical communicator characteristics, for ranges of two to two thousand kilometers, via model-line. Phys Lett A 372:978–984
21. Ursin R, Tiefenbacher F, Schmitt-Manderbach T, Weier H, Scheidl T, Lindenthal M, Blauensteiner B, Jennewein T, Perdigues J, Trojek P, Ömer B, Fürst M, Meyenburg M, Rarity J, Sodnik Z, Barbieri C, Weinfurter H, Zeilinger A (2007) Entanglement-based quantum communication over 144 km. Nat Phys (3 June). doi: 10.1038/nphys629

Chapter 10
Denoising and Time-Frequency Analysis of Signals

Patrick J. Loughlin

10.1 Introduction

The extraction of a signal from noisy data is a common problem in many fields of science and engineering. There are many approaches that have been employed. One approach is to average multiple measurements to reduce the noise, which is effective if the noise is random and the signal of interest is consistent from measurement to measurement. Another approach is to fit the data to a model of the signal of interest in some optimal sense, such as by minimizing the squared error, which is effective if an accurate and parsimonious model of the signal is available. Yet another approach is to filter the data, for which there is a variety of approaches, ranging from linear time-invariant filters, which are effective if the noise and signal have distinct frequencies; to time-varying filters, which can be effective if the properties of either the signal or the noise change over time while the other remains stationary over time; to nonlinear filters such as median or order-statistic filters, which are particularly effective for separating transient-type events from more continuous type of events, such as extracting impulsive events from a sinusoidal or slowly varying drift. No single approach, of course, is suitable for all situations.

A particularly challenging problem, in terms of filtering, can be extracting broadband signals from broadband noise, especially when the noise and signal span common frequencies. However, if the *local* spectral properties of the signal and noise are different, then a phase space approach to filtering can be effective for separating the signal from the noise. By "local" spectral properties, we mean the time-conditional spectral moments of a time-frequency phase space distribution, such as the Wigner distribution [4, 10] or the spectrogram [4, 6]. In this chapter,

P.J. Loughlin (✉)
University of Pittsburgh, Department of Bioengineering & ECE, 4200 Fifth Avenue, Pittsburgh, PA, USA
e-mail: loughlin@pitt.edu

L. Cohen et al. (eds.), *Classical, Semi-classical and Quantum Noise*,
DOI 10.1007/978-1-4419-6624-7_10, © Springer Science+Business Media, LLC 2012

we present the basic idea behind this method of filtering, which involves the design of time-varying filters based on the phase space properties of the signal or noise. The approach is demonstrated by application to marine mammal sounds to remove background ocean noise.

10.2 Time-Frequency Analysis

In this section, we give a brief overview of time-frequency analysis, in particular the Wigner distribution and spectrogram [3, 4]. We also discuss the global and local means and standard deviations of these distributions, and their relationship to the signal, especially the local spectral characteristics of the signal as described by the instantaneous frequency and instantaneous bandwidth. We refer the reader to the referenced text and review article by Cohen for further information and details.

10.2.1 Wigner Distribution and Spectrogram

The time-frequency Wigner distribution and spectrogram of a signal $s(t)$ are given, respectively, by

$$W_s(t,\omega) = \frac{1}{2\pi} \int s^*\left(t - \tfrac{1}{2}\tau\right) s\left(t + \tfrac{1}{2}\tau\right) \mathrm{e}^{-\mathrm{j}\tau\omega}\, \mathrm{d}\tau \tag{10.1}$$

$$SP_s(t,\omega) = \frac{1}{2\pi} \left| \int s(\tau)\, h^*(\tau - t)\, \mathrm{e}^{-\mathrm{j}\omega\tau}\, \mathrm{d}\tau \right|^2, \tag{10.2}$$

where $h(t)$ in the spectrogram formulation is called a "window" and is a function that generally tapers to zero, with a peak at $h(0)$, and is of shorter duration than the signal $s(t)$. The window serves to isolate the signal about successive local regions of time, which are then Fourier analyzed. Clearly, "the" spectrogram of a signal depends on the choice of $h(t)$ and especially its duration. Also clear is that the properties of the spectrogram of a signal, such as its moments, will depend on the signal and window (sometimes in very complicated ways). Often one computes several spectrograms with various window lengths, as a single window is generally insufficient to spectrally and temporally resolve both transient and continuous types of signal components. A "short" duration window is needed to temporally resolve components closely spaced in time, while a "long" duration window is needed to spectrally resolve components closely space in frequency. Resolving components jointly in time and frequency, such as two closely spaced chirps, can require signal-dependent windows [5].

The Wigner distribution, as is clear from its formulation, does not involve a window, and as such the properties of the distribution, such as its moments, depend only on the signal, unlike the spectrogram. One often reads that the Wigner distribution has "high resolution" in comparison with the spectrogram. Discussions of "resolution" and "resolving" signal components in time-frequency are sometimes ambiguous, and it is left to the reader to interpret what is meant. If we are talking about resolving events in time, then the best we can do, without resorting to parametric or other model-based approaches, is to maintain the time resolution of the measured signal. Likewise, if we are talking about resolving events in frequency, then we need to preserve the resolution of the spectrum of the signal. The Wigner distribution does this because it satisfies the marginal properties,

$$\int W_s(t,\omega)\,\mathrm{d}\omega = |\,s(t)\,|^2; \qquad \int W_s(t,\omega)\,\mathrm{d}t = |\,S(\omega)\,|^2, \tag{10.3}$$

where $S(\omega)$ is the Fourier transform of the signal $s(t)$,

$$s(t) = \frac{1}{\sqrt{2\pi}}\int S(\omega)\,\mathrm{e}^{\mathrm{j}\omega t}\mathrm{d}\omega; \qquad S(\omega) = \frac{1}{\sqrt{2\pi}}\int s(t)\,\mathrm{e}^{-\mathrm{j}\omega t}\,\mathrm{d}t. \tag{10.4}$$

The marginals $|\,s(t)\,|^2$ and $|\,S(\omega)\,|^2$ are the energy temporal density or "instantaneous power" and the energy spectral density, respectively. Clearly, any distribution that satisfies the marginals, of which there is an unlimited number [2, 3], has the same time resolution and frequency resolution, in the sense discussed above, as the Wigner distribution. For comparison, the marginals of the spectrogram are

$$\int \mathrm{SP}_s(t,\omega)\,\mathrm{d}\omega = \int |\,s(\tau)\,|^2\,|\,h(\tau-t)\,|^2\mathrm{d}\tau \tag{10.5}$$

$$\int \mathrm{SP}_s(t,\omega)\,\mathrm{d}t = \int |\,S(\theta)\,|^2\,|\,H(\theta-\omega)\,|^2\,\mathrm{d}\theta, \tag{10.6}$$

which makes clear the trade-off in temporal resolution versus spectral resolution imposed by the window $h(t)$ (with Fourier transform $H(\omega)$).

10.2.2 Global Moments: Mean and Standard Deviation

A common measure often associated with resolution is standard deviation, which indicates the spread about the mean. So, for example, for the spectrogram, the standard deviations in time and frequency are given by

$$\sigma_t^{2\,(\mathrm{SP})} = \int\!\!\int \left(t - \langle t\rangle^{(\mathrm{SP})}\right)^2\, SP_s(t,\omega)\,\mathrm{d}t\,\mathrm{d}\omega = \sigma_t^{2\,(s)} + \sigma_t^{2\,(h)} \tag{10.7}$$

$$\sigma_\omega^{2\,(\mathrm{SP})} = \int\!\!\int \left(\omega - \langle \omega\rangle^{(\mathrm{SP})}\right)^2\, SP_s(t,\omega)\,\mathrm{d}t\,\mathrm{d}\omega = \sigma_\omega^{2\,(s)} + \sigma_\omega^{2\,(h)}, \tag{10.8}$$

where the superscripts (SP), (s) and (h) denote the spectrogram, signal, and window, respectively, and $\langle t\rangle^{(\mathrm{SP})}$ and $\langle\omega\rangle^{(\mathrm{SP})}$ are the mean time and mean frequency of the spectrogram,

$$\langle\omega\rangle^{(\mathrm{SP})} = \iint \omega\, \mathrm{SP}_s(t,\omega)\,\mathrm{d}t\,\mathrm{d}\omega = \langle\omega\rangle^{(s)} - \langle\omega\rangle^{(h)} \tag{10.9}$$

$$\langle t\rangle^{(\mathrm{SP})} = \iint t\, \mathrm{SP}_s(t,\omega)\,\mathrm{d}t\,\mathrm{d}\omega = \langle t\rangle^{(s)} - \langle t\rangle^{(h)}. \tag{10.10}$$

Hence, we see that the window increases the spread in time and in frequency, and there is a trade-off in that reducing the spread in one domain increases the spread in the other, because of the uncertainty principle (or time-bandwidth product) of the window

$$\sigma_t^{2\,(h)}\,\sigma_\omega^{2\,(h)} \geq \frac{1}{4}. \tag{10.11}$$

For the Wigner distribution, on the other hand, because it satisfies the marginals, the standard deviations in time and in frequency are equal to those of the signal

$$\sigma_t^{2\,(W)} = \iint (t - \langle t\rangle)^2\, W_s(t,\omega)\,\mathrm{d}t\,\mathrm{d}\omega$$

$$= \int (t - \langle t\rangle)^2\, |\, s(t)\,|^2 = \sigma_t^{2\,(s)} \tag{10.12}$$

$$\sigma_\omega^{2\,(W)} = \iint (\omega - \langle\omega\rangle)^2\, W(t,\omega)\,\mathrm{d}t\,\mathrm{d}\omega$$

$$= \int (\omega - \langle\omega\rangle)^2\, |\, S(\omega)\,|^2 = \sigma_\omega^{2\,(s)}, \tag{10.13}$$

where the superscript (W) and (s) denote the Wigner distribution and signal, respectively, and $\langle t\rangle$ and $\langle\omega\rangle$ are the mean time and mean frequency

$$\langle\omega\rangle = \iint \omega\, W(t,\omega)\,\mathrm{d}t\,\mathrm{d}\omega = \int \omega |\, S(\omega)\,|^2\,\mathrm{d}\omega \tag{10.14}$$

$$\langle t\rangle = \iint t\, W(t,\omega)\,\mathrm{d}t\,\mathrm{d}\omega = \int t |\, s(t)\,|^2\,\mathrm{d}t. \tag{10.15}$$

Clearly, the Wigner distribution satisfies the uncertainty principle (time-bandwidth product) of the signal

$$\sigma_t^{2\,(W)}\,\sigma_\omega^{2\,(W)} = \sigma_t^{2\,(s)}\,\sigma_\omega^{2\,(s)} \geq \frac{1}{4}. \tag{10.16}$$

For the spectrogram, it follows from above that its time-bandwidth product is greater than that of the signal,

$$\sigma_t^{2\,(SP)}\,\sigma_\omega^{2\,(SP)} > \sigma_t^{2\,(s)}\,\sigma_\omega^{2\,(s)} + \sigma_t^{2\,(h)}\,\sigma_\omega^{2\,(h)}. \tag{10.17}$$

In fact, it can be shown that for the spectrogram, $\sigma_t^{2\,(SP)}\,\sigma_\omega^{2\,(SP)} \geq 1$ [4, 11].

10.2.3 Local Moments: Conditional Mean and Standard Deviation

The above calculations quantify the spread in time and the spread in frequency of the distribution, which may indicate something about the resolution of the distribution. Certainly, for the spectrogram, we see clearly how the window comes into play and the trade-offs it forces. However, the standard deviations above are *global* quantities, in that they depend only on the marginals and quantify the spread or resolution in time, and the spread or resolution in frequency. They do not provide any insight into spread or resolution *jointly* in time-frequency. One way to do this, analogous to the global case, is to use the *conditional*, or "local," standard deviations of time and frequency. For arbitrary time-frequency distribution $P(t,\omega)$, these are given by

$$\sigma_{\omega|t}^2 = \int (\omega - \langle\,\omega\,\rangle_t)^2\; P(\omega|t)\,\mathrm{d}\omega \tag{10.18}$$

$$\sigma_{t|\omega}^2 = \int (t - \langle\,t\,\rangle_\omega)^2\; P(t|\omega)\,\mathrm{d}t, \tag{10.19}$$

where $P(\omega\,|t)$ and $P(t\;\omega)$ are the conditional distributions given by

$$P(\omega\,|\,t) = \frac{P(t,\omega)}{P(t)}; \qquad P(t\,|\,\omega) = \frac{P(t,\omega)}{P(\omega)}, \tag{10.20}$$

where $P(t)$ and $P(\omega)$ are the marginal distributions of $P(t,\omega)$, and the quantities $\langle\,\omega\,\rangle_t$ and $\langle\,t\,\rangle_\omega$ are the mean frequency at a given time and the mean time at a given frequency, respectively,

$$\langle\,\omega\,\rangle_t = \int \omega\; P(\omega|t)\,\mathrm{d}\omega; \qquad \langle\,t\,\rangle_\omega = \int t\; P(t|\omega)\,\mathrm{d}t. \tag{10.21}$$

For many time-frequency distributions, including the Wigner distribution but not the spectrogram, the conditional means are equal to the derivative of the phase of the signal and its spectrum. Specifically, if we write the signal and its Fourier transform in terms of amplitude and phase as

$$s(t) = A(t)\mathrm{e}^{\mathrm{j}\varphi(t)}; \qquad S(\omega) = B(\omega)\mathrm{e}^{\mathrm{j}\psi(\omega)}, \tag{10.22}$$

then for the Wigner and many other time-frequency distributions, the conditional means above are given by

$$\langle\omega\rangle_t = \int \omega\, P(\omega|t)\, \mathrm{d}\omega = \varphi'(t); \qquad \langle t\rangle_\omega = \int t\, P(t|\omega)\, \mathrm{d}\omega = -\psi'(\omega), \tag{10.23}$$

where $\varphi'(t)$ is the instantaneous frequency of the signal, and $-\psi'(\omega)$ is the group delay.

The relationship of the conditional standard deviations to the signal is more complicated and varies from distribution to distribution. For many distributions, including the Wigner distribution (but not the spectrogram), the local or conditional spectral variance is given by

$$\sigma^2_{\omega|t} = \frac{1}{2}\left[(1+a)\left(\frac{A'(t)}{A(t)}\right)^2 - (1-a)\frac{A''(t)}{A(t)}\right], \tag{10.24}$$

where the constant a depends on the particular distribution; for example, for the Wigner distribution, $a = 0$, while for the Margenau–Hill distribution, $a = -1$, and for the Cohen–Lee distribution, $a = 1$ [3, 4]. Note that the conditional spectral variance given by (10.24) can be negative, which is problematic from the standpoint of interpreting it in the usual sense, namely as the spread about the mean. However, the *average* of these local variances is nonnegative. In particular, if we define the average local spectral and temporal variances by

$$\langle\, \sigma^2_{\omega|t}\, \rangle = \int \sigma^2_{\omega|t}\, P(t)\, \mathrm{d}t; \qquad \langle\, \sigma^2_{t|\omega}\, \rangle = \int \sigma^2_{t|\omega}\, P(\omega)\, \mathrm{d}\omega, \tag{10.25}$$

then for the Wigner and many other time-frequency distributions, we have that these quantities are related to the signal and spectrum via [7]

$$\langle\sigma^2_{\omega|t}\,\rangle^{(W)} = \int \sigma^2_{\omega|t}\, P(t)\mathrm{d}t = \int \left|\frac{A'(t)}{A(t)}\right|^2 |s(t)|^2\, \mathrm{d}t = \left\langle \left|\frac{A'(t)}{A(t)}\right|^2 \right\rangle \tag{10.26}$$

$$\langle\sigma^2_{t|\omega}\rangle^{(W)} = \int \sigma^2_{t|\omega}\, P(\omega)\mathrm{d}\omega = \int \left|\frac{B'(\omega)}{B(\omega)}\right|^2 |S(\omega)|^2\, \mathrm{d}\omega = \left\langle \left|\frac{B'(\omega)}{B(\omega)}\right|^2 \right\rangle. \tag{10.27}$$

In contrast, for the spectrogram, it can be shown that the average local variances are given by [7]

$$\langle\, \sigma^2_{\omega|t}\, \rangle^{(\mathrm{SP})} \geq \left\langle \left|\frac{A'(t)}{A(t)}\right|^2 \right\rangle + \left\langle \left|\frac{A_h'(t)}{A_h(t)}\right|^2 \right\rangle > \langle\, \sigma^2_{\omega|t}\, \rangle^{(W)} \tag{10.28}$$

$$\langle\, \sigma^2_{t|\omega}\, \rangle^{(\mathrm{SP})} \geq \left\langle \left|\frac{B'(\omega)}{B(\omega)}\right|^2 \right\rangle + \left\langle \left|\frac{B_h'(\omega)}{B_h(\omega)}\right|^2 \right\rangle > \langle\, \sigma^2_{t|\omega}\, \rangle^{(W)}, \tag{10.29}$$

where $A_h(t)$ and $B_h(\omega)$ denote the amplitude of the window and its Fourier transform, respectively. Hence, we see that, as with the global variances, the local variances of the spectrogram are larger than those of the Wigner and other distributions, because of the window. Because of this, the "local uncertainty product" of the spectrogram is larger than that of the Wigner and other distributions, which satisfy the relation [7],

$$0 \leq \langle \sigma^2_{\omega|t} \rangle \langle \sigma^2_{t|\omega} \rangle \leq \sigma^2_\omega \sigma^2_t. \tag{10.30}$$

In the following section, we show how ideas based on the local spectral moments, such as the conditional mean frequency, $\langle \omega \rangle_t$, and the spread about the mean, $\sigma_{\omega|t}$, discussed above, can be used to extract time-varying signals from noise.

10.3 Denoising Based on Local Spectral Moments

As discussed in the Introduction, extracting broadband signals from broadband noise can be particularly challenging when the noise and signal spectra span common frequencies. However, many broadband signals are locally narrowband, meaning that the bandwidth at any given time (the so-called instantaneous bandwidth as described by the conditional standard deviation $\sigma_{\omega|t}$) is small compared to the total (global) bandwidth (given by σ_ω). This situation occurs in particular for signals with significant frequency modulation (FM) but moderate to little amplitude modulation (AM) [4]. This characteristic of broadband signals that are locally narrowband can be exploited to improve noise suppression for such signals. Specifically, in the time-frequency plane, broadband noise is typically distributed across the plane, whereas locally narrowband signal components are concentrated to specific frequencies at particular times. Thus, while the total SNR (which is a global attribute of the signal) may be small, in the time- frequency plane the SNR over a particular time-frequency region can be much greater than the global SNR, due to the dispersement of broadband noise across the plane and the concentration of locally narrowband signal components to specific time-frequency regions. This feature of locally narrowband signals in broadband noise allows one to accurately estimate the AM and FM of the signal components, and extract them from the noise far more effectively than a conventional linear time-invariant filtering approach would allow. We outline this approach next.

10.3.1 AM-FM Signal Model

We model the signal as the real part of a complex signal with amplitude $A(t)$ and phase $\phi(t)$,

$$s(t) = Re\left\{A(t)e^{j[\phi_A(t)+\phi_F(t)]}\right\} = Re\left\{(A_I(t) + jA_Q(t))e^{j\phi_F(t)}\right\}, \tag{10.31}$$

where the signal phase is divided into two parts [8], $\phi(t) = \phi_A(t) + \phi_F(t)$, one of which ($\phi_F(t)$) determines the frequency modulations (FM) of the signal, and the other ($\phi_A(t)$) is coupled with the amplitude to produce quadrature amplitude modulation (QAM)

$$A(t)\mathrm{e}^{j\phi_A(t)} = A_I(t) + jA_Q(t). \tag{10.32}$$

Note that we can reconstruct a signal from its FM and QAM. As shown next, time-frequency distributions allow us to obtain noise-robust estimates of these quantities, in particular for locally narrow-band mono-component signals in broadband noise, thereby resulting in enhanced noise-removal compared to LTI methods.

10.3.2 Algorithm for Noise Removal via TFD-Based AM-FM Decomposition

Let the observed signal be corrupted by additive broadband noise, $x(t) = s(t) + \eta(t)$. The underlying assumption of the proposed method is that in the time-frequency plane, the signal $s(t)$ will stand out above the noise floor caused by $\eta(t)$. This assumption is often accurate, particularly for signals that are locally narrow-band. For such signals, we can obtain from a TFD $P(t,\omega)$ of the signal a robust estimate of the FM part of the phase, $\phi_F(t)$, via peak picking as

$$\hat{\phi}'_F(t) = \max_{\omega} \ \hat{P}(t,\omega), \quad \hat{\phi}_F(t) = \int_0^t \hat{\phi}'_F(\tau)\mathrm{d}\tau, \tag{10.33}$$

where we have assumed the signal starts at $t = 0$ in integrating the instantaneous frequency to get the phase $\hat{\phi}_F(t)$. After estimating the FM part of the phase, we then obtain estimates of the in-phase and quadrature amplitudes via time-varying coherent demodulation

$$\hat{A}_I(t) = \int x(\tau)\cos\{\hat{\phi}_F(\tau)\}h_{lp}(t,\tau)\mathrm{d}\tau \tag{10.34}$$

$$\hat{A}_Q(t) = \int x(\tau)\sin\{\hat{\phi}_F(\tau)\}h_{lp}(t,\tau)\mathrm{d}\tau, \tag{10.35}$$

where $h_{lp}(t,\tau)$ is the impulse response of a linear time-varying low-pass filter. The corner frequency of the filter is set to 2–3 times the instantaneous bandwidth of the signal, which is obtained from the conditional spectral variance of the TFD of the observed signal,

$$\sigma^2_{\omega|t} = \int (\omega - \phi'_F(t))^2 P(\omega|t)\mathrm{d}\omega. \tag{10.36}$$

In practice, the estimates are obtained iteratively to improve accuracy [9].

10.3.3 Example

In this section, we illustrate the method above via application to a whale sound recorded in ambient ocean noise. The time-series of the recorded whale sound is shown on the left in Fig. 10.1a. A time-frequency distribution [1] of this signal is shown on the right. We emphasize that the noise in this signal was not simulated or added to a clean recording; the noise was genuine and was part of the original recording.

In Fig. 10.1c, we show the reconstructed signal and its TFD, using a linear time-invariant filter to remove as much noise outside the signal band as possible. In Fig. 10.1e, the reconstructed signal obtained by the AM-FM time-varying coherent demodulation procedure is shown, along with its TFD. The difference between this reconstructed signal and the original signal is shown in Fig. 10.1f. Note that this difference signal looks like random noise, while the reconstructed signal is virtually noise free. Thus, it is clear that the time-varying approach has far more effectively separated the signal from noise than the conventional LTI approach.

To quantify the improvement in noise removal, we generated synthetic filtered white noise with similar power and spectral characteristics as the ocean noise (as determined from the residual signal in Fig. 10.1g), and added this noise to the AM-FM reconstructed signal in Fig. 10.1e. The SNR of this noisy signal was −2 dB (where we took the reconstructed signal to be the "noise-free" signal in the calculation of the SNR). We then processed this noisy signal as above, i.e., with an LTI filter (passband 2250-4375 Hz), and with the proposed TFD-based AM-FM demodulation method. The LTI filter improved the SNR by just over 3 dB (to an SNR of 1.1 dB) and the TFD-based AM-FM method improved the SNR by more than 15 dB (to an SNR of 13.2 dB). Hence, there is a 12 dB gain in signal-to-noise power enhancement by the time-varying AM-FM denoising procedure compared to the conventional LTI filtering approach.

10.4 Conclusion

In this chapter, we reviewed two common methods for time-frequency analysis, namely the spectrogram and Wigner distribution. We discussed low-order global and local (or conditional) moments of these distributions, in particular the spectral mean and variance, and showed how the window of the spectrogram distorts these moments. The bearing of the global and local variances on the issue of resolution, and the uncertainty principle, were also discussed. Finally, we showed how one can remove noise from broadband signals that are locally narrowband, based on the ideas of the conditional mean and variance and the local spectral properties of the signal, as obtained from time-frequency analysis of the signal.

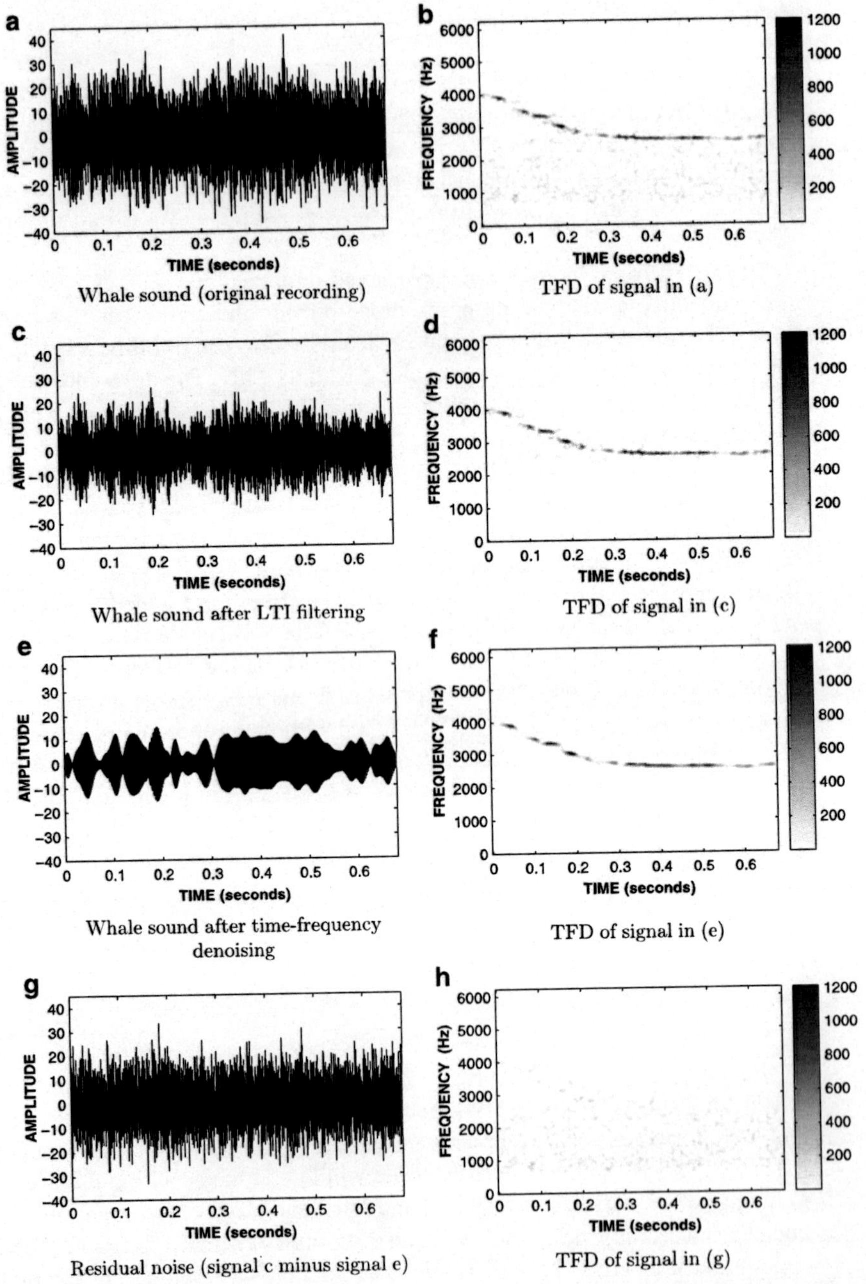

Fig. 10.1 Time series (*left column*) and corresponding time-frequency distribution (*right column*) for a whale sound recorded in the ocean (**a**, **b**) and after various stages of filtering (**c**–**h**). (**a**) Whale sound (original recording), (**b**) TFD of signal in (**a**), (**c**) Whale sound after LTI filtering, (**d**) TFD of signal in (**c**), (**e**) Whale sound after time-frequency denoising, (**f**) TFD of signal in (**e**), (**g**) Residual noise (signal **c** minus signal **e**), (**h**) TFD of signal in (**g**)

Acknowledgements It is a pleasure to dedicate this chapter to David Middleton. The author's research was supported by a grant from the Office of Naval Research (N00014-06-1-0009).

References

1. Çakrak F, Loughlin P (2001) Multiple window time-varying spectral analysis. IEEE Trans Sig Process 49(2):448–453
2. Cohen L (1966) Generalized phase–space distribution functions. J Math Phys 7:781–786
3. Cohen L (1989) Time-frequency distributions-A review. Proc IEEE 77(7):941–981
4. Cohen L (1995) Time-frequency analysis. Prentice-Hall
5. Jones D, Parks T (1992) A resolution comparison of several time-frequency representations. IEEE Trans Signal Process 40(2):413–420
6. Koenig W, Dunn H, Lacy L (1946) The sound spectrograph. J Acoust Soc Am 18(1):19–49
7. Loughlin P, Cohen L (2004) The uncertainty principle: global, local or both? IEEE Trans Signal Process 52(5):1218–1227
8. Loughlin P, Tacer B (1996) On the amplitude- and frequency-modulation decomposition of signals. J Acoust Soc Am 100(3):1594–1601
9. Nelson D, Cristobal G, Kober V, Çakrak F, Loughlin P, Cohen L (1999) Denoising using time-frequency and image processing methods. Proc SPIE Adv Signal Process Algorithm Architect Implementations IX 3807:564–581
10. Wigner EP (1932) On the quantum correction for thermodynamic equilibrium. Phys Rev 40:749–759
11. Wodkiewicz K (1984) Phys Rev Lett 52:1064

Chapter 11
Electromagnetically Induced Transparency with Fields Spectrally Broadened by Phase Noise

Eugeniy Mikhailov, Yuri V. Rostovtsev, and George R. Welch

11.1 Introduction

Electromagnetically induced transparency [1, 2] or EIT is an important nonlinear optical effect with wide ranging applications from slow light [3, 4] to sub-diffraction-limited optical imaging [5]. The most striking feature is the presence of zero (or low) absorption of a probe laser when it is exactly on resonance. In addition to the greatly reduced absorption is greatly enhanced normal dispersion on resonance. It is this steep dispersion coupled with low absorption that makes the system so rich.

There are many techniques for implementing this scheme, but the most basic (often called "canonical") scheme is shown in Fig. 11.1. Here, a weak laser, called the "probe" laser couples the upper level a with lower level b. However, at the same time, a stronger laser called the "drive" or "coupling" laser couples the upper level with another lower level c. We denote the natural decay rate of the upper level by γ and we let γ_{bc} denote the decay rate of coherence between the two lower levels. (This later rate might be due, for example, to the time-of-flight of the atoms through the laser field.) The arrangement of these electromagnetic fields on the energy level diagram gives rise to the name Λ-scheme for this setup. The usual absorptive and dispersive Lorentzian lineshapes that would exist for the probe laser in the absence of the coupling laser are modified as shown in Fig. 11.1.

E. Mikhailov (✉)
College of William and Mary, Williamsburg, VA 23185, USA
e-mail: eemikh@wm.edu

Y.V. Rostovtsev
Department of Physics, University of North Texas, Denton, TX 76203, USA
e-mail: rost@unt.edu

G.R. Welch
Texas A&M University, College Station, TX 77843, USA
e-mail: grw@tamu.edu

L. Cohen et al. (eds.), *Classical, Semi-classical and Quantum Noise*,
DOI 10.1007/978-1-4419-6624-7_11, © Springer Science+Business Media, LLC 2012

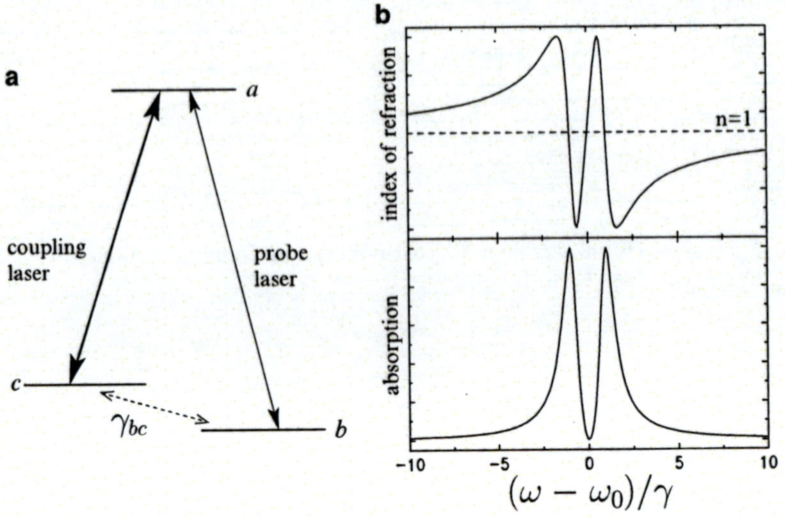

Fig. 11.1 (**a**) Three-level "Λ-scheme" for realizing EIT. (**b**) Absorption and dispersion profile for probe laser propagating under condition of EIT. For this picture, γ_{bc} is taken to be 0, resulting in perfect transmission on resonance

The interaction of an electromagnetic field modulated by phase noise (PN) with media possessing a resonance-like absorption or transmission spectrum has been experimentally and theoretically studied. Armstrong [6] pointed out the conversion of PN to amplitude noise in an optical interferometer, then experiments with broad PN radiation from a diode laser passing through an interferometer [7] supported Armstrong's argument, and showed the conversion of intensity fluctuation statistics from Poissonian to Bose-Einstein (i.e., exponential intensity distribution). Several experiments were done with a laser field and PN, studying the properties of such a field transmitted through a cell with alkali atomic vapor, namely the study of PN conversion to amplitude noise in dense Cs [8] and Rb [9]. There are experimental studies of intensity fluctuations and correlations between drive and probe fields in the EIT regime [10]; however, we are not aware of any study of the spectral density modification after passing through the medium in the EIT regime.

It is usually assumed that the width of the EIT resonance is solely determined by the finite interaction time of atoms with laser fields, and we treat both the probe and drive electromagnetic fields as monochromatic. In reality, the width of both these fields and their relative frequency fluctuations broaden the EIT resonance. That is why the observation of a narrow EIT resonance usually requires the use of a phase-lock device which significantly narrows the spectrum of the beat note between the probe and drive fields (in typical experiments, this beat-note width can be less than 100 Hz, which is usually less than the ground-state dephasing rate). Other ways to get coherent probe and drive fields are use of acousto-optic modulator (AOM), electro-optic modulator (EOM) or modulation of injection current in the case of diode laser. However, these methods have downsides such as the additional expense of buying AOM or EOM, or building a phase-lock device. Also, an EOM produces an extra (Stokes) sideband which complicates the analysis of the system.

Experimental observations with non-phase-locked lasers leads to a broad EIT resonance width of several megahertz, associated with laser linewidth. This effect was theoretically studied by Zubairy et al. [11] and experimentally demonstrated by Kim et al. [12]. They show that the EIT resonance width is limited by linewidths of the drive and probe fields.

Below we present a way of using incoherent probe or drive fields to observe narrow (less than beat-note spectrum width) EIT resonance.

11.2 Experiment

The setup for this experiment is described in [13]. We drive our ^{87}Rb cell with a monochromatic drive field and use a PN broadened probe field with spectral width ∼1 MHz. The output of a Stanford function generator which produces a white-noise signal with spectral width of 10 MHz is used to modulate our probe laser, so the output frequency becomes time dependent

$$f(t) = f_0 + \Delta f(t), \tag{11.1}$$

where f_0 is the carrier frequency (6.835 GHz) and $\Delta f(t)$ is the noise-driven time-dependent frequency shift. Since the phase of such oscillation is equal to

$$\varphi(t) = \varphi_0 + f_0 t + \int_{-\infty}^{t} \Delta f(t)\,\mathrm{d}t, \tag{11.2}$$

we conclude that we have PN broadened signal. This signal drives an EOM which splits part of the drive field to sidebands corresponding to the probe field, so we obtain the PN broadened probe field instead of the monochromatic probe field which is observed without the PN modulation. The spectral density of the probe signal does not coincide with the distribution of the driving frequency because of the limited bandwidth of the EOM response to the modulation frequency.

In canonical EIT resonance observation, the drive field is fixed in frequency and one scans the probe field in the vicinity of the two-photon resonance, observing the transmission signal with respect to the change of this two-photon detuning.

If one observes the beating frequency of the drive and probe field, one would also measure the beat signal of the drive and second side band which is symmetrically placed with respect to the drive laser frequency. To isolate the probe signal, we introduce an additional field which is shifted by an AOM at 60 MHz with respect to the drive field. This field does not propagate through the ^{87}Rb medium. We observe the beating signal of this additional field and probe field, which occurs at a different frequency from other beat notes. Thus, we separate the observation of the probe field from the drive and additional sideband fields. In our PN experiment, we record the spectral density of a PN broadened transmitted probe field with a spectrum analyzer

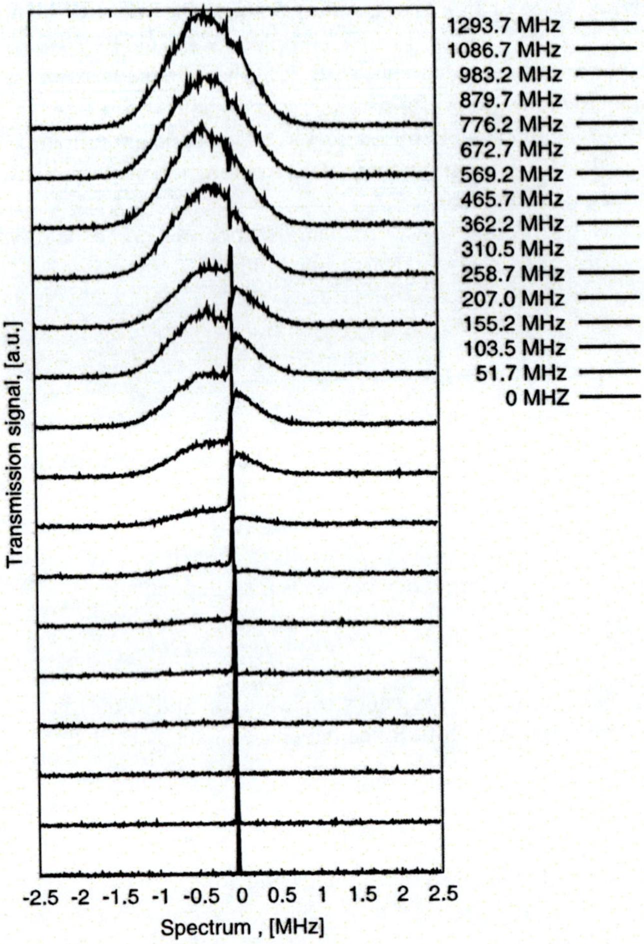

Fig. 11.2 Various transmission spectral density distributions for a noise broadened probe signal with different one-photon detunings of the drive field. Detuning grows from zero for bottom curve up to 1.3 GHz for the top curve. The upper trace can be considered as a spectrum of the input field

tuned in the vicinity of the beatnote frequency of the probe and additional fields. This scheme is described in [4].

A typical spectral density spectrum is depicted in Fig. 11.2. Before the cell, the spectral full width half maximum (FWHM) of the probe field is 980 kHz, and after the cell we see significant narrowing of the spectrum, the FWHM is 4.6 kHz (see also Fig. 11.3). Also, the spectral distribution function is changed; before the cell, we have the Gaussian distribution of the spectral density

$$f_{\mathrm{in}}(\omega) = \mathrm{e}^{-((\omega-\omega_0)^2/\omega_w^2)}, \tag{11.3}$$

where ω_0 is the average probe field frequency and ω_w is the width of the spectral density spectrum. After the cell the probe field has a Lorentz distribution

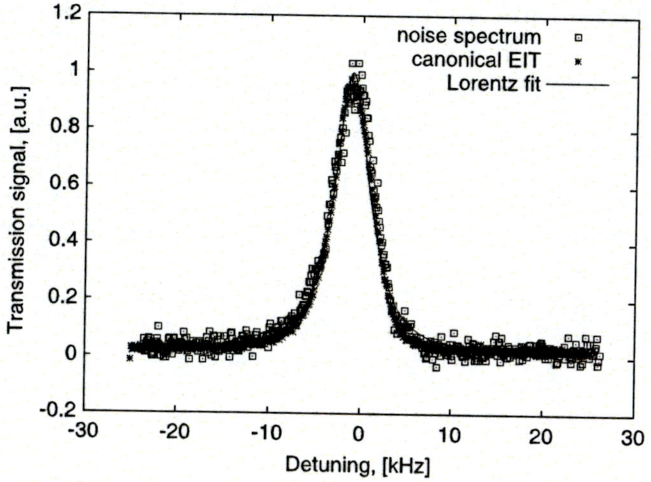

Fig. 11.3 Comparison of EIT resonance as a function of transmission vs two-photon detuning obtained with monochromatic and phase broadened probe field

$$f_{\text{out}} = \frac{\gamma_n^2}{(\omega - \omega_0)^2 + \gamma_n^2}, \tag{11.4}$$

where γ_n is width of spectral density spectrum.

In Fig. 11.3, we compare transmission spectra taken in experimental conditions similar to the canonical EIT resonance observation (with the coherent drive and probe field) and for the PN broadened probe. Note that the measured spectra after normalization coincide within our experimental accuracy, which means that $\gamma_n = \gamma_{\text{EIT}}$.

11.3 Discussion

Even though we modulate the phase of our EOM with a source of white noise, due to limited bandwidth of our electronics and the EOM, we did not obtain infinitely broad probe field spectra (which is unphysical anyway). Instead, we get a spectral density distribution of the probe electromagnetic field that is limited in width.

11.3.1 Very Simple Model

The transition from phase broadened field to spectral distribution is obtained by the following consideration:

$$E_p(t) = E_{p0}\mathrm{e}^{-\mathrm{i}(\omega_p t + \varphi_p(t))}, \tag{11.5}$$

where E_{p0} is the amplitude of the probe field, ω_p is the probe field average frequency, and $\varphi_p(t)$ is a random time-dependent phase of the electromagnetic field. Taking the Fourier transformation of $E_p(t)$, we obtain the spectral distribution of the probe field

$$E_p(t) = \int E_p(\omega)\mathrm{e}^{-\mathrm{i}\omega t}\,\mathrm{d}\omega, \tag{11.6}$$

where $E_p(\omega)$ is the intensity of the spectral component of the probe field with frequency ω.

Now we assume that the drive field is monochromatic. We assume that we can write density matrix and susceptibility equations for each Fourier component independently from other components. Since the susceptibility does not depend on the power of the probe beam for weak probe power, we conclude that the spectral density signal of light transmitted through an optically thin cell is given by the following expression:

$$f_{\mathrm{out}}(\omega) = E_p(\omega)\chi'(\omega, \omega_d, \Omega_d), \tag{11.7}$$

where $\chi'(\omega, \omega_d, \Omega_d)$ must eventually be averaged over the Doppler velocity distribution.

In the case when the input probe field spectra is much broader than the EIT resonance, we can neglect dependence of $E_p(\omega)$ on ω and just take $E_p(\omega) = E_p(\omega_p) = E_p$. This explains the coincidence of spectra in Fig. 11.3.

11.3.2 More General Approach

Let us write the PN broadened probe and drive field as

$$E_p(t) = E_{p0}\mathrm{e}^{-\mathrm{i}(\omega_p t + \varphi_p(t))}, \quad E_d(t) = E_{d0}\mathrm{e}^{-\mathrm{i}(\omega_d t + \varphi_d(t))}. \tag{11.8}$$

Then

$$\dot{\rho}_{ab} = -(\Gamma_{ab} + \dot{\varphi}_p)\rho_{ab} + \mathrm{i}\Omega_p(\rho_{bb} - \rho_{aa}) + \mathrm{i}\Omega_d\rho_{cb}, \tag{11.9}$$

$$\dot{\rho}_{ca} = -(\Gamma_{ca} - \dot{\varphi}_d)\rho_{ca} + \mathrm{i}\Omega_d^*(\rho_{aa} - \rho_{cc}) - \mathrm{i}\Omega_p^*\rho_{cb}, \tag{11.10}$$

$$\dot{\rho}_{cb} = -(\Gamma_{cb} + \dot{\varphi}_p - \dot{\varphi}_d)\rho_{cb} - \mathrm{i}\Omega_p\rho_{ca} + \mathrm{i}\Omega_d^*\rho_{ab}, \tag{11.11}$$

where $\Gamma_{ab} = \gamma + \mathrm{i}(\Delta + \delta)$, $\Gamma_{ca} = \gamma - \mathrm{i}\Delta$, $\Gamma_{cb} = \gamma_{bc} + \mathrm{i}\delta$, γ, γ_{bc} are decays of upper and the ground levels, Δ is the coupling laser detuning, and δ is the two-photon detuning.

In the case when the probe and drive fields are phase correlated ($\dot{\varphi}_d = \dot{\varphi}_p$), then the ρ_{cb} term remains the same as in the case of monochromatic probe and drive fields (this is the essence of a phase-lock technique).

It can be shown that if the phase fluctuation are Brownian such that $\langle \varphi_d \rangle = 0$, $\langle \varphi_p \rangle = 0$, and $\langle \dot{\varphi_d}(t)\dot{\varphi_d}(t') \rangle = 2D_d\delta(t'-t)$, $\langle \dot{\varphi_p}(t)\dot{\varphi_p}(t') \rangle = 2D_p\delta(t'-t)$, then [11, 12]

$$\dot{\rho}_{ab} = -(\Gamma_{ab} + D_p)\rho_{ab} + \mathrm{i}\Omega_p(\rho_{bb} - \rho_{aa}) + \mathrm{i}\Omega_d\rho_{cb}, \tag{11.12}$$

$$\dot{\rho}_{ca} = -(\Gamma_{ca} + D_d)\rho_{ca} + \mathrm{i}\Omega_d^*(\rho_{aa} - \rho_{cc}) - \mathrm{i}\Omega_p^*\rho_{cb}, \tag{11.13}$$

$$\dot{\rho}_{cb} = -(\Gamma_{cb} + D_p + D_d)\rho_{cb} - \mathrm{i}\Omega_p\rho_{ca} + \mathrm{i}\Omega_d^*\rho_{ab}, \tag{11.14}$$

so we see that the solution remains the same with proper modification of Γ_{ab}, Γ_{ca}, and Γ_{cb}. But the main modification is that the EIT resonance is now broadened by the $D_p + D_d$ term.

But let us have a look at propagation of the following signal:

$$S(t) = \Omega_p(t)\Omega_d^*(t), \tag{11.15}$$

where the signal (S) corresponds to the amplitude of the measured beatnote signal.

$$\frac{\partial S}{\partial z} = -\mathrm{i}\eta\rho_{ab}\Omega_d^* + \mathrm{i}\eta\rho_{ca}\Omega_p, \tag{11.16}$$

We substitute ρ_{ab} and ρ_{ca} as solutions of the standard three-level density matrix equations, assuming $\dot{\rho}_{ab} = \dot{\rho}_{ca} = 0$ (adiabatic approximation). We do this because the bandwidth of the PN is less than γ. However, we have to keep a term corresponding to $\dot{\rho}_{cb}$ since Γ_{cb} is smaller than the bandwidth of the PN . Then

$$\frac{\partial S}{\partial z} = \eta\left(\frac{n_{ab}}{\Gamma_{ab}} - \frac{n_{ca}}{\Gamma_{ca}}\right) S - \eta\left(\frac{|\Omega_d|^2}{\Gamma_{ab}} + \frac{|\Omega_p|^2}{\Gamma_{ca}}\right)\rho_{cb}, \tag{11.17}$$

$$\dot{\rho}_{cb} = -\left(\Gamma_{cb} + \frac{|\Omega_d|^2}{\Gamma_{ab}} + \frac{|\Omega_p|^2}{\Gamma_{ca}}\right)\rho_{cb} + \left(\frac{n_{ab}}{\Gamma_{ab}} - \frac{n_{ca}}{\Gamma_{ca}}\right) S. \tag{11.18}$$

At the next step, let us see how the correlation signal ($R(\tau) = \langle S S_\tau \rangle$) propagates

$$\frac{\partial}{\partial z}\langle S S_\tau \rangle = 2\eta\left(\frac{n_{ab}}{\Gamma_{ab}} - \frac{n_{ca}}{\Gamma_{ca}}\right)\langle S S_\tau \rangle - 2\eta\left(\frac{|\Omega_d|^2}{\Gamma_{ab}} + \frac{|\Omega_p|^2}{\Gamma_{ca}}\right)\langle S\rho_{cb}(\tau)\rangle, \tag{11.19}$$

$$\frac{\partial}{\partial \tau}\langle S\rho_{cb}(\tau)\rangle = -\left(\Gamma_{cb} + \frac{|\Omega_d|^2}{\Gamma_{ab}} + \frac{|\Omega_p|^2}{\Gamma_{ca}}\right)\langle S\rho_{cb}(\tau)\rangle + \left(\frac{n_{ab}}{\Gamma_{ab}} - \frac{n_{ca}}{\Gamma_{ca}}\right)\langle S S_\tau \rangle. \tag{11.20}$$

Now let us present spectral density of the correlation function of the signal (I_ω)

$$\langle S(t,z)S(t+\tau,z)\rangle = \int I_\omega \mathrm{e}^{-\mathrm{i}\omega\tau}\,\mathrm{d}\omega, \tag{11.21}$$

and assuming that the signal and coherence correlation expressed as following:

$$\langle S(t,z)\rho_{cb}(t+\tau,z)\rangle = \int \rho_\omega e^{-i\omega\tau}\, d\omega, \tag{11.22}$$

we obtain

$$\rho_\omega = \frac{\left(\frac{n_{ab}}{\Gamma_{ab}} - \frac{n_{ca}}{\Gamma_{ca}}\right) I_\omega}{\Gamma_{cb} + \frac{|\Omega_d|^2}{\Gamma_{ab}} + \frac{|\Omega_p|^2}{\Gamma_{ca}} - i\omega}, \tag{11.23}$$

$$\frac{\partial}{\partial z} I_\omega = 2\eta \frac{(\Gamma_{cb} - i\omega)\left(\frac{n_{ab}}{\Gamma_{ab}} - \frac{n_{ca}}{\Gamma_{ca}}\right)}{\Gamma_{cb} + \frac{|\Omega_d|^2}{\Gamma_{ab}} + \frac{|\Omega_p|^2}{\Gamma_{ca}} - i\omega} I_\omega. \tag{11.24}$$

Assuming that $\Omega_p \ll \Omega_d$ all population is in state $|b\rangle$

$$\frac{\partial}{\partial z} I_\omega = 2\eta \frac{(\Gamma_{cb} - i\omega)}{\Gamma_{cb} + \frac{|\Omega_d|^2}{\Gamma_{ab}} + \frac{|\Omega_p|^2}{\Gamma_{ca}} - i\omega} \frac{1}{\Gamma_{ab}} I_\omega. \tag{11.25}$$

The term in the front of I_ω on the right-hand side of the above equation is just the susceptibility, so all formalism applicable to the probe field propagation is applicable for the spectral density of the beatnote signal correlation function propagation. For example, the expression for the width of the EIT signal is identical to the expression for the width of the spectral density of the correlation function.

11.4 Applications

As we have just shown, all formal apparatus developed for calculation of EIT properties is valid for the spectral density of the beatnote signal of the incoherent probe as well. Thus, we can apply our general knowledge about EIT properties for the case of the incoherent probe and see what possible application are.

11.4.1 *Modification of Correlation Function*

Let us suppose that our spectrally broadened probe field has spectral distribution $S_p(\omega)$, then

$$\langle |S_p|^2 \rangle = I_p = \int |S_p(\omega)|^2\, d\omega = \int I_\omega\, d\omega, \tag{11.26}$$

where I_ω is the spectral distributions of the probe electromagnetic field intensity. Then the correlation function $R(\tau)$ defined as

$$R(\tau) = \langle S_p(t) S_p(t+\tau) \rangle \tag{11.27}$$

can be expressed through the spectral density of intensity as [14]

$$R(\tau) = \int I_\omega e^{-i\omega\tau}\, d\omega \tag{11.28}$$

and, on the other hand,

$$I_\omega = \frac{1}{2\pi} \int R(\tau) e^{i\omega\tau}\, d\tau. \tag{11.29}$$

Using (11.3) and (11.4), we obtain

$$R_{\text{in}}(\tau) = \frac{1}{2\pi} \int e^{-((\omega-\omega_p)^2/\omega_w^2)} e^{-i\omega\tau}\, d\omega = \frac{\omega_w}{2\sqrt{\pi}} e^{-(\omega_w^2\tau^2/4)-i\omega_p\tau}, \tag{11.30}$$

$$R_{\text{out}}(\tau) = \frac{1}{2\pi} \int \frac{\gamma_n^2}{(\omega-\omega_0)^2+\gamma_n^2} e^{-i\omega\tau}\, d\omega = \frac{\gamma_n}{\pi} e^{-\gamma_n|\tau|-i\omega_0\tau}. \tag{11.31}$$

From (11.28), we see that by changing the spectral density we change the correlation function I_ω as well. As we can see in Fig. 11.2, we modify the input spectrum from a broad Gaussian to a narrow Lorentzian at the output of the cell. Thus, we significantly increase the correlation time of the output probe field τ_{out} with respect to the correlation time of the ingoing probe field τ_{in}. The characteristic coherence time $\tau_{\text{in}} = 2/\omega_w$ for the input probe field is much smaller than the characteristic coherence time of the output radiation $\tau_{\text{out}} = 2/\gamma_n = 2/\gamma_{\text{EIT}}$, since $\omega_w \gg \gamma_{\text{EIT}}$.

If one desires a source of radiation with long coherence time, the use of coherent media under condition of EIT resonance would be a good tool.

Let us note that the narrow EIT resonance behaves as a narrow spectral filter. We supply a wide spectrally broadened probe field and obtain in the output of the cell a filtered signal with the spectral width of the order of several kHz (see, for example, Fig. 11.2). Also, this filter has *variable* spectral width; we can change the spectral width of the filter by varying several experimental parameters such as a density of ^{87}Rb atoms and applied power of the probe or drive field. We know that the width of EIT changes with the density of atoms

$$\gamma_{\text{EIT}} = \frac{\gamma_{bc} + \frac{\Omega^2}{\gamma}}{\sqrt{\eta L}}, \tag{11.32}$$

where γ_{bc} is the decay rate of the ground levels coherence, Ω is the Rabi frequency of the driving field, γ is the decay rate of upper level, L is the cell length, N is the density of atoms, λ is the probe field wave length. Thus, higher density results in narrower EIT resonance width. Since the density of ^{87}Rb atoms can be easily changed by changing the temperature of the cell, we have a nice control over the width of EIT. This can also be achieved if we change the power of the drive or probe laser which can be done very quickly by changing the current driving the lasers, or more slowly by changing a filter in the laser field. This will change the power broadening of the EIT resonance width and thus the spectral filter bandwidth.

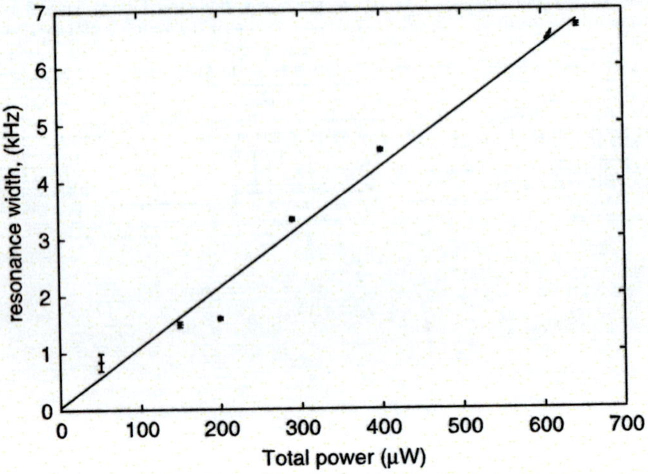

Fig. 11.4 Dependence of spectral width of transmitted trough the cell probe field vs total power under condition of EIT resonance

This point is illustrated in Fig. 11.4, where we measure the dependence of the transmitted spectral width of the signal on the power of the drive and probe lasers (the power of both lasers is changed with the same proportionality constant by varying the angle between two crossed polarizers). All of the above is applied to the correlation function modification. This means we can modify the correlation time in a controllable fashion.

We can also tune the position of the EIT resonance, and thus change the central frequency of our spectral signal filter. This may be done, for example, by applying a magnetic field and splitting the EIT resonance, where the position of the EIT resonance depends on the magnetic field (see Fig. 11.5). Other mechanisms can be used for this as well, for example, the Stark shift of the EIT resonance depends on laser power. Also, the position of the two photon resonance is related to the ground level splitting and only slightly changes with one-photon detuning. Thus, changing the one-photon detuning of the drive laser, we will change the frequency of transmitted field or central frequency of the filter according to the two-photon resonance condition

$$\omega_p = \omega_d + \omega_{gs}, \tag{11.33}$$

where ω_{gs} is the ground level splitting frequency (≈6.835 GHz for ^{87}Rb).

11.4.2 Magnetometer

A simple magnetometer (similar to that suggested by Wynands et al. [15]) is also possible with this technique. Since the position of different magnetic sublevels is

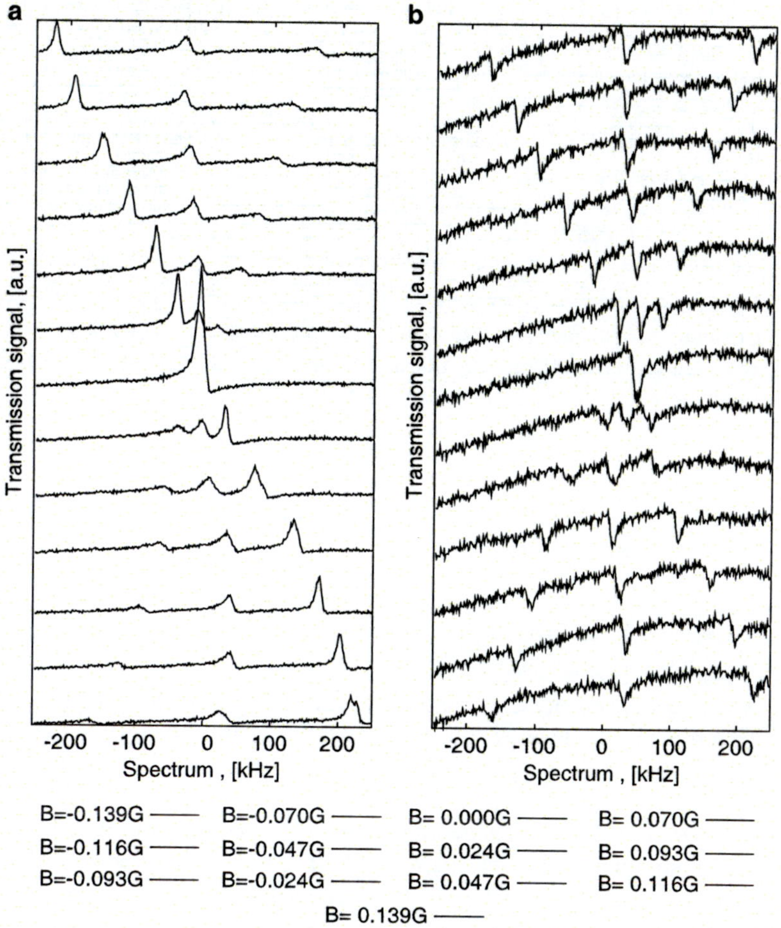

Fig. 11.5 (**a**) Transmission spectra of noise broadened probe field for different magnetic field (voltages applied to internal solenoid) taken in EIT regime. (**b**) Transmission spectra of noise broadened probe field for different magnetic field (voltages applied to internal solenoid) taken in buffer gas induced absorption (BGIA) regime

dependent on magnetic field (Zeeman splitting), we see that a nonzero magnetic field we result in several EIT resonances corresponding to the transitions between different magnetic sublevels.

The difference in the position of these levels, and hence the difference in frequency of the EIT two-photon resonance position, is well known and given by following expression ([16], p 37):

$$\Delta\omega_{\mathrm{EIT}}^{\pm}(m, m' = m - 1) = 2\pi(\Gamma^{\pm} B \mp (2m - 1)GB^2), \tag{11.34}$$

where B is the magnetic field, m and m' are the quantum number of different Zeeman sublevels, $\Gamma^{\pm}$ and G are constants that are different for different atoms. Sign "+" corresponds to magnetic sublevels with quantum number $F = I + 1/2$ and sign "–" corresponds to $F = I - 1/2$, where I is nuclear spin of the atom. For ^{87}Rb, $\Gamma^{+} = 699\,605$ Hz/G, $\Gamma^{-} = 702\,399$ Hz/G, and $G = 71.8$ Hz/G^2.

If we use a probe field with spectral width larger than $\omega_{\mathrm{EIT}}(m, m')$, we will find after the cell the filtered probe field with frequencies corresponding to two photon resonances for different Zeeman sublevels. This can be detected with a spectrum analyzer (see Fig. 11.5a). Similar results based on the buffer gas induced absorption (BGIA) resonance discussed in [17] are shown in Fig. 11.5b. Also, we can detect the beat note frequency corresponding to the frequency equal to $\Delta\omega_{\mathrm{EIT}}(m, m')$ on a frequency counter. From the measurement of this frequency, we can easily calculate the magnetic field B since the other parameters are known.

11.4.3 Modification of the Spectral Density Distribution and Spectral Hole Burning

11.5 Conclusion

We have shown that the spectral width of a broad-band probe field is greatly reduced (more than 1,000 times) by propagation through a medium exhibiting EIT. The experimental results agree with an analysis of the modification of a broad emission spectral line after transmission through a coherently prepared resonant medium. The final spectral lineshape is defined by the spectral shape of the EIT resonance. This may find applications for spectral filtering, as well as other possibilities such as sensing external effects that shift energy levels.

References

1. Marangos JP (1998) Topical review electromagnetically induced transparency. J Mod Opt 45:471
2. Harris SE (1997) Electromagnetically induced transparency. Phys Today 50(7):36
3. Hau LV, Harris SE, Dutton Z, Behroozi CH (1999) Light speed reduction to 17 metres per second in an ultracold atomic gas. Nature 397(6720):594–598
4. Kash MM, Sautenkov VA, Zibrov AS, Hollberg L, Welch GR, Lukin MD, Rostovtsev Y, Fry ES, Scully MO (1999) Ultraslow group velocity and enhanced nonlinear optical effects in a coherently driven hot atomic gas. Phys Rev Lett 82(26):5229–5232
5. Li H, Sautenkov VA, Kash MM, Sokolov AV, Welch GR, Rostovtsev YV, Zubairy MS, Scully MO (2008) Optical imaging beyond the diffraction limit via dark states. Phys Rev A 78:013803
6. Armstrong JA (1966) J Opt Soc Am 56:1024
7. Centeno NR, Boersma DM, Nienhuis G, van Exter MP, Woerdman JP (1992) Spectral filtering within the Schawlow-Townes linewidth of a semiconductor laser. Phys Rev Lett 69(4):593

8. Bahoura M, Andre Clairon A (2001) Diode-laser noise conversion in an optically dense atomic sample. Opt Lett 26(12):926
9. See, for example, materials of the 14th International conference on atomic physics. Diode laser noise conversion and reduction in rubidium, vol 1A-7, 1994, where many examples of noise conversion were reported
10. Garrido Alzar CL, Cruz LS, Aguirre Gomez JG, Franca Santos M, Nussenzveig P (2003) Super-poissonian intensity fluctuations and correlations between pump and probe fields in electromagnetically induced transparency. Europhys Lett 61(4):485–491
11. Sultana S, Zubairy MS (1994) Effect of finite bandwidth on refractive-index enhancement and lasing without inversion. Phys Rev A 49(1):438–448
12. Kim HA, Kwon KA, Kim JB (1997) Effects of laser linewidth, rabi frequency and detunings on electromagnetically induced. J Kor Phys Soc 30(2):407–412
13. Mikhailov EE, Sautenkov VA, Rostovtsev YV, Zhang A, Zubairy MS, Scully MO, Welch GR (2006) Spectral narrowing of a phase broadened optical field in a coherently prepared medium. Phys Rev A 74:013807
14. Ahmanov SA, Dyakov YE, Chirkin AS (1981) Vvedenie v Statisticheskuyu Radiofiziku i Optiku. Nauka, Moscow
15. Wynands R, Nagel A (1998) Precision spectroscopy with coherent dark states. Appl Phys B 68:1
16. Pomerantsev NM, Rizhkov VM, Skrotskiy GV (1972) Fizicheskie Osnovi Kvantovoy Optiki. Nauka, Moscow
17. Mikhailov EE, Novikova I, Rostovtsev YV, Welch GR (2004) Buffer-gas induced absorption resonances in Rb vapor. Phys Rev A 70:033806

Chapter 12
Multiple-Access Interference

H. Vincent Poor

12.1 Introduction

Dramatic growth rates in capacity demands in wireless and other broadband systems have resulted in a rise in the use of communication networks in which multiple users share common communication resources. A significant consequence of this trend is the increasing presence of multiple-access interference (MAI), which arises in communication systems employing non-orthogonal multiplexing, that is, in multiple-access systems. This issue arises naturally, for example, in code-division multiple-access (CDMA) communication systems using nonorthogonal spreading codes. It also arises in orthogonally multiplexed wireless channels, such as time-division multiple-access (TDMA) and orthogonal frequency division multiple-access (OFDMA) channels, due to effects such as multipath or nonideal frequency channelization, and in wireline channels such as those arising in digital subscriber line (DSL) systems or powerline communications (PLCs) in which crosstalk and other types of interference are major impairments. MAI also arises in optical wave-division multiplexing (WDM) systems due to mode interactions caused by nonlinearities.

Unlike ambient noise from natural sources, MAI is structured, and as such can be modeled and mitigated using appropriate receiver techniques. Such techniques, of which there are a great variety, are collectively termed multiuser detection (MUD). The original formulation of MUD, and much of its subsequent development for wireless systems, was cast within the context of classical channel models, in which ambient noise is modeled as being additive, white and Gaussian. Reviews of these techniques can be found in [11, 24, 27]. However, the modeling of MAI and the

H.V. Poor (✉)
School of Engineering and Applied Science, Princeton University, Princeton, NJ 08544, USA
e-mail: poor@princeton.edu

L. Cohen et al. (eds.), *Classical, Semi-classical and Quantum Noise*,
DOI 10.1007/978-1-4419-6624-7_12, © Springer Science+Business Media, LLC 2012

development of consequent MUD techniques have also been considered in semi-classical models with Poisson observations or impulsive noise, and more recently in quantum models. This chapter provides an overview of those developments.

12.2 Classical Channels

In this section, we discuss the classical case, in which ambient noise is assumed to be additive, white and Gaussian. To model this situation, we can consider the reception of a multiple-access communication signal as illustrated in Fig. 12.1.

Such a signal can be written in the following form:

$$r(t) = \sum_{k=1}^{K}\sum_{i=1}^{B} b_{k,i}\, p_k(t - iT - \tau_k) + n(t), \; -\infty < t < \infty, \tag{12.1}$$

where:

- K is the number of users active in the channel.
- B is the number of symbols per user in a received frame to be processed.
- T is the per-user symbol interval (so, $1/T$ is the per-user signaling rate).
- $b_{k,i}$ denotes the ith symbol of the kth user.
- $p_k(\cdot)$ is the received signaling waveform of user k.
- $\tau_k \in [0, T]$ is the delay with which user k's signal is received.
- $n(\cdot)$ is a white Gaussian process with spectral height $\frac{N_0}{2}$.

It is assumed that the symbols are all drawn from the same finite alphabet $\mathcal{M}$. For the sake of exposition, we further assume that these observations are real valued, although the extensions to complex or vector-valued waveforms (the latter of which correspond to the case of multiple receive antennas) are straightforward (e.g., [8]).

The essential problem of interest is to make decisions about the set of symbols $\{b_{k,i}\}$ in (12.1), of which there are KB. Singling out a given one of these, say $b_{k,i}$, we can rewrite (12.1) as

$$r(t) = b_{k,i}\, p_k(t - iT - \tau_k) + I_{k,i}(t) + n(t), \; -\infty < t < \infty, \tag{12.2}$$

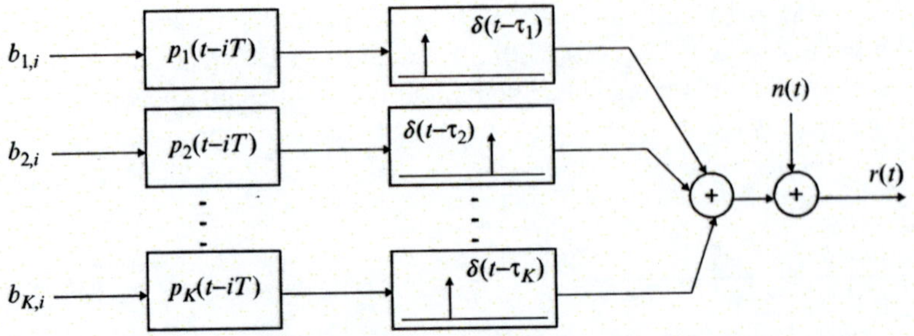

Fig. 12.1 Illustration of a classical multiple-access communication signal

where $I_{k,i}(\cdot)$ denotes the MAI affecting the symbol $b_{k,i}$, given by

$$I_{k,i}(t) = \sum_{(\ell,j)\neq(k,i)} b_{\ell,j}\, p_k(t - jT - \tau\ell), \; -\infty < t < \infty. \tag{12.3}$$

Note that if the waveforms $\{p_k(\cdot)\}_{k=1,2,\dots,K}$ are orthogonal to one another and do not extend beyond a single symbol interval, then the problem of symbol detection in (12.1) is a classical one, solved by the conventional matched filter receiver [17]:

$$\hat{b}_{k,i} = \mathrm{Q}(y_{k,i}), \; k = 1, \dots, K, \; i = 1, \dots, B. \tag{12.4}$$

where Q(·) denotes a quantizer mapping its argument to the symbol alphabet $\mathcal{M}$, and where $y_{k,i}$ denotes the output of a filter matched to the waveform of the kth user in its ith symbol interval; i.e.,

$$y_{k,i} = \int_{-\infty}^{\infty} r(t)\, p_k(t - iT - \tau_k)\, \mathrm{d}t. \tag{12.5}$$

Here, however, we are interested in the more general case, in which these waveforms are not orthogonal so that we have MAI. This situation involves the use of multiuser detection.

In general, a sufficient statistic for making inferences about $\{b_{k,i}\}$ is formed by the set of KB matched-filter outputs (see, e.g., [17]):

$$y_{k,i}, \; k = 1, \dots, K, \; i = 1, \dots, B. \tag{12.6}$$

Organizing these observables into a vector $\mathbf{y} \in \mathbf{R}^{KB}$ (here $\mathbf{R}$ denotes the real line) by sorting them first by symbol number and then by user number, the model (12.1) can be rewritten as a linear model

$$\mathbf{y} = \mathbf{H}\,\mathbf{b} + \mathcal{N}\left(0, \frac{N_0}{2}\mathbf{H}\right), \tag{12.7}$$

where $\mathbf{b} \in \mathcal{M}^{KB}$ denotes a vector containing the symbols $\{b_{k,i}\}$ sorted conformally with $\mathbf{y}$, and where $\mathbf{H}$ denotes a matrix of cross-correlations

$$H_{p,q} = \int_{-\infty}^{\infty} p_k(t - iT - \tau_k)\, p_\ell(t - jT - \tau_k)\, \mathrm{d}t \tag{12.8}$$

with the indices (k, i) and (ℓ, j) corresponding in the model (12.1) to the indices p and q, respectively, in the vector $\mathbf{y}$.[1] The term $\mathcal{N}\left(0, \frac{N_0}{2}\mathbf{H}\right)$ denotes a noise term having the multivariate Gaussian distribution with zero mean and covariance matrix $\frac{N_0}{2}\mathbf{H}$.

---

[1]The relationship between p, q and $(k, i), (\ell, j)$ is

$$k = [p-1]_K, \; i = \left\lfloor \frac{p-1}{K} \right\rfloor, \; \ell = [q-1]_K, \text{ and } j = \left\lfloor \frac{q-1}{K} \right\rfloor, \tag{12.9}$$

where $[\cdot]_K$ denotes reduction modulo K.

Fig. 12.2 Canonical structure of classical multi-user detectors

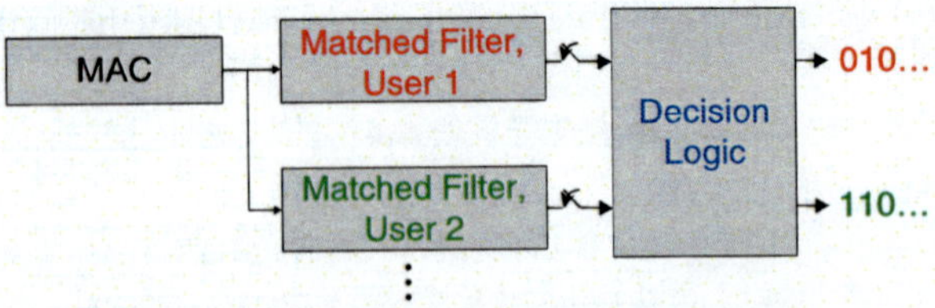

Thus, the basic problem of multiuser detection is, in general, a problem of linear model fitting, which involves choosing a vector $\hat{\mathbf{b}}$ of estimates of the symbols $\mathbf{b}$ in the linear model (12.7). That is, a canonical structure for multiuser detection in classical channels is a bank of filters matched to the various waveforms in the channel (which extracts $\mathbf{y}$), followed by a decision logic (which determines $\hat{\mathbf{b}}$), as illustrated in Fig. 12.2.

The presence of MAI in this problem is manifested in the fact that the matrix $\mathbf{H}$ in (12.7) is not a diagonal matrix; and this is also the source of complexity in the symbol detection problem. This choice can be made to satisfy optimality criteria such as maximum likelihood (ML) or maximum *a posteriori* probability (MAP), or it can be constrained to limit the complexity of the system. The ML and MAP detectors involve dynamic programming [18], which allows the brute-force complexity of $\mathcal{O}\left(M^{KB}\right)$, where $M = |\mathcal{M}|$ is the symbol alphabet size, to be reduced to $\mathcal{O}\left(M^{K\Delta}\right)$ where Δ is the span (in units of symbols) of the signaling waveforms.

Although this complexity can sometimes be managed in practical systems, it is still rather high, and consequently lower-complexity linear or iterative detectors are also of interest and have been studied extensively. Linear detectors perform a linear fit of (12.7), which yields a continuous estimate of the symbols. This linear estimate must then be quantized into the discrete set $\mathcal{M}$. Similarly, iterative detectors solve (12.7) iteratively, and they can be either linear, followed by a final quantization after iteration stops, or nonlinear, allowing intermediate quantizations between iterations. (The interested reader is referred to [11] or [27] for details of these types of algorithms.)

The ML and MAP detectors can achieve a level of performance very close to that which would be observed in an interference-free channel (e.g., [24]), while linear and iterative detectors can significantly limit the effects of the MAI on overall performance. For example, they can assure that the detector performance is not interference limited, which would be the case if the classical matched filter detector of (12.4) were to be misapplied to this problem.

Thus, for the classical model in which the basic multiple-access communication channel is modeled as being corrupted by additive white Gaussian noise, the problems of MAI modeling and mitigation have been well studied.

12.3 Semi-classical Channels

The classical model of (12.1) is sufficiently accurate for many communication systems, such as wireless systems in which the dominant ambient noise is thermal noise generated in the receiver. However, other situations are less well modeled in this way. In this section, we discuss two refinements to this model, one in which the basic additive-noise model is preserved but the noise is impulsive rather than Gaussian, and the other in which the additive-noise model is replaced by a point-process model. Although this latter model is more properly semiclassical in nature, the impulsive noise model can be viewed as a bridge between the white Gaussian noise model and the full semiclassical model.

12.3.1 Impulsive Channels

Middleton and others (e.g., [14]) have demonstrated that many physical noises affecting communication receivers are impulsive, rather than thermal, in nature. Such noises typically arise from phenomena external to the receiver, and are often man-made. The modeling and mitigation of MAI in such channels has been addressed in [19] and elsewhere, albeit to a degree less thorough than the study of the classical model.

The model of (12.1) is useful in considering impulsive channels, in that this situation means primarily a refinement of the statistical model for the noise process $n(\cdot)$ to account for impulsive components. This can be accomplished by replacing the white Gaussian model with a model consisting of a mixture of Gaussian and point-process components. From a receiver design viewpoint, however, this model is not entirely satisfactory, as the development of a discrete-index sufficient statistic is no longer tractable for this more general model. Thus, it is desirable to replace (12.1) with a discrete-time model that allows us to leverage discrete-time impulsive noise models, such as the well-known Middleton models [14].

Thus, we consider a discrete-time counterpart to (12.1), in which we consider observations sampled from the output of a linear filter whose input is the underlying continuous time received process. It is assumed that the filter is of sufficient bandwidth to capture the signals of interest largely undistorted, that the sampling rate is matched to this filter's bandwidth, and that the resulting samples of the noise are independent and identically distributed (i.i.d.). For simplicity, we consider the synchronous case of (12.1), in which can take $\tau_1 = \tau_2 = \cdots = \tau_K = 0$, and we assume that the signaling waveforms span only a single symbol interval; i.e., $s_k(t) = 0$, for $t \notin [0, T]$ and for $k = 1, 2, \ldots, K$. These assumptions, which can be relaxed, allow us to restrict attention to a single symbol interval, say $t \in [0, T]$, and yields a discrete-time model of the form

$$r_t = \sum_{k=1}^{K} b_k \, p_{k,t} + n_t, \; t = 0, 1, \ldots, N-1, \tag{12.10}$$

where we have dropped the symbol index i, which equals 0 here, as the essential problem is the same for each symbol interval under the above assumptions. Here, $p_{k,t}$ and n_t denote the sampled outputs, at sampling time t, of the response of the front-end linear filter to inputs $p_k(\cdot)$ and $n(\cdot)$, respectively, and N is the number of samples taken per symbol interval.

Detection of the data symbols in (12.10) can again be accomplished based on various optimality criteria, including ML and MAP, as discussed in [19]. However, the non-Gaussian nature of the additive noise significantly increases the complexity of ML and MAP detectors, as dynamical programming can no longer be applied. Thus, from a practical viewpoint, other detector structures are much more useful. One such detector structure, described in [26], is similar in nature to the linear and iterative detectors developed for the classical model (12.1)–(12.7). In particular, the detector of [26] is based on first obtaining continuous estimates for the symbols $b_1, b_2, \ldots, b_K$, and then quantizing these continuous estimates into the symbol alphabet $\mathcal{M}$. These estimates are particularly determined by solving the following minimization problem

$$\min_{b_1, b_2, \ldots, b_K \in \mathbb{R}^K} \sum_{t=0}^{N-1} \rho\left(r_t - \sum_{k=1}^{K} b_k \, p_{k,t} \right), \tag{12.11}$$

where $\rho(\cdot)$ is a nonlinear function. Note that (12.11) determines a best fit to the model (12.10) with error-weighting function $\rho(\cdot)$. Detectors that operate by quantizing the solutions to problems of the form (12.11) are known as *M-detectors*.

The choice of $\rho(x) = x^2$ in (12.11) gives a least-squares fit to (12.10), which is equivalent to the linear detectors for this problem described in Sect. 16.1.1. However, this choice of $\rho(\cdot)$ is not generally a good one for impulsive noise, as the resulting linear detectors are overly sensitive to impulses. Under relatively mild assumptions, the best detector to use for a given continuous noise distribution is $\rho(x) = -\log f(x)$, where $f(\cdot)$ is the marginal probability density function of the noise. If the noise were Gaussian, as in (12.1), then the corresponding optimal $\rho(\cdot)$ is quadratic and this problem reduces to linear detection in (12.7), as noted above. More generally, however, this optimal $\rho(\cdot)$ is not quadratic, and typically it is such that larger errors are de-emphasized relative to the quadratic weighting. This has the effect of suppressing impulses in the received signal, thereby improving detector performance. For example, a simplified version of a Middleton model for impulsive noise is a Gaussian mixture model in which $f(\cdot)$ is of the form

$$f = (1-\epsilon)\,\mathcal{N}(0, \sigma^2) + \epsilon\,\mathcal{N}(0, \kappa\sigma^2), \tag{12.12}$$

where $\epsilon \in [0, 1], \sigma > 0$, and $\kappa > 1$, are parameters of the model. Note that this model allows a fraction ϵ of the noise samples to have a variance that is a factor κ larger than the ambient variance σ^2, with the interpretation that these samples are due to impulses. Choosing ρ to be minus the log of this density gives the general effect of de-emphasizing large fitting errors relative to least squares, as described above. It should be noted that (12.11) can be solved in closed form for the quadratic case, but usually requires iterative solution for more general cases, providing a further connection to the iterative detectors developed for the classical channel discussed in Sect. 16.1.1.

This approach can also be extended to asynchronous channels, to situations in which the vectors $(p_{k,0}, p_{k,1}, \ldots, p_{k,N-1})$ are not known for all k, and to situations in which the probability density function $f(\cdot)$ may not be known exactly. The reader is referred to [26] for details of these extensions and of the analysis of all of these detectors. Related approaches for these and other settings are found in [20–22, 28].

12.3.2 Poisson Channels

Another refinement of the classical model is that in which the measurements are point processes, rather than evolving continuously in time as in (12.1). Such models arise in direct-detected optical systems or shot-noise limited systems, and the classical model of (12.1) can be viewed as limiting form of this case in the asymptote as the rate of the point process increases without bound. The multiple-access problem with point process measurements was considered originally in [23], with further developments found in [1, 4, 13, 16], and in other works cited therein.

With shot-noise-limited systems, the additive noise model of (12.1) can be adapted as in the case of impulsive noise, and this can be viewed as a particular case of that formalism. However, for direct-detected optical systems, a more direct point-process model is needed, in which the data symbols are encoded in the rate of the point process, rather than in additive form as in (12.1). Thus, now (12.1) should be replaced by a Poisson counting process whose rate is determined by the intensity of the received (multiuser) electromagnetic wave. That is, the model (12.1) is replaced by

$$r(t) = X(t), \; -\infty < t < \infty, \tag{12.13}$$

where $X(\cdot)$ is a random process with independent increments and having the property that, for all $t < u$, the increment

$$X(t) - X(u), \tag{12.14}$$

has the Poisson distribution with parameter

$$\int_u^t \lambda(\mathbf{b}; s)\, \mathrm{d}s. \tag{12.15}$$

In (12.15), the function $\lambda(\mathbf{b}; \cdot)$ is the rate process of the doubly-stochastic Poisson counting process $X(\cdot)$, which depends explicitly on the series $\mathbf{b}$ of KB symbols transmitted by the K users over a frame B symbols in length. The particular way in which the symbols enter the rate process depends on the physics of the observation process, as is discussed in [23]. Two models of particular interest are *additive-light*, in which the intensity of the sum of the users' electromagnetic signals produces the rate process, and *additive-rate*, in which individual rate processes produced by the intensities of individual users' signals add to produce a composite rate process.

The classical (matched-filter) receiver for such models is known to become MAI-limited with only moderate numbers of users (see, e.g., [1]). Thus, it is of interest to develop multiuser detectors in this context. The problem of optimal multiuser detection in such channels has been treated in [23], in which both additive-light and additive-rate models are considered. In both cases, dynamic programming solutions are derived for ML and minimum-probability-of-error detection. Although similar in result to the dynamic programming solutions for the classical model of (12.1), the decomposition that allows this solution is quite different in the semi-classical case, and further interesting distinctions arise as well[2]. (See [23] for further details.) Performance analyses of multiuser detection in this setting are given in [2] and [16], while lower-complexity multiuser detectors for Poisson channels based on iterative techniques are described in [3, 12, 15].

12.4 Quantum Channels

We now turn to the quantum case, in which the communication channel can be viewed as a quantum electrodynamical system. This case can be viewed as the most fundamental of models, subsuming all of those discussed above.

As with the other problems considered above, there are two basic problems here: modeling and detector design. Note that, at the outset, it is clear that quantum measurements will present a particular challenge for detector design in the multiple-access case. In particular, the canonical structure of Fig. 12.2, in which the output of the multiple-access channel is processed via parallel physical measurements is problematic for quantum observations. In particular, the quantum measurements required for "matching" to one user's signal may not be compatible with those required for other users' signals. For example, one user's signal may require an accurate measurement of phase, while another requires an accurate measurement of amplitude; the uncertainty relationship between these two quantities may prevent both measurements being taken. Moreover, the so-called "no-cloning theorem" tells us that we cannot make multiple copies of the quantum observations in order to extract parallel incompatible quantum measurements. Thus, unlike the classical and semiclassical cases, optimal quantum multiuser detection requires the joint design of the instrument and the detector. Nevertheless, this joint design is quite worthwhile, as the example below illustrates.

[2]The dynamic programming solutions arising in both the classical and semiclassical models are both special cases of a more general dynamic programming model developed in [25].

While sophisticated modeling of the quantum multiple-access channel is developed in [5], a simple model in which the benefits of active mitigation of MAI in quantum systems can be examined was proposed in [6]. This model considers a simple two-user system in which the users each excite a mode of a quantum electromagnetic field with respective complex amplitudes α_1 and α_2. The transmitted symbol (binary, in this model) are encoded in these amplitudes. The modes are mixed-up in the channel, resulting in MAI. The received quantum state is a coherent state $|\psi\rangle = |\beta_1\beta_2\rangle$, where the output amplitudes β_1 and β_2 depend on α_1, α_2 and a correlation coefficient.

As there are four possible symbol patterns, there are four possible received states: $|\psi_{0,0}\rangle$, $|\psi_{0,1}\rangle$, $|\psi_{1,0}\rangle$, and $|\psi_{1,1}\rangle$. The basic problem of data detection here is thus one of quantum multiple-hypothesis testing, which can be approached using results in [10]. In particular, this observable has four orthonormal eigenvectors, $\{|w_i\rangle\}$, which can be sought by maximizing the objective function

$$J = \sum_{i=00}^{11} |\langle w_i, \psi_i\rangle|^2. \tag{12.16}$$

This maximization can be accomplished by applying a sequence of axis rotations; see [10] for details.[3]

An example of the performance of the resulting multiuser detector and other detectors in given in Fig. 12.3, which shows the performance of multiuser detection applied after photon counting, of quantum multiuser detection as above, and of a detector in a channel without interference. Note that the first of these is essentially the detector described in Sect. 12.3.2 for Poisson channels. Figure 12.3 clearly shows the advantages of quantum modeling in this problem. In particular, the performance of the quantum multiuser detector coincides for the single-user bound for correlation coefficients up to almost 0.45. And, moreover, the gain over a multiuser detector based on photon counting is quite substantial (several orders of magnitude in error probability) even for relatively small cross-correlations. Such levels of cross-correlation are not unrealistic, for example, for those arising from mode interactions in WDM systems.

Lower complexity detectors, in the spirit of the linear multiuser detectors used in the classical case, are developed in [5] and [7].

12.5 Conclusion

In this chapter, we have briefly reviewed the modeling and mitigation of multiple-access interference (MAI) in classical, semiclassical, and quantum channels. A fundamental message of the work discussed herein is that the effects of MAI can be

[3]Detection when the states defining the various hypotheses are uncertain is treated in [9].

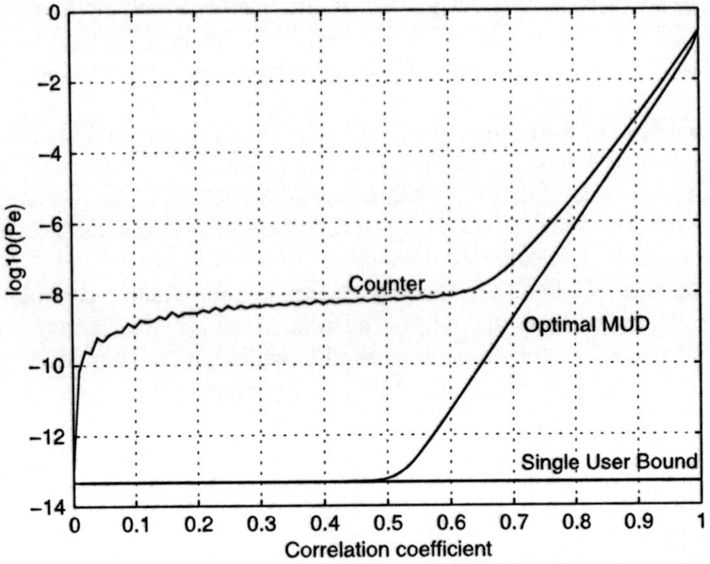

Fig. 12.3 Comparison of quantum and semi-classical multiple-access models: Two-user case; 30 photons (average) per user

ameliorated in all of these cases, provided that proper models of the interference are used to derive optimal detectors. A second important message is that such optimal detectors are often of relatively high complexity, but that MAI can still be mitigated to a large extent with simpler, more practical, detectors based on the same models.

The intent of this chapter has been to give a general overview of how these problems have been treated, rather than to elaborate on the details of those treatments. However, it is hoped that this discussion has given the interested reader a useful entry to the field and its literature that will allow for delving more deeply into the subject.

Acknowledgments This work was prepared under the support of the National Science Foundation under Grant CNS-09-05398.

References

1. Brady D, Verdú S (1991) A semiclassical analysis of optical code division multiple access. IEEE Trans Commun 39(1):85–93
2. Brandt-Pearce M, Aazhang B (1994) Performance analysis of single-user and multiuser detectors for optical code-division multiple-access communication systems. IEEE Trans Commun 42(2-4):434–444
3. Brandt-Pearce M, Aazhang B (1994) Multiuser detection for optical code-division multiple-access systems. IEEE Trans Commun 42(2-4):1801–1810

4. Bross SI, Burnashev M, Shamai S (2001) Error exponents for the two-user Poisson multiple-access channel. IEEE Trans Inform Theor 46(5):1999–2016
5. Concha JI, Poor HV (2004) Multiaccess quantum channels. IEEE Trans Inform Theor 50(5):725–747
6. Concha JI, Poor HV (2001) Least-squares detectors in quantum channels. Proceedings of the 39th Annual Allerton Conference on Communication, Control and Computing, University of Illinois at Urbana-Champaign, Monticello, IL, October 3–5
7. Concha JI, Poor HV (2002) An optimality property of the square-root measurement for mixed states. Proceedings of the Sixth International Conference on Quantum Communication, Measurement and Computing, MIT, Cambridge, MA, July 22–26
8. Dai H et al. (2007) Multi-user Receiver Design. In: Bilgieri E, et al. (eds), MIMO wireless communications, Chap. 6. Cambridge University Press, Cambridge, UK, pp. 230–292
9. Elron N, Eldar YC (2005) Quantum detection with uncertain states. Phys Rev A 72(3):032338
10. Helstrom CW (1976) Quantum detection and estimation theory. Academic, New York
11. Honig M (ed) (2009) Advances in multiuser detection. Wiley, New York
12. Karabulut GZ, Kurt T, Yongacoglu A (2005) Optical CDMA detection by basis selection. J Lightwave Technol 23(11):3708–3715
13. Lapidoth A, Shamai S (1998) The Poisson multiple-access channel. IEEE Trans Inform Theor 44:488–501
14. Middleton D (1977) Statistical-physical models of electromagnetic interference. IEEE Trans Electromagn C EMC-19(3):106–127
15. Motahari AS, Nasiri-Kenari M (2004) Multiuser detection for optical CDMA networks based on expectation-maximization algorithm. IEEE Trans Commun 52(4):652–660
16. Nelson LB, Poor HV (1995) Performance of multiuser detection for optical CDMA - Part I: Error probabilities. IEEE Trans Commun 43:2803–2811
17. Poor HV (1994) An introduction to signal detection and estimation, 2nd edn. Springer, New York
18. Poor HV (2002) Dynamic programming in digital communications: Viterbi decoding to turbo multiuser detection. J Optim Theor Appl 115(3):629–657
19. Poor HV, Tanda M (1999) Multiuser detection in impulsive channels. Annales des Telecommun 54(7–8):392–400
20. Poor HV, Tanda M (2002) Multiuser detection in flat fading non-Gaussian channels. IEEE Trans Commun 50(11):1769–1777
21. Seyfe B, Sharafat AR (2005) Signed-rank nonparametric multiuser detection in non-Gaussian channels. IEEE Trans Inform Theor 51(4):1478–1486
22. Seyfe B, Sharafat AR (2006) Nonparametric multiuser detection in non-Gaussian channels. IEEE Trans Signal Process 54(1):22–33
23. Verdú S (1986) Multiple-access channels with point-process observations: Optimum demodulation. IEEE Trans Inform Theor IT-32:642–651
24. Verdú S (1998) Multiuser detection. Cambridge University Press, Cambridge, UK
25. Verdú S, Poor HV (1987) Abstract dynamic programming models under commutativity conditions. SIAM J Contr Optim 25(4):990–1006
26. Wang X, Poor HV (1999) Robust multiuser detection in non-Gaussian channels. IEEE Trans Signal Process 47(2):289–305
27. Wang X, Poor HV (2004) Wireless communication systems: Advanced techniques for signal reception. Prentice-Hall, Upper Saddle River, NJ
28. Yao YW, Poor HV (2004) Blind detection of synchronous CDMA in non-Gaussian channels. IEEE Trans Signal Process 52(1):271–279

Chapter 13
Classical Capacities of Bosonic Channels

Jeffrey H. Shapiro

13.1 Introduction

Ubiquitous, reliable, high data rate communication – carried by electromagnetic waves at microwave to optical frequencies – is an essential ingredient of our technological age. Information theory seeks to delineate the ultimate limits on reliable communication that arise from the presence of noise and other disturbances, and to establish means by which these limits can be approached in practical systems. The mathematical foundation for this assessment of limits is Shannon's Noisy Channel Coding Theorem [1], which introduced the notion of channel capacity – the maximum mutual information between a channel's input and output – as the highest rate at which error-free communication could be maintained. Textbook treatments of channel capacity [2, 3] study channel models – ranging from the binary symmetric channel's digital abstraction to the additive white-Gaussian-noise channel's idealization of thermal-noise-limited waveform transmission – for which classical physics is the underlying paradigm. Fundamentally, however, electromagnetic waves are quantum mechanical; they are boson fields [4, 5]. Moreover, high-sensitivity photodetection systems have long been limited by noises of quantum mechanical origin [6]. Thus, it would seem that determining the ultimate limits on optical communication would necessarily involve an explicitly quantum analysis, but such has not been the case. Nearly all work on the communication theory of optical channels – work done for systems with laser transmitters and either coherent-detection or direct-detection receivers – uses semiclassical (shot-noise) models (see, e.g., [7,8]). Here, electromagnetic waves are taken to be classical entities, and the fundamental noise is due to the random release of discrete charge carriers in the process of photodetection. Inasmuch as the quantitative results obtained from shot-noise analyses of such systems are known to coincide

J.H. Shapiro (✉)
Research Laboratory of Electronics, Massachusetts Institute of Technology,
Cambridge, MA 02139, USA
e-mail: jhs@mit.edu

L. Cohen et al. (eds.), *Classical, Semi-classical and Quantum Noise*,
DOI 10.1007/978-1-4419-6624-7_13,

with those derived in rigorous quantum-mechanical treatments [9], it might be hoped that the semiclassical approach would suffice. But, Helstrom's derivation [10, 11] of the optimum quantum receiver for binary coherent-state (laser light) signaling demonstrated that the lowest error probability, at constant average photon number, required a receiver that was neither coherent detection nor direct detection. That Dolinar [12] was able to show how Helstrom's optimum receiver could be realized with a photodetection feedback system, which admits to a semiclassical analysis, did not alleviate the need for a fully quantum-mechanical theory of optical communication, as Shapiro et al. [13] soon proved that even better binary-communication performance could be obtained by use of two-photon coherent state (now known as squeezed state) light, for which semiclassical photodetection theory did not apply.

In quantum mechanics, the state of a physical system together with the measurement that is made on that system determines the statistics of the outcome of that measurement, see, e.g., [14]. Thus, in seeking the classical information capacity of a bosonic channel, we must allow for optimization over both the transmitted quantum states and the receiver's quantum measurement. In particular, it is not appropriate to immediately restrict consideration to coherent-state transmitters and coherent-detection or direct-detection receivers. What is needed for deducing the fundamental limits on optical communication is the analog of Shannon's Noisy Channel Coding Theorem – unfettered by unjustified structural constraints on the transmitter and receiver – that applies to transmission of classical information over a noisy quantum channel, viz., the Holevo–Schumacher–Westmoreland (HSW) Theorem [15–17]. We shall review recent progress that has been made in applying the HSW Theorem to bosonic channels, and we shall treat both single-user and multi-user (multiple-access and broadcast) communications.

13.2 Single-User Bosonic Channels and the HSW Theorem

We are interested in the classical communication capacity of the lossy bosonic channel with isotropic quantum Gaussian noise. For brevity, we will only present the single-mode channel model, although results for multi-mode (temporal and spatio-temporal, scalar field and vector field) channels, are easily obtained by tensoring up a collection of single-mode models. For the single-mode channel, the input is an electromagnetic-field mode with annihilation operator $\hat{a}$ and the output is another field mode with annihilation operator $\hat{a}'$. The lossy, noisy bosonic channel is not a unitary evolution, so it is governed by a trace-preserving completely-positive (TPCP) map [18] that relates its output density operator, $\hat{\rho}'$, to its input density operator, $\hat{\rho}$.

The TPCP map $\mathscr{E}_\eta^N(\cdot)$ for the single-mode lossy channel can be derived from the commutator-preserving beam splitter relation

$$\hat{a}' = \sqrt{\eta}\,\hat{a} + \sqrt{1-\eta}\,\hat{e}, \tag{13.1}$$

in which the annihilation operator $\hat{e}$ is associated with an environmental (noise) mode, and $0 \leq \eta \leq 1$ is the channel transmissivity. (See [19] for how the quantum version of the Huygens–Fresnel diffraction integral leads to a collection of single-mode beam splitter relations, and for a quantum characteristic function specification of the associated TPCP map.) For the pure-loss channel, the $\hat{e}$ mode is in its vacuum state; for the thermal-noise channel, this mode is in a thermal state, i.e., an isotropic-Gaussian mixture of coherent states with average photon number $N > 0$,

$$\hat{\rho}_e = \int d^2\mu \frac{\exp(-|\mu|^2/N)}{\pi N} |\mu\rangle\langle\mu|. \tag{13.2}$$

The classical capacity of a quantum channel is established by random coding arguments akin to those employed in classical information theory. A set of symbols $\{j\}$ is represented by a collection of input states $\{\hat{\rho}_j\}$ that are selected according to some prior distribution $\{p_j\}$. The output states $\{\hat{\rho}'_j\}$ are obtained by applying the channel's TPCP map $\mathscr{E}_\eta^N(\cdot)$ to these input symbols. The Holevo information associated with priors $\{p_j\}$ and states $\{\hat{\sigma}_j\}$ is given by

$$\chi(p_j, \hat{\sigma}_j) = S\left(\sum_j p_j \hat{\sigma}_j\right) - \sum_j p_j S(\hat{\sigma}_j), \tag{13.3}$$

where $S(\hat{\sigma}) \equiv -\text{tr}(\hat{\sigma}\ln(\hat{\sigma}))$ is the von Neumann entropy, which is the Shannon entropy of the density operator's eigenvalues. According to the HSW Theorem, the capacity of this channel, in nats per use, is

$$C = \sup_n(C_n/n) = \sup_n\left\{\max_{\{p_j,\hat{\rho}_j\}}\left[\chi(p_j, \mathscr{E}_\eta^{N\otimes n}(\hat{\rho}_j))/n\right]\right\}, \tag{13.4}$$

where C_n is the capacity achieved when coding is performed over n-channel-use symbols and the supremum over n is necessitated by the fact that channel capacity may be superadditive, viz., $C_n > nC_1$ is possible for quantum channels, whereas such is not the case for classical channels. The HSW Theorem tells us that the Holevo information plays the role for classical information transmission over a quantum channel that Shannon's mutual information does for a classical channel.

Until a few years ago, the only bosonic channel whose classical information capacity had been determined was the lossless channel [20, 21], in which the field mode controlled by the transmitter is available for measurement – without loss, hence without additional quantum noise – at the receiver. This channel, which corresponds to (13.1) with $\eta = 1$, has little relevance to optical communication owing to its lossless, noiseless character. Its classical information capacity, when the transmitter has its average photon number constrained to satisfy $\langle\hat{a}^\dagger\hat{a}\rangle \leq N_S$, is

$$C = g(N_S) \text{ nats/use}, \tag{13.5}$$

where $g(x) \equiv (x+1)\ln(x+1) - x\ln(x)$ is the Shannon entropy of the Bose–Einstein distribution with mean value x. This capacity was shown to be achievable

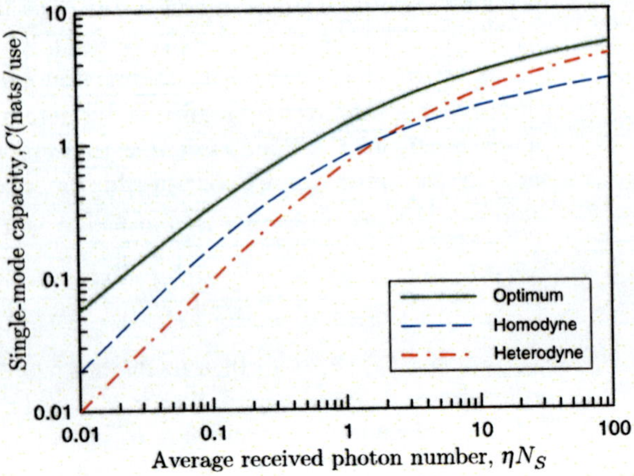

Fig. 13.1 Comparison of channel capacities, in nats per channel use, for the pure-loss channel. The *solid curve* is the optimum capacity from the HSW Theorem. The *dashed curve* is the capacity for coherent-state encoding with homodyne detection. The *dot-dashed curve* is the capacity for coherent-state encoding with heterodyne detection

with random coding over photon-number eigenstates, $|n\rangle$, using the Bose–Einstein probability distribution with mean N_S. Thus, photon-number measurements by the receiver perfectly distinguish all distinct codewords. Moreover, the capacity for this channel is additive, i.e., $C = C_1$, so that single-letter encoding suffices.

More than a decade after the capacity proof for the lossless bosonic channel, we derived the capacity for the pure-loss channel [22], which is the beam splitter relation (13.1) with the environmental mode $\hat{e}$ in its vacuum state. Under the same average photon number constraint at the transmitter, we found that the pure-loss channel's capacity is additive and given by

$$C = g(\eta N_S) \text{ nats/use for all } 0 \leq \eta \leq 1, \tag{13.6}$$

where capacity is achieved by random coding over coherent states, $|\alpha\rangle$, using the isotropic-Gaussian probability density function

$$p(\alpha) = \frac{e^{-|\alpha|^2/N_S}}{\pi N_S}. \tag{13.7}$$

Because our demonstration included the lossless $\eta = 1$ case, it showed that the capacity in (13.5) could also be achieved with coherent states. High-quality laser sources produce excellent approximations to coherent-state light, thus it is pleasing that these sources can achieve the capacity of the pure-loss channel. Figure 13.1 illustrates the capacity advantage afforded by the HSW Theorem, for the pure-loss channel, by comparing the quantum result from (13.6) with the capacities achieved over this bosonic channel when optical homodyne or heterodyne detection are employed in conjunction with coherent-state encoding, namely,

$$C_{\text{hom}} = \frac{\ln(1 + 4\eta N_S)}{2} \quad \text{and} \quad C_{\text{het}} = \ln(1 + \eta N_S) \text{ nats/use}. \tag{13.8}$$

In subsequent work [23], we addressed the capacity of the thermal-noise channel. There, we showed that the rate achieved with single-letter, coherent-state encoding using the same probability density that achieved capacity for the pure-loss channel is given by

$$R = g(\eta N_S + (1-\eta)N) - g((1-\eta)N) \text{ nats/use}, \tag{13.9}$$

and is a lower bound on C. We also conjectured that this lower bound is in fact the capacity. Our capacity conjecture is intimately related to our work [24–26] on the minimum output entropy for the thermal-noise channel. Indeed, the following strong-form minimum output entropy conjecture suffices to prove that $C = R$ for the thermal-noise channel.

Strong Conjecture 1. Let the environmental modes $\hat{\boldsymbol{e}} = [\hat{e}_1\ \hat{e}_2 \cdots \hat{e}_n]^{\mathrm{T}}$ be in a tensor product of n thermal states with total von Neumann entropy $ng(K)$. Then the von Neumann entropy of the output modes $\hat{\boldsymbol{a}}' = [\hat{a}'_1\ \hat{a}'_2 \cdots \hat{a}'_n]^{\mathrm{T}} = \sqrt{\eta}\,\hat{\boldsymbol{a}} + \sqrt{1-\eta}\,\hat{\boldsymbol{e}}$ is minimized when the input modes $\hat{\boldsymbol{a}} = [\hat{a}_1\ \hat{a}_2 \cdots \hat{a}_n]^{\mathrm{T}}$ are in their vacuum states. The resulting minimum von Neumann output entropy is $ng((1-\eta)K)$.

Moreover, proving the weak ($n = 1$) version of this conjecture suffices to prove $C_1 = R$ for the thermal-noise channel.

The output entropy from a bosonic channel, such as the thermal-noise channel, when one of its inputs is in a pure state is a measure of the degree to which the channel output is entangled with its environment. It seems eminently reasonable, therefore, that the input state which will result in the least entanglement with the environment is the vacuum state. Nevertheless, despite considerable effort, neither Strong Conjecture 1 nor its weak form has been proven. It is known, for example, that the conjecture is true in the limits of low and high noise, and it is known that the vacuum-state input yields a local minimum of the output entropy. For other evidence supportive of this conjecture, see [24–26].

13.3 The Bosonic Multiple-Access Channel

The lossy single-user channel provides a natural basis for the two-user multiple-access channel (MAC). Suppose that Alice and Bob control electromagnetic field modes with annihilation operators $\hat{a}$ and $\hat{b}$, respectively, and that Charlie – their common intended receiver – has access to the mode whose annihilation operator $\hat{c}$ satisfies

$$\hat{c} = \sqrt{\eta_A}\,\hat{a} + \sqrt{\eta_B}\,\hat{b} + \sqrt{1-\eta_A-\eta_B}\,\hat{e}, \tag{13.10}$$

where the transmissivities obey $0 \le \eta_A, \eta_B, 1-\eta_A-\eta_B$ and $\hat{e}$ is the annihilation operator of an environmental (noise) mode. Charlie's density operator, $\hat{\rho}_C$, when Alice and Bob transmit the product state $\hat{\rho}_A \otimes \hat{\rho}_B$ is then governed by the TPCP map $\mathscr{M}^N_{\eta_A,\eta_B}(\cdot)$ that is associated with the preceding three-input beam splitter relation.

Lossless multiple access corresponds to $\eta_A + \eta_B = 1$. Lossy multiple access corresponds to $\eta_A + \eta_B < 1$ with the $\hat{e}$ mode being in its vacuum state ($N = 0$) for pure loss, and in the isotropic-Gaussian coherent-state mixture (13.2) when there is excess noise. Extending this two-user model to $M > 2$ users is straightforward – an M-input beam splitter relation will generate the required TPCP – as is generalizing to multi-mode operation. It is also possible to incorporate optical amplification into the inputs or output of the preceding beam splitter relation to compensate for low-transmissivity couplings at the expense of additional quantum noise.

For the two-user quantum MAC, Winter has shown [27] that reliable communication from Alice and Bob to Charlie is possible if and only if their respective rates, R_A and R_B, lie within a capacity region that is the closure of the convex hull of all (R_A, R_B) satisfying

$$R_A < H_A - \sum_i \sum_j p_i^A p_j^B S(\hat{\rho}_{ij}), \tag{13.11}$$

$$R_B < H_B - \sum_i \sum_j p_i^A p_j^B S(\hat{\rho}_{ij}), \tag{13.12}$$

$$R_A + R_B < S(\hat{\rho}) - \sum_i \sum_j p_i^A p_j^B S(\hat{\rho}_{ij}), \tag{13.13}$$

for some product distribution $p_i^A p_j^B$ over input-symbol states $\{\hat{\rho}_{ij}\}$. Here,

$$\hat{\rho} \equiv \sum_i \sum_j p_i^A p_j^B \hat{\rho}_{ij}, \quad \hat{\rho}_i^A \equiv \sum_j p_j^B \hat{\rho}_{ij}, \quad \hat{\rho}_j^B \equiv \sum_i p_i^A \hat{\rho}_{ij} \tag{13.14}$$

specifies the joint and conditional density operators for Alice and Bob and

$$H_A \equiv \sum_j p_j^B S(\hat{\rho}_j^B), \quad H_B \equiv \sum_i p_i^A S(\hat{\rho}_i^A) \tag{13.15}$$

gives their conditional von Neumann entropies. Winter's results apply to finite-dimensional (qudit) channels, but can be extended (via an appropriate limiting argument) to the bosonic case of interest to us. As stated in (13.11)–(13.15), the coding is over single channel uses, thus, the superadditivity issue mentioned in conjunction with single-user links also arises – and must be dealt with – in the multiple-access case.

We have studied the lossless bosonic MAC, in which $\eta_A = \eta = 1 - \eta_B$ and Alice and Bob's transmitters are subject to the average photon-number constraints $\langle \hat{a}^\dagger \hat{a} \rangle \le N_A$ and $\langle \hat{b}^\dagger \hat{b} \rangle \le N_B$ [28]. We showed that single-letter, independent random coding over coherent states ($|\alpha\rangle$ for Alice and $|\beta\rangle$ for Bob) using the isotropic-Gaussian probability densities,

$$p_A(\alpha) = \frac{e^{-|\alpha|^2/N_A}}{\pi N_A} \quad \text{and} \quad p_B(\beta) = \frac{e^{-|\beta|^2/N_B}}{\pi N_B} \tag{13.16}$$

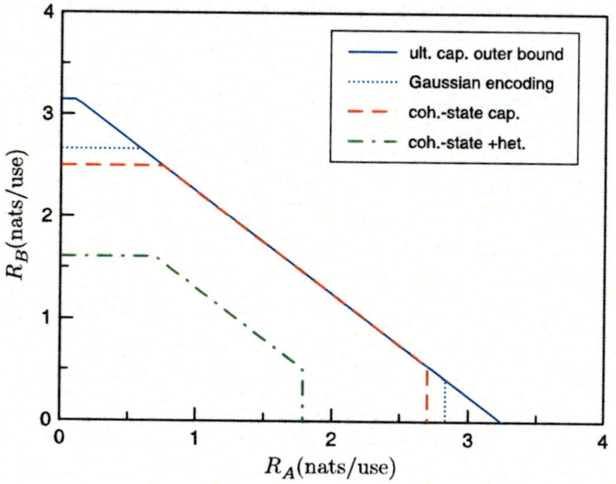

Fig. 13.2 Capacity region for the two-user, lossless bosonic MAC. Inner bounds from single-letter, coherent-state and Gaussian-state encodings [28] and outer bounds are plotted, along with the capacity regions for single-letter coherent-state encoding with optimum quantum reception and for single-letter coherent-state encoding with heterodyne detection. The figure assumes $\eta = 1/2$, $N_A = 10$, and $N_B = 8$. In all cases, the capacity region lies below and to the left of the lines in the figure

achieves the following inner bound on the capacity region,

$$R_A \le g(\eta N_A), \quad R_B \le g((1-\eta)N_B), \quad \text{and}$$
$$R_A + R_B \le g(\eta N_A + (1-\eta)N_B) \text{ nats/use.} \tag{13.17}$$

Furthermore, we were able to show that $R_A + R_B \le g(\eta N_A + (1-\eta)N_B)$ is the sum capacity for this channel and that the sum capacity is additive. On the other hand, we demonstrated that $R_A \le g(\eta N_A)$ and $R_B \le g((1-\eta)N_B)$ do not achieve the single-letter limit on single-user capacity. We did this by exhibiting more general Gaussian-state encodings that exceed these coherent-state rates. See Fig. 13.2 for an example, in which we have also plotted our outer bound,

$$R_A \le g(N_A), \quad R_B \le g(N_B), \quad \text{and} \quad R_A + R_B \le g(\eta N_A + (1-\eta)N_B) \text{ nats/use} \tag{13.18}$$

on the capacity region, and the capacity region for single-letter, coherent-state encoding with heterodyne detection,

$$R_A \le \ln(1 + \eta N_A), \quad R_B \le \ln(1 + (1-\eta)N_B),$$
$$R_A + R_B \le \ln(1 + \eta N_A + (1-\eta)N_B) \text{ nats/use.} \tag{13.19}$$

13.4 The Bosonic Broadcast Channel

The lossy single-user channel also provides a natural basis for the two-user bosonic broadcast channel. In this case, Alice controls an electromagnetic field mode with annihilation operator $\hat{a}$ with which she wants to send different information to Bob and to Charlie, whose receivers can measure the electromagnetic modes with annihilation operators $\hat{b}$ and $\hat{c}$, where

$$\hat{b} = \sqrt{\eta}\,\hat{a} + \sqrt{1-\eta}\,\hat{e} \quad \text{and} \quad \hat{c} = \sqrt{1-\eta}\,\hat{a} - \sqrt{\eta}\,\hat{e}. \tag{13.20}$$

Here, $\hat{e}$ is the annihilation operator for the environment and we have assumed a lossless situation in which Alice's photons must be received by either Bob or Charlie.

For the two-user degraded quantum broadcast channel, Yard et al. [29] have obtained the capacity region for the finite-dimensional (qudit) case, where once again superadditivity may occur, and we have extended their work to the infinite-dimensional bosonic channel [30]. When Alice's transmission is limited to an average photon number obeying $\langle \hat{a}^\dagger \hat{a} \rangle \le N_S$, on a channel with $\eta > 1/2$ and an environment that is in its vacuum state, we have found the following inner bound on the capacity region:

$$R_B \le g(\eta\beta N_S) \quad \text{and} \quad R_C \le g((1-\eta)N_S) - g((1-\eta)\beta N_S) \text{ nats/use,} \tag{13.21}$$

as β is varied from 0 to 1. This region is achievable with single-letter coherent-state encoding (see [30] for details). Moreover, we have shown that this region is the capacity region if the following strong-form minimum output entropy conjecture is true.

Strong Conjecture 2. Let the input modes $\hat{\boldsymbol{a}} = [\hat{a}_1\ \hat{a}_2 \cdots \hat{a}_n]^{\mathrm{T}}$ be in their vacuum states, and let the von Neumann entropy of the input modes $\hat{\boldsymbol{e}} = [\hat{e}_1\ \hat{e}_2 \cdots \hat{e}_n]^{\mathrm{T}}$ be $ng(K)$. Then, putting the $\hat{\boldsymbol{e}}$ modes in a tensor product of mean-photon-number K thermal states minimizes the von Neumann entropy of the output modes $\hat{\boldsymbol{b}} = \sqrt{\eta}\,\hat{\boldsymbol{a}} + \sqrt{1-\eta}\,\hat{\boldsymbol{e}}$. The resulting minimum von Neumann output entropy is $ng((1-\eta)K)$.

Proving the weak ($n = 1$) version of this conjecture suffices to prove that our coherent-state encoding achieves the single-letter capacity region for bosonic broadcast channel.

Strong Conjecture 2 appears to be the dual of Strong Conjecture 1, and likewise for their weak forms. As with Strong Conjecture 1, we have not been able to prove Strong Conjecture 2. We do know that the conjectures are equivalent when the input states are restricted to be Gaussian, and that their weak forms are both true under this Gaussian-input restriction.

Figure 13.3 shows an example of our inner bound on the two-user bosonic broadcast channel's capacity region. Also included in this figure are the capacity

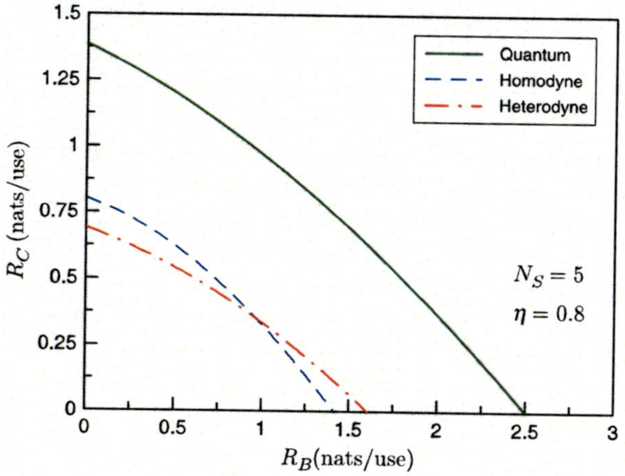

Fig. 13.3 Comparison of capacity regions for the two-user, lossless bosonic broadcast channel when the environmental mode is in its vacuum state. The *solid curve* is the inner bound on the quantum optimum capacity region. The *dashed curve* is the capacity region for coherent-state encoding with homodyne detection. The *dot-dashed curve* is the capacity region for coherent-state encoding with heterodyne detection. The figure assumes $N_S = 5$ and $\eta = 0.8$. In all cases, the capacity region lies below and to the left of the curves in the figure

regions for single-letter coherent-state encoding when homodyne or heterodyne detection is employed,

$$R_B^{\text{hom}} \leq \frac{\ln(1 + 4\eta\beta N_S)}{2},$$
$$R_C^{\text{hom}} \leq \frac{\ln\left[1 + \dfrac{4(1-\eta)(1-\beta)N_S}{1 + 4(1-\eta)\beta N_S}\right]}{2} \text{ nats/use,} \tag{13.22}$$

and

$$R_B^{\text{het}} \leq \ln(1 + \eta\beta N_S),$$
$$R_C^{\text{het}} \leq \ln\left[1 + \frac{(1-\eta)(1-\beta)N_S}{1 + (1-\eta)\beta N_S}\right] \text{ nats/use,} \tag{13.23}$$

as β varies from 0 to 1.

13.5 The Entropy Photon-Number Inequality

The Entropy Power Inequality (EPI) from classical information theory [31] is widely used in coding theorem converse proofs for Gaussian channels, see, e.g., [32]. By analogy with the EPI, we have postulated the Entropy Photon-Number Inequality

(EPnI) [33]. This inequality has yet to be proven. However, both Strong Conjecture 1 and Strong Conjecture 2 have been shown to be simple consequences of the EPnI. Moreover, existing proof techniques for the classical EPI may guide us to a proof for the EPnI.

Let $\hat{\boldsymbol{a}} = [\hat{a}_1\ \hat{a}_2\ \cdots\ \hat{a}_n\,]^{\mathrm{T}}$ and $\hat{\boldsymbol{b}} = [\hat{b}_1\ \hat{b}_2\ \cdots\ \hat{b}_n\,]^{\mathrm{T}}$ be vectors of photon annihilation operators for a collection of $2n$ different electromagnetic field modes. The joint state of the modes associated with $\hat{\boldsymbol{a}}$ and $\hat{\boldsymbol{b}}$ is given by the product-state density operator $\hat{\rho}_{\boldsymbol{ab}} = \hat{\rho}_{\boldsymbol{a}} \otimes \hat{\rho}_{\boldsymbol{b}}$, where $\hat{\rho}_{\boldsymbol{a}}$ and $\hat{\rho}_{\boldsymbol{b}}$ are the density operators associated with the $\hat{\boldsymbol{a}}$ and $\hat{\boldsymbol{b}}$ modes, respectively. The von Neumann entropies of the $\hat{\boldsymbol{a}}$ and $\hat{\boldsymbol{b}}$ modes are $S(\hat{\rho}_{\boldsymbol{a}}) = -\mathrm{tr}[\hat{\rho}_{\boldsymbol{a}} \ln(\hat{\rho}_{\boldsymbol{a}})]$ and $S(\hat{\rho}_{\boldsymbol{b}}) = -\mathrm{tr}[\hat{\rho}_{\boldsymbol{b}} \ln(\hat{\rho}_{\boldsymbol{b}})]$. The entropy photon-numbers of the density operators $\hat{\rho}_{\boldsymbol{a}}$ and $\hat{\rho}_{\boldsymbol{b}}$ are defined as follows:

$$N(\hat{\rho}_{\boldsymbol{a}}) \equiv g^{-1}(S(\hat{\rho}_{\boldsymbol{a}})/n) \text{ and } N(\hat{\rho}_{\boldsymbol{b}}) \equiv g^{-1}(S(\hat{\rho}_{\boldsymbol{b}})/n). \tag{13.24}$$

Thus, if $\hat{\rho}_{\tilde{\boldsymbol{a}}} \equiv \bigotimes_{i=1}^{n} \hat{\rho}_{T_{a_i}}$ and $\hat{\rho}_{\tilde{\boldsymbol{b}}} \equiv \bigotimes_{i=1}^{n} \hat{\rho}_{T_{b_i}}$, where $\hat{\rho}_{T_{a_i}}$ is the thermal state of average photon number $N(\hat{\rho}_{\boldsymbol{a}})$ for the $\hat{a}_i$ mode and $\hat{\rho}_{T_{b_i}}$ is the thermal state of average photon number $N(\hat{\rho}_{\boldsymbol{b}})$ for the $\hat{b}_i$ mode, then we have $S(\hat{\rho}_{\tilde{\boldsymbol{a}}}) = S(\hat{\rho}_{\boldsymbol{a}})$ and $S(\hat{\rho}_{\tilde{\boldsymbol{b}}}) = S(\hat{\rho}_{\boldsymbol{b}})$. We define a new vector of photon annihilation operators, $\hat{\boldsymbol{c}} = [\hat{c}_1\ \hat{c}_2\ \cdots\ \hat{c}_n\,]^{\mathrm{T}}$, by the beam splitter relation

$$\hat{\boldsymbol{c}} \equiv \sqrt{\eta}\,\hat{\boldsymbol{a}} + \sqrt{1-\eta}\,\hat{\boldsymbol{b}} \quad \text{for } 0 \leq \eta \leq 1, \tag{13.25}$$

and use $\hat{\rho}_{\boldsymbol{c}}$ to denote its density operator. We can now state two equivalent forms of the EPnI:

$$N(\hat{\rho}_{\boldsymbol{c}}) \geq \eta N(\hat{\rho}_{\boldsymbol{a}}) + (1-\eta) N(\hat{\rho}_{\boldsymbol{b}}) \tag{13.26}$$

$$S(\hat{\rho}_{\boldsymbol{c}}) \geq S(\hat{\rho}_{\tilde{\boldsymbol{c}}}), \tag{13.27}$$

where $\hat{\rho}_{\tilde{\boldsymbol{c}}} \equiv \bigotimes_{i=1}^{n} \hat{\rho}_{T_{c_i}}$ with $\hat{\rho}_{T_{c_i}}$ being the thermal state of average photon number $\eta N(\hat{\rho}_{\boldsymbol{a}}) + (1-\eta) N(\hat{\rho}_{\boldsymbol{b}})$ for $\hat{c}_i$.

13.6 Conclusions

We have reviewed recent results for the classical information capacities of single-user and multi-user bosonic channels. Two minimum output entropy conjectures figure prominently in closing open questions regarding the capacity of the single-user thermal-noise channel and the multi-user broadcast channel, respectively. Both are subsumed by the Entropy Photon-Number Inequality, which is our conjectured quantum analog of the EPI from classical information theory.

Acknowledgments This research was supported in part by the W.M. Keck Foundation Center for Extreme Quantum Information Theory. The author gratefully acknowledges the contributions to this research by B.I. Erkmen, V. Giovannetti, S. Guha, S. Lloyd, L. Maccone, B.J. Yen, and H.P. Yuen.

References

1. Shannon CE (1948) The mathematical theory of communication. Bell Syst Tech J 27:379–423, 623–656
2. Gallager RG (1968) Information theory and reliable communication. Wiley, New York (Chaps. 4, 5 and 8)
3. Cover TM, Thomas JA (1991) Elements of information theory. Wiley, New York (Chaps. 8 and 10)
4. Louisell WH (1973) Quantum statistical properties of radiation. Wiley, New York (Sects. 4.3 and 4.4)
5. Mandel L, Wolf E (1995) Optical coherence and quantum optics. Cambridge University Press, Cambridge (Sects. 10.1–10.3)
6. Kingston RH (1978) Detection of optical and infrared radiation. Springer, Berlin
7. Gagliardi RM, Karp S (1976) Optical communications. Wiley, New York
8. Gowar J (1984) Optical communication systems. Prentice-Hall, Englewood Cliffs
9. Yuen HP, Shapiro JH (1980) Optical communication with two-photon coherent states—part III: quantum measurements realizable with photoemissive detectors. IEEE Trans Inform Theor 26:78–92
10. Helstrom CW (1964) Detection theory and quantum mechanics. Inform Contr 10:254–291
11. Helstrom CW (1976) Quantum detection and estimation theory. Academic Press, New York (Chaps. 4 and 6)
12. Dolinar SJ (1973) An optimum receiver for the binary coherent state quantum channel. MIT Res Lab Electron Quart Progr Rep 111:115–120
13. Shapiro JH, Yuen HP, Machado Mata JA (1979) Optical communication with two-photon coherent states—part II: photoemissive detection and structured receiver performance. IEEE Trans Inform Theor 25:179–192
14. Louisell WH (1973) Quantum statistical properties of radiation. Wiley, New York (Chaps. 1 and 2)
15. Holevo AS (1998) The capacity of a quantum channel with general signal states. IEEE Trans Inform Theor 44:269–273
16. Hausladen P, Jozsa R, Schumacher B, Westmoreland M, Wootters WK (1996) Classical information capacity of a quantum channel. Phys Rev A 54:1869–1876
17. Schumacher B, Westmoreland MD (1997) Sending classical information via noisy quantum channels. Phys Rev A 56:131–138
18. Nielsen MA, Chuang IL (2000) Quantum computation and quantum information. Cambridge University Press, Cambridge (Sect. 8.2.4)
19. Yuen HP, Shapiro JH (1978) Optical communication with two-photon coherent states—part I: quantum state propagation and quantum noise reduction. IEEE Trans Inform Theor 24: 657–668
20. Yuen HP, Ozawa M (1993) Ultimate information carrying limit of quantum systems. Phys Rev Lett 70:363–366
21. Caves CM, Drummond PD (1994) Quantum limits on bosonic communication rates. Rev Mod Phys 66:481–537
22. Giovannetti V, Guha S, Lloyd S, Maccone L, Shapiro JH, Yuen HP (2004) Classical capacity of lossy bosonic channels: the exact solution. Phys Rev Lett 92:027902

23. Shapiro JH, Giovannetti V, Guha S, Lloyd S, Maccone L, Yen BJ (2004) Capacity of bosonic communications. In: Barnett SM, Anderson E, Jeffers J, Öhberg P, Hirota O (eds) Proceedings of the 7th international conference on quantum communication, measurement and computing. American Institute of Physics, New York, pp 15–20
24. Giovannetti V, Guha S, Lloyd S, Maccone L, Shapiro JH (2004) Minimum output entropy of bosonic channels: a conjecture. Phys Rev A 70:032315
25. Giovannetti V, Lloyd S, Maccone L, Shapiro JH, Yen BJ (2004) Minimum Rényi and Wehrl entropies at the output of bosonic channels. Phys Rev A 70:022328
26. Giovannetti V, Lloyd S (2004) Additivity properties of a Gaussian channel. Phys Rev A 69:062307
27. Winter A (2001) The capacity of the quantum multiple-access channel. IEEE Trans Inform Theor 47:3059–3065
28. Yen BJ, Shapiro JH (2005) Multiple-access bosonic communications. Phys Rev A 72:062312
29. Yard J, Hayden P, Devetak I (2006) Quantum broadcast channels, e-print arXiv:quant-ph/0603098
30. Guha S, Shapiro JH, Erkmen BI (2007) Classical capacity of bosonic broadcast communication and a minimum output entropy conjecture. Phys Rev A 76:032303
31. Cover TM, Thomas JA (1991) Elements of information theory. Wiley, New York (Sect. 16.7)
32. Bergmans P (1973) Random coding theorem for broadcast channels with degraded components. IEEE Trans Inform Theor 19:197–207
33. Guha S, Shapiro JH, Erkmen BI (2008) Capacity of the bosonic wiretap channel and the entropy photon-number inequality. Proceedings of the 2008 IEEE International Symposium on Information Theory, IEEE, New York, pp 91–95

Chapter 14
The Physics of Ghost Imaging

Yanhua Shih

14.1 Introduction

Assuming an object that is either self-luminous or externally illuminated, imagining each point on the object surface as a point radiation sub-source, each point sub-source will emit spherical waves to all possible directions. How much chance do we expect to observe a spherical wave collapsing into a point or a "speckle" by free propagation? Obviously, the chance is zero unless an imaging system is applied. The concept of optical imaging was well developed in classical optics for this purpose. Figure 14.1 schematically illustrates a standard imaging setup. In this setup an object is illuminated by a radiation source, an imaging lens is used to focus the scattered and reflected light from the object onto an image plane, which is defined by the "Gaussian thin lens equation"

$$\frac{1}{s_i} + \frac{1}{s_o} = \frac{1}{f}, \tag{14.1}$$

where s_o is the distance between the object and the imaging lens, s_i the distance between the imaging lens and the image plane, and f the focal length of the imaging lens. Basically this equation defines a point-to-point relationship between the object plane and the image plane: any radiation starting from a point on the object plane will "collapse" to a unique point on the image plane. It is not difficult to see from Fig. 14.1 that the point-to-point relationship is the result of *constructive–destructive interference*. The radiation fields coming from a point on the object plane will experience equal distance propagation to superpose constructively at one unique point on the image plane, and experience unequal distance propagations

Y. Shih (✉)
Department of Physics, University of Maryland, Baltimore County,
Baltimore, MD 21250, USA
e-mail: shih@umbc.edu

L. Cohen et al. (eds.), *Classical, Semi-classical and Quantum Noise*,
DOI 10.1007/978-1-4419-6624-7_14, © Springer Science+Business Media, LLC 2012

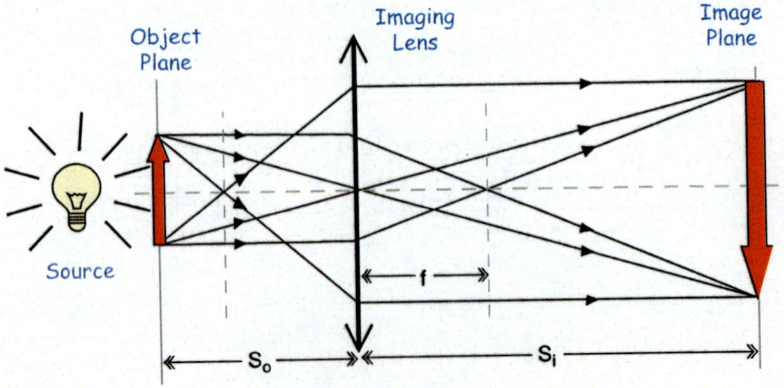

Fig. 14.1 Optical imaging: a lens produces an *image* of an object in the plane defined by the Gaussian thin-lens equation $1/s_i + 1/s_o = 1/f$. Image formation is based on a point-to-point relationship between the object plane and the image plane. All radiations emitted from a point on the object plane will "collapse" to a unique point on the image plane

to superpose destructively at all other points on the image plane. The use of the imaging lens makes this constructive–destructive interference possible.

A perfect point-to-point image-forming relationship between the object and image planes produces a perfect image. The observed image is a reproduction, either magnified or demagnified, of the illuminated object, mathematically corresponding to a convolution between the object distribution function $|A(\vec{\rho}_o)|^2$ (aperture function) and a δ-function, which characterizes the perfect point-to-point relationship between the object and image planes:

$$I(\vec{\rho}_i) = \int_{\text{obj}} d\vec{\rho}_o \left|A(\vec{\rho}_o)\right|^2 \delta\left(\vec{\rho}_o + \frac{\vec{\rho}_i}{m}\right), \tag{14.2}$$

where $I(\vec{\rho}_i)$ is the intensity in the image plane, $\vec{\rho}_o$ and $\vec{\rho}_i$ are 2-D vectors of the transverse coordinates in the object and image planes, respectively, and $m = s_i/s_o$ is the image magnification factor.

In reality, limited by the finite size of the imaging system, we may never obtain a perfect point-to-point correspondence. The incomplete constructive–destructive interference turns the point-to-point correspondence into a point-to-"spot" relationship. The δ-function in the convolution of (14.2) will be replaced by a point-spread function:

$$I(\vec{\rho}_i) = \int_{\text{obj}} d\vec{\rho}_o \left|A(\vec{\rho}_o)\right|^2 somb^2\left[\frac{R}{s_o}\frac{\omega}{c}\left|\vec{\rho}_o + \frac{\vec{\rho}_i}{m}\right|\right], \tag{14.3}$$

where the sombrero-like function, or the Airy disk, is defined as

$$somb(x) = \frac{2J_1(x)}{x},$$

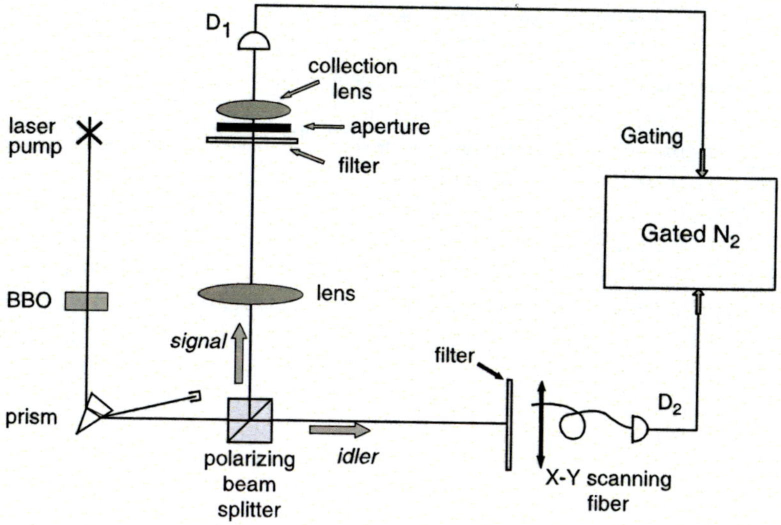

Fig. 14.2 Schematic setup of the first "ghost" image experiment. The experimental demonstrations of ghost imaging and ghost interference [6] in 1995 together stimulated the foundation of quantum imaging in terms of geometrical and physical optics

and $J_1(x)$ is the first-order Bessel function, and R the radius of the imaging lens, and R/s_o is known as the numerical aperture of the imaging system. The sombrero-like point-spread function, or the Airy disk, defines the spot size on the image plane that is produced by the radiation coming from point $\vec{\rho}_o$. It is clear from (14.3) that a larger imaging lens and shorter wavelength will result in a narrower point-spread function, and thus a higher spatial resolution of the image. The finite size of the spot determines the spatial resolution of the imaging system.

Type-one and type-two ghost imaging, in certain aspects, exhibit a similar point-to-point imaging-forming function as that of classical imaging except the ghost image is reproducible only in the joint-detection between two independent photodetectors, and the point-to-point imaging-forming function is in the form of second-order correlation,

$$R_{12}(\vec{\rho}_i) = \int_{obj} d\vec{\rho}_o \left|A(\vec{\rho}_o)\right|^2 G^{(2)}(\vec{\rho}_o, \vec{\rho}_i), \tag{14.4}$$

where $R_{12}(\vec{\rho}_i)$ is the joint-detection counting rate between photodetectors D_1 and D_2. Mathematically, the convolution is taken between the aperture function of the object $\left|A(\vec{\rho}_o)\right|^2$ and a nontrivial point-to-point second-order correlation function $G^{(2)}(\vec{\rho}_o, \vec{\rho}_i)$, corresponding to the probability of observing a joint photo-detection event at coordinates $\vec{\rho}_o$ and $\vec{\rho}_i$. It is the physics behind this point-to-point correlation function of $G^{(2)}(\vec{\rho}_o, \vec{\rho}_i)$ made ghost imaging so special.

The first type-one ghost imaging experiment was demonstrated by Pittman et al. in 1995 [1] enlightened by the theoretical work of Klyshko [2–4]. The schematic setup of the experiment is shown in Fig. 14.2. A continuous wave (CW) laser is used to pump a nonlinear crystal to produce an entangled pair of orthogonally polarized signal (e-ray of the crystal) and idler (o-ray of the crystal) photons in the nonlinear optical process of spontaneous parametric down-conversion (SPDC). The pair emerges from the crystal collinearly with $\omega_s \cong \omega_i \cong \omega_p/2$ (degenerate SPDC). The pump is then separated from the signal-idler pair by a dispersion prism, and the signal and idler are sent in different directions by a polarization beam splitting Thompson prism. The signal photon passes through a convex lens of 400 mm focal length and illuminates a chosen aperture (mask). As an example, one of the demonstrations used the letters "UMBC" for the object mask. Behind the aperture is the "bucket" detector package D_1, which is made by an avalanche photodiode placed at the focus of a short focal length collection lens. During the experiment, D_1 is kept in a fixed position. The idler photon is captured by detector package D_2, which consists of an optical fiber coupled to another avalanche photodiode. The input tip of the fiber is scannable in the transverse plane by two step motors (along orthogonal directions). The output pulses of D_1 and D_2, both operate in the photon counting mode, are independently counted as the counting rate of D_1 and D_2, respectively, and simultaneously, sent to a coincidence circuit for counting the joint-detection events of the pair. The single detector counting rates of D_1 and D_2 are both monitored to be constants during the measurement. Surprisingly, a ghost image of the chosen aperture is observed in coincidences during the scanning of the fiber tip, when the following two experimental conditions are satisfactory: (1) D_1 and D_2 always measure a pair; (2) the distances s_o, which is the optical distance between the aperture to the lens, s_i, which is the optical distance from the imaging lens going backward along the signal photon path to the two-photon source of SPDC then going forward along the idler photon path to the fiber tip, and the focal length of the imaging lens f satisfy the Gaussian thin lens equation of (14.1).

Figure 14.3 shows a typical measured ghost image. It is interesting to note that while the size of the "UMBC" aperture inserted in the signal path is only about 3.5×7 mm, the observed image measures 7×14 mm. The image is therefore magnified by a factor of 2 which equals the expected magnification $m = s_i/s_o$. In this measurement, $s_o = 600$ mm and $s_i = 1{,}200$ mm. When D_2 was scanned on transverse planes other than the ghost image plane the images blurred out.

The experiment was immediately given the name "ghost imaging" by the physics community due to its nonlocal feature. In the language of Einstein–Podolsky–Rosen (EPR) [5], the non-factorizable[1] point-to-point image-forming correlation

$$G^{(2)}(\vec{\rho}_o, \vec{\rho}_i) \sim \delta(\vec{\rho}_o + \vec{\rho}_i/m) \tag{14.5}$$

observed in this experiment represents a nonlocal behavior of a measured pair of photons: neither the signal photon nor idler photon "knows" precisely where to go when the pair is created at the source. However, if one of them is observed at a

[1] A factorizable correlation function $G^{(2)}(\mathbf{r}_1, t_1; \mathbf{r}_2, t_2) = G^{(1)}(\mathbf{r}_1, t_1)\, G^{(1)}(\mathbf{r}_2, t_2)$ characters independent radiations at space-time $(\mathbf{r}_1, t_1)$ and $(\mathbf{r}_2, t_2)$. In ghost imaging, the light on the object plane and the light at the CCD array is described by a nonfactorizeable point-to-point image-forming function, indicating nontrivial statistical correlation between the two measured intensities.

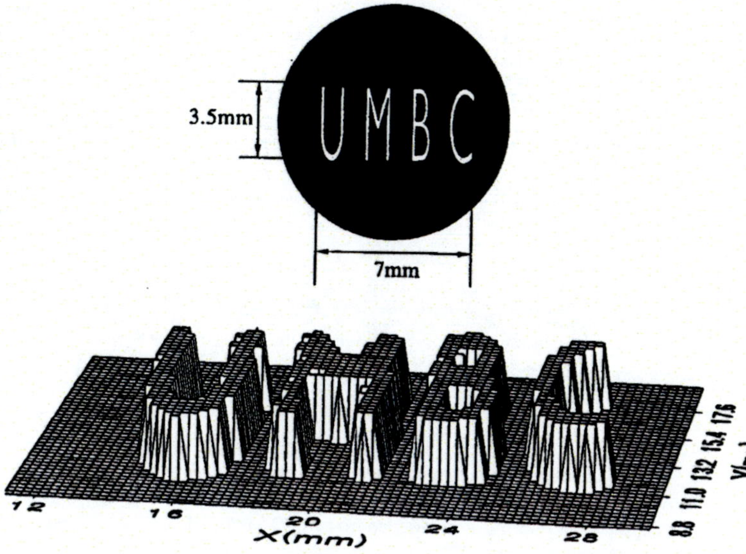

Fig. 14.3 Upper: A reproduction of the actual aperture "UMBC" placed in the signal beam. Lower: The image of "UMBC": coincidence counts as a function of the fiber tip's transverse coordinates in the image plane. The step size is 0.25 mm. The image shown is a "slice" at the half maximum value

point on the object plane, the other one must arrive at a unique corresponding point on the image plane.[2] Although questions regarding fundamental issues of quantum theory still exist, the experimental demonstration of ghost imaging [1] and ghost interference [6] in 1995 together stimulated the foundation of quantum imaging in terms of geometrical and physical optics.

Type-two ghost imaging uses chaotic-thermal radiation sources. Different from type-one, the nonfactorizable point-to-point image-forming correlation between the object and image planes is only partial with at least 50% constant background,

$$G^{(2)}(\vec{\rho}_o, \vec{\rho}_i) \sim 1 + \delta(\vec{\rho}_o - \vec{\rho}_i). \tag{14.6}$$

The first Fresnel near-field lensless ghost imaging experiment, which realized this nonfactorizable point-to-point image-forming correlation, was demonstrated by D'Angelo et al. and Scarcelli et al. in 2005 and 2006 [7, 8] after their experimental demonstration of two-photon interference of chaotic-thermal light in 2004 [9]. Figure 14.4 illustrates an improved setup of the type-two ghost imaging experiment by Meyers et al.. [10]. The thermal radiation of a chaotic source, which has a fairly large size in its transverse dimension, is split into two by a 50%–50% beamsplitter.

[2]The ghost imaging experiment is thus considered a demonstration of the historical Einstein–Podolsky–Rosen (EPR) experiment.

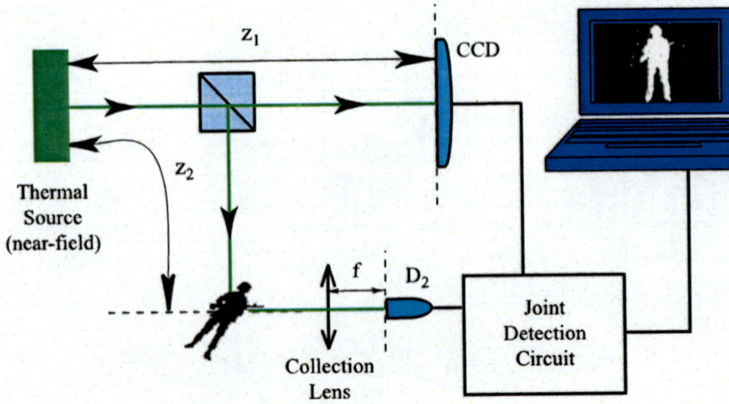

Fig. 14.4 Fresnel near-field lensless ghost imaging of chaotic light demonstrated by Meyers et al. D_2 is a "bucket" photon counting detector that is used to collect and count all random scattered and reflected photons from the object. The joint-detection between D_2 and the CCD array is realized by a photon-counting-coincidence circuit. D_2 is fixed in space. The counting rate of D_2 and the ungated output of the CCD are both monitored to be constants during the measurement. Surprisingly, a 1:1 ghost image of the object is captured in joint-detection between D_2 and the CCD, when taking $z_1 = z_2$. The images "blurred out" when the CCD is moved away from $z_1 = z_2$, in the direction of either $z_1 > z_2$ or $z_1 < z_2$

One of the beams illuminates a toy soldier as shown in Fig. 14.4. The scattered and reflected photons from the solider (object) are collected and counted by a "bucket" detector D_2. In the other beam a high resolution CCD array, operated at the photon counting regime, is placed toward the radiation source for joint-detection with the "bucket" detector D_2. The counting rate of D_2 and the ungated output of the CCD are both monitored to be constants during the measurement. Surprisingly, a 1:1 ghost image of the toy soldier is captured in the joint-detection between D_2 and the CCD, when taking $z_1 = z_2$. The 1:1 ghost image of the toy soldier is shown in Fig. 14.5. The images "blurred out" when the CCD is moved away from $z_1 = z_2$ to the side of either $z_1 > z_2$ or $z_1 < z_2$.

There is no doubt that chaotic-thermal radiations propagate to any transverse plane in a random and chaotic manner. A brief discussion for Fresnel free-propagation is given in the appendix. In the lensless ghost imaging experiment, a large transverse sized chaotic light source, as shown in Fig. 14.4, is usually used for achieving better spatial resolution. The source consists a large number of independent point subsources randomly distributed on the source plane. Each point subsource may randomly radiate independent spherical waves to the object and image planes. Due to the chaotic nature of the source, these independent subintensities simply add together yielding a constant total intensity in space and in time on any transverse plane. In the lensless ghost imaging setup, there is no lens applied to force these spherical waves collapsing to a point or a "speckle", and there is no chance to have two identical copies of any "speckle" of the source onto the object and image planes. What is the physical cause of the point-to-point

Fig. 14.5 Ghost image of a toy soldier model

image-forming correlation? Although the nonfactorizable point-to-point correlation between the object and image planes is only partial, the type-two ghost imaging looks more surprising than type-one because of the nature of the light source. Unlike the signal-idler photon pair, the jointly measured photons in type-two ghost imaging are just two independent photons that fall into the coincidence time window by chance only. Nevertheless, analogous to EPR, the nonfactorizable partial point-to-point correlation represents a nonlocal behavior of a measured pair of independent photons: neither photon-one nor photon-two "knows" precisely where to go when they are created at each independent subsources; however, if one of them is observed at a point on the object plane, the other one has twice greater probability of arriving at a unique corresponding point on the image plane.[3]

We have concluded and will show that the partial point-to-point correlation between the object and image planes in type-two ghost imaging is the result of *two-photon interference*. Similar to that of type-one, it involves the nonlocal superposition of two-photon amplitudes, a nonclassical entity corresponding to different yet indistinguishable alternative ways of triggering a joint-detection event [11]. Different from that of type-one, the joint-detection events observed in type-two ghost imaging are triggered by two randomly distributed independent photons. It is interesting to see that the quantum mechanical concept of *two-photon interference* is

[3] Similar to the HBT correlation, the contrast of the near-field partial point-to-point image-forming function is 50%, i.e., two to one ratio between the maximum value and the constant background, see (14.33).

applicable to "classical" thermal light.[4] In fact, this is not the first time in the history of physics we apply quantum mechanical concepts to thermal light. We should not forget Planck's theory of blackbody radiation originated the quantum physics. The radiation Planck dealt with was thermal radiation. Although the concept of "two-photon interference" comes from the study of entangled biphoton states [11], the concept should not be restricted to entangled systems. The concept is generally true and applicable to any radiation, including "classical" thermal light. The partial point-to-point correlation of thermal radiation is not a new discovery either. The first set of temporal and spatial far-field intensity-intensity correlations of thermal light was demonstrated by Hanbury Brown and Twiss (HBT) in 1956 [12, 13]. The HBT experiment created quite a surprise in the physics community and lead to a debate about the classical or quantum nature of the phenomenon [13–15]. Although the discovery of HBT initiated a number of key concepts of modern quantum optics, the HBT phenomenon itself was finally interpreted as statistical correlation of intensity fluctuations and considered as a classical effect. It is then reasonable to ask: Is the Fresnel near-field type-two ghost imaging with thermal light a simple classical effect similar to that of HBT? Is it possible that the ghost imaging phenomenon itself, including the type-one ghost imaging of 1995, is merely a simple classical effect of intensity fluctuation correlation? [16–19]. This article will address these important questions and explore the multiphoton interference nature of ghost imaging.

To explore the two-photon interference nature, we will analyze the physics of type-one and type-two ghost imaging in five steps. (1) Review the physics of coherent and incoherent light propagation; (2) Review classical imaging as the result of constructive–destructive interference among electromagnetic waves; (3) analyze type-one ghost imaging in terms of constructive–destructive interference between the biphoton amplitudes of an entangled photon-pair; (4) analyze type-two ghost imaging in terms of two-photon interference between chaotic subfields; and (5) discuss the physics of the phenomenon: whether it is a quantum interference or a classical intensity fluctuation correlation.

14.2 Classical Imaging

To understand the multiphoton interference nature of ghost imaging, it might be helpful to see the constructive–destructive interference nature of classical imaging first. We start from a typical classical imaging setup of Fig. 14.6 and ask a simple question: how does the radiation field propagate from the object plane to the image plane? In classical optics, such propagation is usually described by an optical transfer function $h(\mathbf{r}-\mathbf{r}_0, t-t_0)$. We prefer to work with the single-mode propagator, namely the Green's function, $g(\mathbf{k}, \mathbf{r}-\mathbf{r}_0, t-t_0)$ [20, 21], which propagates each

[4]There exist a number of definitions for classical light and for quantum light. One of the commonly accepted definitions considers thermal light classical because its positive P-function.

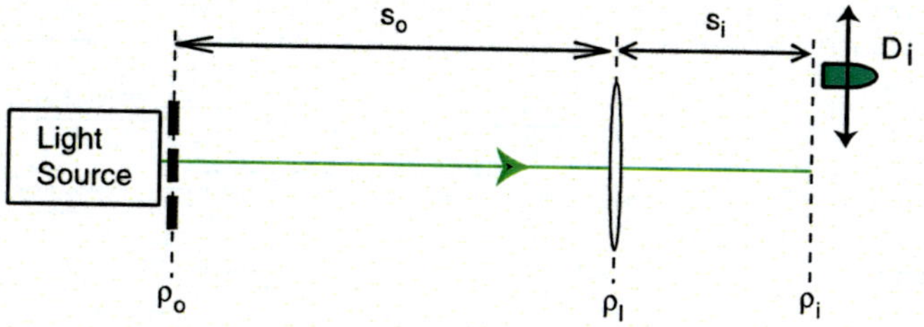

Fig. 14.6 Typical imaging setup. A lens of finite size is used to produce a magnified or demagnified image of an object with limited spatial resolution

mode of the radiation from space-time point $(\mathbf{r}_0, t_0)$ to space-time point $(\mathbf{r}, t)$. We treat the field $E(\mathbf{r}, t)$ as a superposition of these modes. A detailed discussion about $g(\mathbf{k}, \mathbf{r}-\mathbf{r}_0, t-t_0)$ is given in the appendix. It is convenient to write the field $E(\mathbf{r}, t)$ as a superposition of its longitudinal and transverse modes under the Fresnel paraxial approximation,

$$E(\vec{\rho}, z, t) = \int \mathrm{d}\vec{\kappa}\, \mathrm{d}\omega\, \tilde{E}(\vec{\kappa}, \omega)\, g(\vec{\kappa}, \omega; \vec{\rho}, z)\, \mathrm{e}^{-\mathrm{i}\omega t}, \tag{14.7}$$

where $\tilde{E}(\vec{\kappa}, \omega)$ is the complex amplitude for the mode of frequency ω and transverse wave-vector $\vec{\kappa}$. In (14.7), we have taken $z_0 = 0$ and $t_0 = 0$ at the object plane as usual. To simplify the notation, we have assumed one polarization.

Based on the experimental setup of Fig. 14.6 and following the Appendix, $g(\vec{\kappa}, \omega; \vec{\rho}, z)$ is found to be

$$\begin{aligned} g(\vec{\kappa}, \omega; \vec{\rho}_i, s_o + s_i) = &\int_{\text{obj}} \mathrm{d}\vec{\rho}_o \int_{\text{lens}} \mathrm{d}\vec{\rho}_l \left\{ A(\vec{\rho}_o)\, \mathrm{e}^{\mathrm{i}\vec{\kappa}\cdot\vec{\rho}_o} \right\} \\ &\times \left\{ \frac{-\mathrm{i}\omega}{2\pi c} \frac{\mathrm{e}^{\mathrm{i}\frac{\omega}{c}s_o}}{s_o} \mathrm{e}^{\mathrm{i}\frac{\omega}{2cs_o}|\vec{\rho}_l - \vec{\rho}_o|^2} \right\} \left\{ \mathrm{e}^{-\mathrm{i}\frac{\omega}{2cf}|\vec{\rho}_l|^2} \right\} \\ &\times \left\{ \frac{-\mathrm{i}\omega}{2\pi c} \frac{\mathrm{e}^{\mathrm{i}\frac{\omega}{c}s_i}}{s_i} \mathrm{e}^{\mathrm{i}\frac{\omega}{2cs_i}|\vec{\rho}_i - \vec{\rho}_l|^2} \right\}, \end{aligned} \tag{14.8}$$

where $\vec{\rho}_o$, $\vec{\rho}_l$, and $\vec{\rho}_i$ are two-dimensional vectors defined, respectively, on the object, lens, and image planes. The first curly bracket includes the aperture function $A(\vec{\rho}_o)$ of the object and the phase factor $\mathrm{e}^{\mathrm{i}\vec{\kappa}\cdot\vec{\rho}_o}$ contributed at the object plane by each transverse mode $\vec{\kappa}$. The terms in the second and fourth curly brackets describe free-space Fresnel propagation-diffraction from the source/object plane to the imaging lens, and from the imaging lens to the detection plane, respectively. The Fresnel propagator includes a spherical wave function $\mathrm{e}^{\mathrm{i}\frac{\omega}{c}(z_j - z_k)}/(z_j - z_k)$ and a Fresnel phase factor $\mathrm{e}^{\mathrm{i}\omega|\vec{\rho}_j - \vec{\rho}_k|^2/2c(z_j - z_k)}$. The third curly bracket adds the phase factor introduced by the imaging lens.

We now rewrite (14.8) into the following form

$$g(\vec{\kappa},\omega;\vec{\rho}_i, z = s_o + s_i) = \frac{-\omega^2}{(2\pi c)^2 s_o s_i}\, e^{i\frac{\omega}{c}(s_o+s_i)}\, e^{i\frac{\omega}{2cs_i}|\vec{\rho}_i|^2}$$
$$\times \int_{obj} d\vec{\rho}_o\, A(\vec{\rho}_o)\, e^{i\frac{\omega}{2cs_o}|\vec{\rho}_o|^2}\, e^{i\vec{\kappa}\cdot\vec{\rho}_o} \int_{lens} d\vec{\rho}_l\, e^{i\frac{\omega}{2c}\left[\frac{1}{s_o}+\frac{1}{s_i}-\frac{1}{f}\right]|\vec{\rho}_l|^2}\, e^{-i\frac{\omega}{c}\left(\frac{\vec{\rho}_o}{s_o}+\frac{\vec{\rho}_i}{s_i}\right)\cdot\vec{\rho}_l}. \tag{14.9}$$

The image plane is defined by the Gaussian thin-lens equation of (14.1). Hence, the second integral in (14.9) reduces to, for a finite sized lens of radius R, the so-called point-spread function, or the Airy disk, of the imaging system:

$$\int_{lens} d\vec{\rho}_l\, e^{-i\frac{\omega}{c}\left(\frac{\vec{\rho}_o}{s_o}+\frac{\vec{\rho}_i}{s_i}\right)\cdot\vec{\rho}_l} = \frac{2J_1(x)}{x} = somb(x), \tag{14.10}$$

where the sombrero-like function $somb(x) = 2J_1(x)/x$ with argument $x = \left[\frac{R}{s_o}\frac{\omega}{c}|\vec{\rho}_o + \rho_i/m|\right]$ has been defined in (14.3). Equation (14.10) indicates the result of an interference.

Substituting (14.9) and (14.10) into (14.7) enables one to obtain the classical self-correlation function of the field, or, equivalently, the intensity on the image plane

$$I(\vec{\rho}_i, z_i, t_i) = \langle\, E^*(\vec{\rho}_i, z_i, t_i)\, E(\vec{\rho}_i, z_i, t_i)\,\rangle, \tag{14.11}$$

where $\langle \ldots \rangle$ denotes an ensemble average. To simplify the mathematics, monochromatic light is assumed as usual.

Case (I): Incoherent imaging. The ensemble average yields zeros except when $\vec{\kappa} = \vec{\kappa}'$. The image is thus

$$I(\vec{\rho}_i) \propto \int d\vec{\rho}_o\, \left|A(\vec{\rho}_o)\right|^2 somb^2\left[\frac{R}{s_o}\frac{\omega}{c}\left|\vec{\rho}_o + \frac{\vec{\rho}_i}{m}\right|\right]. \tag{14.12}$$

An incoherent image, magnified by a factor of m, is thus given by the convolution between the modulus square of the object aperture function and the point-spread function. The spatial resolution of the image is determined by the finite width of the $|somb|^2$-function.

Case (II): Coherent imaging. The coherent superposition of the $\vec{\kappa}$ modes in both $E^*(\vec{\rho}_i, \tau)$ and $E(\vec{\rho}_i, \tau)$ results in a wavepacket. The image, or the intensity distribution on the image plane, is

$$I(\vec{\rho}_i) \propto \left| \int_{obj} d\vec{\rho}_o\, A(\vec{\rho}_o)\, e^{i\frac{\omega}{2cs_o}|\vec{\rho}_o|^2} somb\left[\frac{R}{s_o}\frac{\omega}{c}\left|\vec{\rho}_o + \frac{\vec{\rho}_i}{m}\right|\right]\right|^2. \tag{14.13}$$

A coherent image, magnified by a factor of m, is thus given by the modulus square of the convolution between the object aperture function (multiplied by a Fresnel phase factor) and the point-spread function.

For $s_i < s_o$ and $s_o > f$, both (14.12) and (14.13) describe a real demagnified inverted image. In both cases, a narrower *somb*-function yields a higher spatial resolution. Therefore, the use of a larger imaging lens and shorter wavelengths will improve the spatial resolution of an imaging system.

14.3 Biphoton and Type-One Ghost Imaging

In this section, we analyze type-one ghost imaging. Type-one ghost imaging uses entangled photon pairs such as the signal-idler biphoton pairs of SPDC [11, 22]. The nearly collinear signal-idler system generated by SPDC can be described, in the ideal case, by the following entangled biphoton state [11]:

$$|\,\Psi\,\rangle = \Psi_0 \int \mathrm{d}\vec{\kappa}_s\, \mathrm{d}\vec{\kappa}_i\, \delta(\vec{\kappa}_s + \vec{\kappa}_i) \int \mathrm{d}\omega_s\, \mathrm{d}\omega_i\, \delta(\omega_s + \omega_i - \omega_p)$$
$$a^\dagger(\vec{\kappa}_s, \omega_s)\, a^\dagger(\vec{\kappa}_i, \omega_i)\,|\,0\,\rangle, \qquad (14.14)$$

where $\omega_j, \vec{\kappa}_j$ $(j = s, i, p)$ are the frequency and transverse wavevector of the signal, idler, and pump, respectively. For simplicity, a CW single mode pump with $\vec{\kappa}_p = 0$ is assumed. Equation (14.14) indicates that the biphoton state of the signal-idler pair is an entangled state. The single-photon state of the signal and the idler can be evaluated by taking a partial trace of its twin,

$$\hat{\rho}_s = tr_i\,|\,\Psi\,\rangle\langle\,\Psi\,| = \int \mathrm{d}\vec{\kappa}_s\, \mathrm{d}\omega_s\, a^\dagger(\vec{\kappa}_s, \omega_s)|\,0\,\rangle\langle\,0\,|a(\vec{\kappa}_s, \omega_s),$$
$$\hat{\rho}_i = tr_s\,|\,\Psi\,\rangle\langle\,\Psi\,| = \int \mathrm{d}\vec{\kappa}_i\, \mathrm{d}\omega_i\, a^\dagger(\vec{\kappa}_i, \omega_i)|\,0\,\rangle\langle\,0\,|a(\vec{\kappa}_i, \omega_i). \qquad (14.15)$$

Although the signal-idler system is in a pure state, the state of the signal photon and the idler photon, respectively, are both mixed states.

Let us imagine a measurement in which two point-like photon counting detectors (D_1 and D_2) are placed at the output plane of an SPDC source for the detection of the signal photon and the idler photon, respectively, and for the joint-detection of the signal-idler pair. The probability of observing a photo-detection event in the SPDC output plane $\vec{\rho}_j$ at time t_j, $j = s, i$, is calculated from the first-order photo-detection theory of Glauber [23]

$$G^{(1)}(\vec{\rho}_j, t_j) = tr\,\hat{\rho}\,E^{(-)}(\vec{\rho}_j, t_j)E^{(+)}(\vec{\rho}_j, t_j), \qquad (14.16)$$

where we have chosen $z_j = 0$ for the SPDC output plane as usual. It is easy to find that

$$G^{(1)}(\vec{\rho}_s, t_s) = \text{constant}, \quad G^{(1)}(\vec{\rho}_i, t_i) = \text{constant}, \qquad (14.17)$$

which means that the signal photon and the idler photon both have equal probability to be observed at any position in the output plane of the SPDC at any time. The probability of observing a joint-detection event between D_1 and D_2 located at $\vec{\rho}_s$ and $\vec{\rho}_i$ in the SPDC output plane of $z_s = z_i = 0$ is calculated from the second-order photo-detection theory of Glauber [23]:

$$\begin{aligned} G^{(2)}(\vec{\rho}_s, t_s; \vec{\rho}_i, t_i) &= \langle \Psi \,|\, E_s^{(-)}(\vec{\rho}_s, t_s) E_i^{(-)}(\vec{\rho}_i, t_i) E_i^{(+)}(\vec{\rho}_i, t_i) E_s^{(+)}(\vec{\rho}_s, t_s) \,|\, \Psi \rangle \\ &= |\langle 0 \,|\, E_i^{(+)}(\vec{\rho}_i, t_i) E_s^{(+)}(\vec{\rho}_s, t_s) \,|\, \Psi \rangle|^2 \\ &\equiv |\, \Psi(\vec{\rho}_s, t_s; \vec{\rho}_i, t_i) \,|^2, \end{aligned} \tag{14.18}$$

where $\Psi(\vec{\rho}_s, t_s; \vec{\rho}_i, t_i)$ is defined as the effective biphoton wavefunction. The transverse spatial part of the effective biphoton wavefunction is easily calculated to be:

$$\Psi(\vec{\rho}_s, \vec{\rho}_i) \simeq \delta(\vec{\rho}_s - \vec{\rho}_i), \tag{14.19}$$

under the condition $t_s \simeq t_i$. Equations (14.14), (14.17), and (14.19) suggest that the entangled signal-idler photon pair is characterized by the EPR correlation [5] in transverse momentum and transverse position; hence, similar to the original EPR state, we have [24]:

$$\begin{gathered} \Delta(\vec{\kappa}_s + \vec{\kappa}_i) = 0 \;\;\&\;\; \Delta(\vec{\rho}_s - \vec{\rho}_i) = 0 \\ \text{with } \Delta\vec{\kappa}_s \sim \infty, \;\; \Delta\vec{\kappa}_i \sim \infty, \;\; \Delta\vec{\rho}_s \sim \infty, \;\; \Delta\vec{\rho}_i \sim \infty. \end{gathered} \tag{14.20}$$

In EPR's language, the signal photon and the idler photon may come from any point in the output plane of the SPDC. However, if the signal (idler) is found in a certain position, the idler (signal) must be observed in the same position, with certainty (100%). Simultaneously, the signal photon and the idler photon may have any transverse momentum. However, if a certain value and direction of the transverse momentum of the signal (idler) is observed, the transverse momentum of the idler (signal) will be uniquely determined with equal value and opposite direction.

The EPR δ-functions, $\delta(\vec{\rho}_s - \vec{\rho}_i)$ and $\delta(\vec{\kappa}_s + \vec{\kappa}_i)$ in transverse position and momentum, are the keys to understanding the ghost imaging experiment of Pittman et al.. of 1995. $\delta(\vec{\rho}_s - \vec{\rho}_i)$ indicates that the signal-idler pair is always emitted from the same point on the output plane of the biphoton source. Simultaneously, $\delta(\vec{\kappa}_s + \vec{\kappa}_i)$ defines the angular correlation of the pair: the signal-idler pair always exists at roughly equal but opposite angles relative to the pump for degenerate SPDC. This then allows for a simple explanation of the experiment in terms of "usual" geometrical optics in the following manner: we envision the nonlinear crystal as a "hinge point" and "unfold" the schematic of Fig. 14.2 into the Klyshko picture [2–4] of Fig. 14.7. The signal-idler biphoton amplitudes can then be represented by straight lines (but keep in mind the different propagation directions) and therefore the image is reproduced in coincidences when the aperture, lens, and fiber tip are located according to the Gaussian thin lens equation of (14.1). The image is exactly the same as that one would observe on a screen placed at the fiber tip if detector D_1

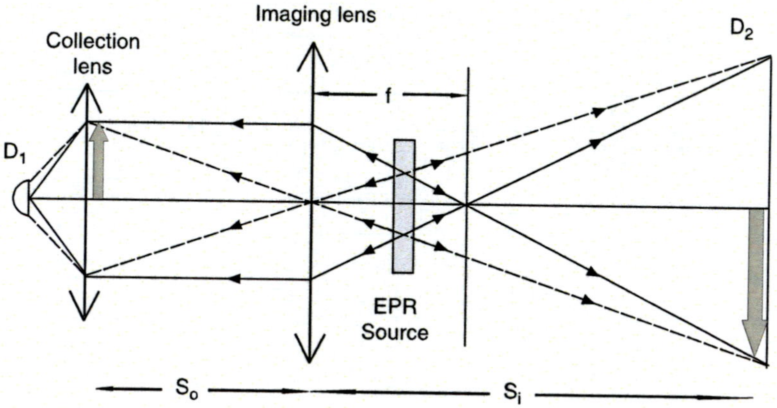

Fig. 14.7 An unfolded schematic of the 1995 ghost imaging experiment, which is helpful for understanding the physics. Since the biphoton "light" propagates along "straight lines", it is obvious that any point on the object plane corresponds to a unique point on the image plane. Although the placement of the lens, the object, and detector D_2 obeys the Gaussian thin lens equation, it is important to notice that the geometric rays in the figure actually represent the biphoton amplitudes of an entangled photon pair. The point-to-point correspondence is the result of a constructive–destructive interference of these biphoton amplitudes

were replaced by a point-like light source and the nonlinear crystal by a reflecting mirror.

Comparing the "unfolded" schematic of the ghost imaging experiment with that of the classical imaging setup of Fig. 14.1, it is not difficult to find that any "light point" on the object plane has a unique corresponding "light point" on the image plane. This point-to-point correspondence is the result of the constructive–destructive interference among these biphoton amplitudes that are illustrated as the geometrical rays in Fig. 14.7. Similar to the situation in classical imaging, these biphoton amplitudes which experience equal optical path propagation will superpose constructively at each pair of one-to-one points of the object plane and the image plane for a joint-detection event, while these that experience unequal distance propagation will superpose destructively at all other points on the object and image planes. The use of the imaging lens makes this constructive-destructive interference possible. It is this unique point-to-point EPR correlation that makes the "ghost" image of the object-aperture function possible. Despite the completely different physics from classical optics, the remarkable feature is that the relationship between the focal length f of the lens, the aperture's optical distance s_o, and the image's optical distance s_i, satisfies the Gaussian thin lens equation of (14.1). It is worth emphasizing again that the geometric rays in Fig. 14.7 represent the biphoton amplitudes of a signal-idler photon pair, and the point-to-point correspondence is the result of the constructive-destructive interference of these biphoton amplitudes.

We now calculate $G^{(2)}(\vec{\rho}_1, \vec{\rho}_2)$ for the "ghost" imaging experiment in detail, where $\vec{\rho}_1$ and $\vec{\rho}_2$ are the transverse coordinates of the point-like photodetector D_1

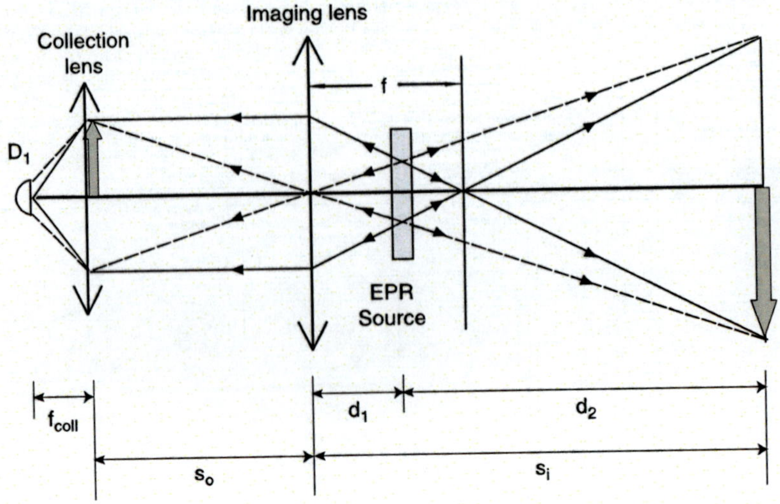

Fig. 14.8 In arm-1, the signal propagates freely over a distance d_1 from the output plane of the source to the imaging lens, passes an object aperture at distance s_o, and then is focused onto photon-counting detector D_1 by a collection lens. In arm-2, the idler propagates freely over a distance d_2 from the output plane of the source to a point-like photon counting detector D_2

and D_2, on the object and image planes, respectively. We will show that there exists a δ-function-like point-to-point correlation between the object and image planes, $\delta(\vec{\rho}_1 - \vec{\rho}_2/m)$. We will then show how the object function of $A(\vec{\rho}_o)$ is transferred to the image plane as a magnified image $A(\vec{\rho}_2/m)$.

We first calculate the effective biphoton wavefunction $\Psi(\vec{\rho}_1, z_1, t_1; \vec{\rho}_2, z_2, t_2)$, as defined in (14.18). By inserting the field operators into $\Psi(\vec{\rho}_1, z_1, t_1; \vec{\rho}_2, z_2, t_2)$, and considering the commutation relations of the field operators, the effective biphoton wavefunction is calculated to be

$$\Psi(\vec{\rho}_1, z_1, t_1; \vec{\rho}_2, z_2, t_2) = \Psi_0 \int d\vec{\kappa}_s \, d\vec{\kappa}_i \, \delta(\vec{\kappa}_s + \vec{\kappa}_i) \int d\omega_s \, d\omega_i \, \delta(\omega_s + \omega_i - \omega_p) \times g(\vec{\kappa}_s, \omega_s; \vec{\rho}_1, z_1) \, e^{-i\omega_s t_1} \, g(\vec{\kappa}_i, \omega_i; \vec{\rho}_2, z_2) \, e^{-i\omega_i t_2}. \quad (14.21)$$

Equation (14.21) indicates a coherent superposition of all the biphoton amplitudes shown in Fig. 14.7. Next, we follow the unfolded experimental setup of Fig. 14.8 to establish the Green's functions $g(\vec{\kappa}_s, \omega_s, \vec{\rho}_1, z_1)$ and $g(\vec{\kappa}_i, \omega_i, \vec{\rho}_2, z_2)$. In arm-1, the signal propagates freely over a distance d_1 from the output plane of the source to the imaging lens, passes an object aperture at distance s_o, and then is focused onto photon-counting detector D_1 by a collection lens. We will evaluate $g(\vec{\kappa}_s, \omega_s, \vec{\rho}_1, z_1)$ by propagating the field from the output plane of the biphoton source to the object plane. In arm-2, the idler propagates freely over a distance d_2 from the output plane of the biphoton source to a point-like detector D_2. $g(\vec{\kappa}_i, \omega_i, \vec{\rho}_2, z_2)$ is thus a free propagator.

(I) Arm-1 (source to object):

The optical transfer function or Green's function in arm-1, which propagates the field from the source plane to the object plane, is given by:

$$g(\vec{\kappa}_s, \omega_s; \vec{\rho}_1, z_1 = d_1 + s_o) = \mathrm{e}^{\mathrm{i}\frac{\omega_s}{c} z_1} \int_{\text{lens}} \mathrm{d}\vec{\rho}_l \int_{\text{source}} \mathrm{d}\vec{\rho}_s \left\{ \frac{-\mathrm{i}\omega_s}{2\pi c d_1} \mathrm{e}^{\mathrm{i}\vec{\kappa}_s \cdot \vec{\rho}_s} \mathrm{e}^{\mathrm{i}\frac{\omega_s}{2cd_1} |\vec{\rho}_s - \vec{\rho}_l|^2} \right\} \times \mathrm{e}^{-\mathrm{i}\frac{\omega}{2cf} |\vec{\rho}_l|^2} \left\{ \frac{-\mathrm{i}\omega_s}{2\pi c s_o} \mathrm{e}^{\mathrm{i}\frac{\omega_s}{2cs_o} |\vec{\rho}_l - \vec{\rho}_1|^2} \right\}, \tag{14.22}$$

where $\vec{\rho}_s$ and $\vec{\rho}_l$ are the transverse vectors defined, respectively, on the output plane of the source and on the plane of the imaging lens. The terms in the first and second curly brackets in (14.22) describe free space propagation from the output plane of the source to the imaging lens and from the imaging lens to the object plane, respectively. Again, $\mathrm{e}^{\mathrm{i}\frac{\omega_s}{2cd_1} |\vec{\rho}_s - \vec{\rho}_l|^2}$ and $\mathrm{e}^{\mathrm{i}\frac{\omega_s}{2cs_o} |\vec{\rho}_l - \vec{\rho}_1|^2}$ are the Fresnel phases as defined in the Appendix. Here, the imaging lens is treated as a thin-lens, and the transformation function of the imaging lens is approximated as a Gaussian, $l(|\vec{\rho}_l|, f) \cong \mathrm{e}^{-\mathrm{i}\frac{\omega}{2cf} |\vec{\rho}_l|^2}$.

(II) Arm-2 (from source to image):

In arm-2, the idler propagates freely from the source to the plane of D_2, which is also the plane of the image. The Green's function is

$$g(\vec{\kappa}_i, \omega_i; \vec{\rho}_2, z_2 = d_2) = \frac{-\mathrm{i}\omega_i}{2\pi c d_2} \mathrm{e}^{\mathrm{i}\frac{\omega_i}{c} d_2} \int_{\text{source}} \mathrm{d}\vec{\rho}_s' \, \mathrm{e}^{\mathrm{i}\frac{\omega_i}{2cd_2} |\vec{\rho}_s' - \vec{\rho}_2|^2} \mathrm{e}^{\mathrm{i}\vec{\kappa}_i \cdot \vec{\rho}_s'}, \tag{14.23}$$

where $\vec{\rho}_s'$ and $\vec{\rho}_2$ are the transverse vectors defined, respectively, on the output plane of the source and the plane of photodetector D_2.

(III) $\Psi(\vec{\rho}_1, \vec{\rho}_2)$ and $G^{(2)}(\vec{\rho}_1, \vec{\rho}_2)$ (object plane - image plane):

For simplicity, in the following calculation we consider degenerate ($\omega_s = \omega_i = \omega$) and collinear SPDC. The effective transverse biphoton wavefunction $\Psi(\vec{\rho}_1, \vec{\rho}_2)$ is then evaluated by substituting the Green's functions $g(\vec{\kappa}_s, \omega; \vec{\rho}_1, z_1)$ and $g(\vec{\kappa}_i, \omega; \vec{\rho}_2, z_2)$ into (14.21),

$$\Psi(\vec{\rho}_1, \vec{\rho}_2) \propto \int \mathrm{d}\vec{\kappa}_s \, \mathrm{d}\vec{\kappa}_i \, \delta(\vec{\kappa}_s + \vec{\kappa}_i) \, g(\vec{\kappa}_s, \omega; \vec{\rho}_1, z_1) \, g(\vec{\kappa}_i, \omega; \vec{\rho}_2, z_2)$$

$$\propto e^{i\frac{\omega}{c}(s_o+s_i)} \int d\vec{\kappa}_s\, d\vec{\kappa}_i\, \delta(\vec{\kappa}_s + \vec{\kappa}_i) \int_{\text{lens}} d\vec{\rho}_l \int_{\text{source}} d\vec{\rho}_s\, e^{i\vec{\kappa}_s\cdot\vec{\rho}_s} e^{i\frac{\omega}{2cd_1}|\vec{\rho}_s-\vec{\rho}_l|^2}$$

$$\times\, e^{-i\frac{\omega}{2cf}|\vec{\rho}_l|^2}\, e^{i\frac{\omega_s}{2cs_o}|\vec{\rho}_l-\vec{\rho}_1|^2} \int_{\text{source}} d\vec{\rho}_s'\, e^{i\vec{\kappa}_i\cdot\vec{\rho}_s'}\, e^{i\frac{\omega_i}{2cd_2}|\vec{\rho}_s'-\vec{\rho}_2|^2}, \tag{14.24}$$

where all the proportionality constants have been ignored. After completing the double integral of $d\vec{\kappa}_s$ and $d\vec{\kappa}_i$

$$\int d\vec{\kappa}_s\, d\vec{\kappa}_i\, \delta(\vec{\kappa}_s + \vec{\kappa}_i)\, e^{i\vec{\kappa}_s\cdot\vec{\rho}_s}\, e^{i\vec{\kappa}_i\cdot\vec{\rho}_s'} \sim \delta(\vec{\rho}_s - \vec{\rho}_s'),$$

(14.24) becomes

$$\Psi(\vec{\rho}_1, \vec{\rho}_2) \propto e^{i\frac{\omega}{c}(s_o+s_i)} \int_{\text{lens}} d\vec{\rho}_l \int_{\text{source}} d\vec{\rho}_s\, e^{i\frac{\omega}{2cd_2}|\vec{\rho}_2-\vec{\rho}_s|^2}\, e^{i\frac{\omega}{2cd_1}|\vec{\rho}_s-\vec{\rho}_l|^2}$$

$$e^{-i\frac{\omega}{2cf}|\vec{\rho}_l|^2}\, e^{i\frac{\omega}{2cs_o}|\vec{\rho}_l-\vec{\rho}_o|^2}.$$

Next, we complete the integral for $d\vec{\rho}_s$,

$$\Psi(\vec{\rho}_1, \vec{\rho}_2) \propto e^{i\frac{\omega}{c}(s_0+s_i)} \int_{\text{lens}} d\vec{\rho}_l\, e^{i\frac{\omega}{2cs_i}|\vec{\rho}_2-\vec{\rho}_l|^2}\, e^{-i\frac{\omega}{2cf}|\vec{\rho}_l|^2} e^{i\frac{\omega}{2cs_o}|\vec{\rho}_l-\vec{\rho}_1|^2}, \tag{14.25}$$

where we have replaced $d_1 + d_2$ with s_i (as depicted in Fig. 14.8). Although the signal and idler propagate in different directions along two optical arms, interestingly, the Green function in (14.25) is equivalent to that of a classical imaging setup, as if the field is originated from a point $\vec{\rho}_1$ on the object plane and propagated the lens and then arrived at point $\vec{\rho}_2$ on the imaging plane. The mathematics is consistent with our previous qualitative analysis of the experiment.

The finite integral on $d\vec{\rho}_l$ yields a point-to-"spot" relationship between the object plane and the image plane that is defined by the Gaussian thin-lens equation

$$\Psi(\vec{\rho}_1, \vec{\rho}_2) \propto \int_{\text{lens}} d\vec{\rho}_l\, e^{i\frac{\omega}{2c}[\frac{1}{s_o}+\frac{1}{s_i}-\frac{1}{f}]|\vec{\rho}_l|^2}\, e^{-i\frac{\omega}{c}(\frac{\vec{\rho}_1}{s_o}+\frac{\vec{\rho}_2}{s_i})\cdot\vec{\rho}_l}$$

$$= somb\left(\frac{R}{s_o}\frac{\omega}{c}\left|\vec{\rho}_1 + \frac{\vec{\rho}_2}{m}\right|\right). \tag{14.26}$$

If the integral is taken to infinity, by imposing the condition of the Gaussian thin-lens equation the effective transverse biphoton wavefunction can be approximated as a δ function

$$\Psi(\vec{\rho}_1, \vec{\rho}_2) \sim \delta(\vec{\rho}_1 + \vec{\rho}_2/m) \sim \delta(\vec{\rho}_o + \vec{\rho}_I/m), \tag{14.27}$$

where we have replaced $\vec{\rho}_1$ and $\vec{\rho}_2$ with $\vec{\rho}_o$ and $\vec{\rho}_I$, respectively, to emphasize the point-to-point EPR correlation between the object and image planes. To avoid confusion with the "idler", we have used $\vec{\rho}_I$ to label the image plane.

We now include an object-aperture function, a collection lens and a photon counting detector D_1 into the optical transfer function of arm-1 as shown in Fig. 14.2. The collection-lens–D_1 package can be simply treated as a "bucket" detector. The "bucket" detector integrates the biphoton amplitudes $\Psi(\vec{\rho}_o, \vec{\rho}_2)$, which are modulated by the object aperture function $A(\vec{\rho}_o)$ into a joint photodetection event. This process is equivalent to the following convolution

$$R_{1,2} \propto \int_{\text{object}} d\vec{\rho}_o \left|A(\vec{\rho}_o)\right|^2 \left|\Psi(\vec{\rho}_o, \vec{\rho}_2)\right|^2 \simeq \left|A(\vec{\rho}_2/m)\right|^2 = \left|A(\vec{\rho}_I/m)\right|^2. \quad (14.28)$$

Again, D_2 is scanned in the image plane ($\vec{\rho}_2 = \vec{\rho}_I$). A ghost image of the object is thus reproduced on the image plane by means of the joint-detection between the point-like-detector D_2 and the bucket detector D_1.

The physical process corresponding to the above convolution can be summarized as follows. Due to the unique point-to-point correlation between the object plane and the image plane, whenever the bucket detector receives a signal photon that is either transmitted, scattered or reflected from a unique point of the object, the scanning point photo-detector D_2 or a CCD element that receives the idler photon identifies the coordinate of $\vec{\rho}_o$ and the value of the aperture function $A(\vec{\rho}_o)$ for that joint-photo-detection event. For instance, at time t, the bucket detector receives a signal photon that is either transmitted, scattered, or reflected from a unique point $\vec{\rho}_o$ of the object plane within a coincidence time window. The joint-detection of the idler photon by the scanning point detector D_2 or a CCD element, with known coordinate $\vec{\rho}_I$, identifies $\vec{\rho}_o$ immediately for that event. At time t', the bucket detector receives another signal photon that is either transmitted, scattered, or reflected from another unique point $\vec{\rho'}_o$ of the object plane within the coincidence window, and the joint-detection of the idler photon by the scanning point detector D_2 or a CCD element, with known coordinate $\vec{\rho'}_I$, identifies $\vec{\rho'}_o$ immediately for that event. The probability of receiving a joint-photo-detection event at $(\vec{\rho}_o, \vec{\rho}_I)$ and at $(\vec{\rho'}_o, \vec{\rho'}_I)$ is proportional to the value of the aperture function $A(\vec{\rho}_o)$ and $A(\vec{\rho'}_o)$, respectively. Accumulating a large number of joint-detection events at each transverse coordinate on the image plane, the aperture function $A(\vec{\rho}_o)$ is thus reproduced in the joint-detection as a function of $\vec{\rho}_I$.

The observation of type-one ghost imaging has demonstrated a nonfactorizable point-to-point EPR correlation between the object and image planes. This point-to-point correlation is the result of a constructive–destructive interference between biphoton amplitudes,

$$\begin{aligned} G^{(2)}(\vec{\rho}_o, \vec{\rho}_I) &= \left| \int d\vec{\kappa}_s \, d\vec{\kappa}_i \, \delta(\vec{\kappa}_s + \vec{\kappa}_i) \, g(\vec{\kappa}_s, \vec{\rho}_o) \, g(\vec{\kappa}_i, \vec{\rho}_I) \right|^2 \\ &= somb^2 \left(\frac{R}{s_o} \frac{\omega}{c} \left| \vec{\rho}_O + \frac{\vec{\rho}_I}{m} \right| \right). \end{aligned} \quad (14.29)$$

In this view, we consider the ghost imaging experiment of Pittman et al.. a realization of the 1935 EPR *gedankenexperiment* [24, 25].

Classical theory has difficulties when facing type-one ghost imaging phenomenon. In the classical theory of light, a joint measurement between two photodetectors D_1 and D_2 measures the statistical correlation of intensity fluctuations,

$$G^{(2)}(\vec{\rho}_1, \vec{\rho}_2) = \langle I(\vec{\rho}_1) I(\vec{\rho}_2) \rangle = \bar{I}(\vec{\rho}_1)\bar{I}(\vec{\rho}_2) + \langle \Delta I(\vec{\rho}_1)\Delta I(\vec{\rho}_2)\rangle. \qquad (14.30)$$

Therefore, the point-to-point image-forming correlation is considered as a result of the statistical correlation of intensity fluctuations between the object and the image planes. Comparing (14.30), which has a constant background $\bar{I}(\vec{\rho}_1)\bar{I}(\vec{\rho}_2)$, with (14.29), which has a zero background, the mean intensities $\bar{I}(\vec{\rho}_1)$ and $\bar{I}(\vec{\rho}_2)$ must be zero; otherwise, the result would lead to nonphysical conclusions. The measurements, however, never yield zero mean values of $\bar{I}(\vec{\rho}_1)$ and $\bar{I}(\vec{\rho}_2)$ under any circumstances. In fact, the individual-detector counting rates of D_1 and D_2 were monitored in the experiment of Pittman et al.. with much greater value than that of the coincidences. It is clear that the classical theory of statistical correlation of intensity fluctuations does not reflect the correct physics behind type-one ghost imaging.

14.4 Type-Two Ghost Imaging with Chaotic-thermal Radiation

In this section, we discuss the physics of type-two ghost imaging. The Fresnel near-field lensless ghost imaging with chaotic-thermal radiation was first demonstrated by D'Angelo et al.. and Scarcelli et al. in the years from 2005 to 2006 [7,8] following their experimental demonstration of two-photon interference of chaotic-thermal light in 2004 [9]. The schematic experimental setup of their 2006 demonstration is shown in Fig. 14.9. Radiation with a narrow spectral bandwidth $\Delta\omega$ of a few millimeters diameter from a chaotic pseudothermal source [26] was equally divided into two by a 50%–50% nonpolarizing beam-splitter. In the reflected arm, a double-slit with slit separation $b = 1.5$ mm and slit width $a = 0.2$ mm, was placed at a distance $z_2 = 139$ mm from the source and a bucket detector D_2 was placed just behind the object. In the transmitted arm, a point detector D_1 was scanned in the transverse plane of $z_1 = z_2$. Scarcelli et al. tested two different joint detection schemes, namely the photon counting coincidence circuit and the standard HBT correlator. In the photon counting regime two Geiger mode avalanche photodiodes were employed for single-photon detection. In the bright light condition, two silicon PIN diodes were used with a standard analog HBT linear multiplier. The bucket detector D_2 was simulated by using a short focal length lens ($f = 25$ mm) to focus the light coming from the object onto the active area of the detector while the point detector D_2 was simulated by a pin-hole like aperture. After a large number of reaped measurements for different experimental schemes and conditions. Scarcelli et al. reported the following observations.

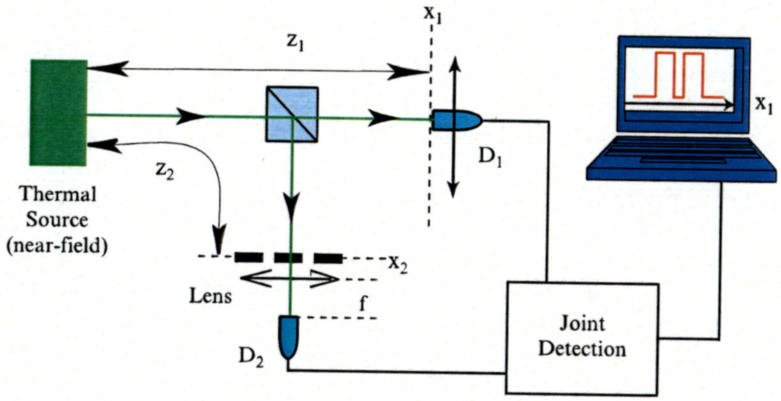

Fig. 14.9 Fresnel near-field lensless ghost imaging with chaotic light demonstrated in 2006 by Scarcelli et al. D_1 is a point-like photodetector that is scannable along the x_1-axis. The joint-detection between D_1 and the bucket detector D_2 is realized either by a photon-counting coincidence counter or by a standard HBT linear multiplier (RF mixer). In this measurement, D_2 is fixed in the focal point of a convex lens, playing the role of a bucket detector. The counting rate or the photocurrent of D_1 and D_2, respectively, is measured to be constants. Surprisingly, an image of the 1-D object is observed in the joint-detection between D_1 and D_2 by scanning D_1 in the plane of $z_1 = z_2$ along the x_1-axis. The image is blurred out when $z_1 \neq z_2$. There is no doubt that thermal radiations propagate to any transverse plane in a random and chaotic manner. There is no lens applied to force the thermal radiation "collapsing" to a point or speckle either. What is the physical cause of the point-to-point image-forming correlation in coincidences?

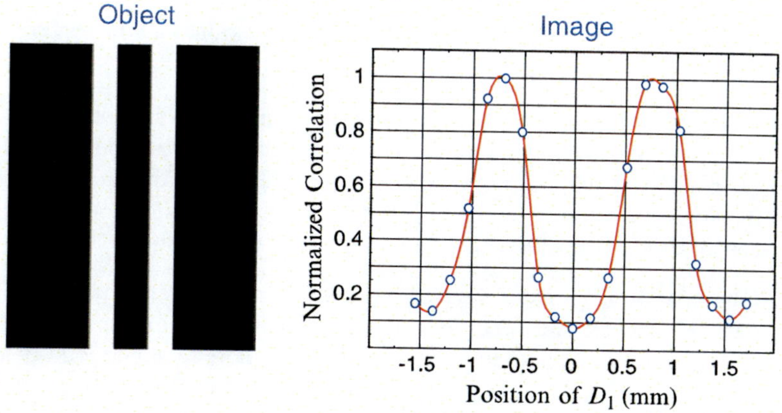

Fig. 14.10 The double-slit and its ghost image. Notice the constant background has been removed from the correlation

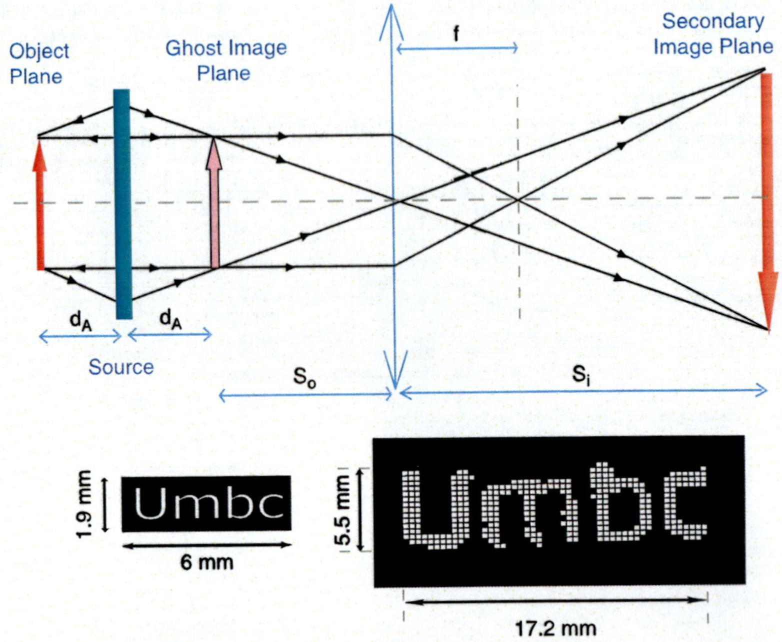

Fig. 14.11 *Upper*: unfolded schematic experimental setup of a secondary image measurement of the ghost image and the measured secondary images. By using a convex lens of focal length f, the ghost image is imaged onto a secondary image plane, which is defined by the Gaussian thin-lens equation, $1/s_o + 1/s_i = 1/f$, with magnification $m = -s_i/s_o$. This setup is useful for distant large-scale ghost imaging applications. *Lower*: a secondary image of the ghost image of "UMBC" with a magnification factor of $m = s_i/s_o \sim 2.9$. Notice the secondary image is "blurred" out quickly when the scanning photodetector D_1 is moved away from the image plane

Observation (1): A typical measured ghost image of the double-slit is shown in Fig. 14.10. The measured curve reports the joint-detection counting rate between D_1 and D_2, or the output current of a HBT linear multiplier, as a function of the transverse position of the point detector D_1 along x_1 axis. Notice in Fig. 14.10 the constant background has been removed from the correlation.

Observation (2): The measured contrasts vary significantly under different experimental schemes and conditions. It was found that the image contrast can achieve $\sim$50% in photon counting measurement if no more than one joint-detection event occurring within the time window of the coincidence circuit. 50% is the maximum image contrast we expect for thermal light ghost imaging.

Observation (3): To achieve less than one joint-detection event per coincidence time window, weak light source is not a necessary condition. It can be easily achieved under bright light sources by using adjustable ND-filters with D_1 and D_2.

To confirm the observation is an image but not a "projection shadow", Scarcelli et al.. constructed a secondary imaging system, which is illustrated in Fig. 14.11.

By using a convex imaging lens of focus length f, the ghost image is imaged onto a secondary image plane, where the scanning photodetector D_1 is placed. The secondary image of the ghost image is recorded in the joint detection between D_1 and D_2 by means of either photon counting coincidences or current–current correlation. Figure 14.11 shows a secondary image of the ghost image of "UMBC" with a magnification factor of $m = s_i/s_o \sim 2.9$. The secondary imaging system is useful in certain experimental conditions, especially when the size of the 1:1 ghost image is either too big or too small to be captured by the CCD (scanning photodetector D_1). A demagnified or enlarged secondary image would be helpful for certain applications. For instance, in a satellite-based ghost imaging system, the spatial resolution of the $1:1$ primary ghost image, which could be as large as kilometers, is determined by the angular size of the thermal source such as the Sun. When capturing the entire view of the $1:1$ ghost image, the spatial resolution of the secondary image will be limited by the classical imaging system; however, the secondary ghost image may achieve the same spatial resolution of the primary ghost image by reducing its field of view and to reconstruct the entire ghost image piece by piece.

In 2008, Meyers et al.. published a modified version of the original near-field lensless ghost imaging experiment. Figure 14.4 schematically illustrates their experimental setup. This experiment improves the ghost imaging experiment of Scarcelli et al.. in two aspects: (1) the bucket detector is not triggered by the transmitted light that passing through a mask but instead trigged by the randomly scattered and reflected photons from the surface of a toy soldier; (2) no scanning point photodetector, but instead a CCD array of 2-D operated at photon counting regime is used for the joint-detection with the bucket detector D_2. A ghost image of the toy soldier is captured by the gated CCD when taking $z_1 = z_2$.

14.4.1 *What Is the Physical Cause of Chaotic-thermal Light Ghost Imaging?*

It is the partial point-to-point correlation between the object plane and the image plane that makes ghost imaging with thermal light possible. Similar but different from classical imaging and type-one ghost imaging, mathematically, type-two ghost imaging is the result of a convolution between the aperture function $|A(\vec{\rho}_2)|^2$ and a δ-function like partial point-to-"spot" correlation function

$$R_{12} \propto \int_{\text{object}} d\vec{\rho}_2 \, \left|A(\vec{\rho}_2)\right|^2 \left[1 + somb^2\left(\frac{\pi \Delta\theta \, |\vec{\rho}_1 - \vec{\rho}_2|}{\lambda}\right)\right] \tag{14.31}$$

in 2-D, where $\Delta\theta$ is the angular diameter of the radiation source viewed from the photodetector, $\vec{\rho}_1$ and $\vec{\rho}_2$ are the transverse coordinates on the object plane and the image plane, respectively, or

$$R_{12} \propto \int_{\text{object}} dx_2 \, \left|A(x_2)\right|^2 \left[1 + \text{sinc}^2\left(\frac{\pi \Delta\theta \, (x_1 - x_2)}{\lambda}\right)\right]. \tag{14.32}$$

in 1-D. For a chosen wavelength, the spatial resolution of the ghost image is determined by the angular diameter of the light source: the larger the size of the source in transverse dimensions, the higher the spatial resolution of the lensless ghost image. The point-to-"spot" image-forming functions in (14.31) and (14.32) have been verified experimentally by Scarcelli et al..

Similar to that of type-one ghost imaging, the physical process corresponding to the above convolution is rather simple if the sombrero-like correlation function can be approximated as a δ-function of $|\vec{\rho}_1 - \vec{\rho}_2|$, which is true for a large angular sized thermal source radiating at visible and shorter wavelengths such as the Sun. Suppose the scanning point detector D_1 or a CCD element is triggered by a photon at a transverse position of $\vec{\rho}_1$ in a joint-detection event with the bucket detector D_2, which is triggered by another photon that is scattered from the object. At $z_1 = z_2$, this photon would have twice the probability to be scattered from $\vec{\rho}_{obj} = \vec{\rho}_1$. After moving D_1 to another transverse position $\vec{\rho'}_1$, and registering another joint-detection event between D_1 and D_2, the photon that triggers D_2 would have twice the probability of being scattered from $\vec{\rho'}_{obj} = \vec{\rho'}_1$. The probability of receiving a joint detection event at $\vec{\rho}_1 = \vec{\rho}_{obj}$ and at $\vec{\rho'}_1 = \vec{\rho'}_{obj}$ is proportional to the value of the aperture function $A(\vec{\rho}_{obj})$ and $A(\vec{\rho'}_{obj})$, respectively. After accumulating a large number of joint-detection events at different transverse coordinates $\vec{\rho}_1$, a 50% contrast aperture function $A(\vec{\rho}_1) = A(\vec{\rho}_{obj})$ is thus reproduced in the joint-detection as a function of $\vec{\rho}_1$.

A necessary condition for observing high-contrast ghost image of thermal light requires single-photon detection of the bucket detector within the coincidence time window. Otherwise, the information of $A(\vec{\rho}_{obj})$ and $A(\vec{\rho'}_{obj})$ will be mixed up by the scanning point photo-detector D_1 or by the CCD elements located at $\vec{\rho}_1 = \vec{\rho}_{obj}$ and $\vec{\rho'}_1 = \vec{\rho'}_{obj}$. With mixed $A(\vec{\rho}_{obj})$ and $A(\vec{\rho'}_{obj})$ at $\vec{\rho}_1$ and $\vec{\rho'}_1$, we will lose the ability to reproduce a joint-detection counting rate that is proportional to $A(\vec{\rho}_{obj})$ at $\vec{\rho}_1$ only, and a different joint-detection counting rate that is proportional to $A(\vec{\rho'}_{obj})$ at $\vec{\rho'}_1$ only, i.e., lose the ability of imaging. However, this requirement does not require single-photon sources. Due to the stochastic nature of the thermal light, the jointly detected photon pair is just two randomly distributed photons that fall into the coincidence time window by chance. The single-photon detection condition can be easily achieved by using adjustable ND-filters with D_1 and D_2. High contrast ghost image of thermal light can be reproduced with either bright or weak light sources, provided that no more than one joint-detection event within the coincidence time window is registered. Observing a ghost image, the natural, nonfactorizeable, point-to-point image-forming correlation of thermal light must be obtainable at quantum level. Thus, the lensless Fresnel near-field ghost imaging of thermal light essentially belongs to quantum imaging.

It is necessary to emphasize that the mathematics of the convolution between the aperture function and the point-to-point image-forming function has no difference in any optical imaging systems, including the traditional classical imaging, the ghost imaging with thermal light, and the ghost imaging with entangled states. Differences between different imaging systems come from different mechanisms

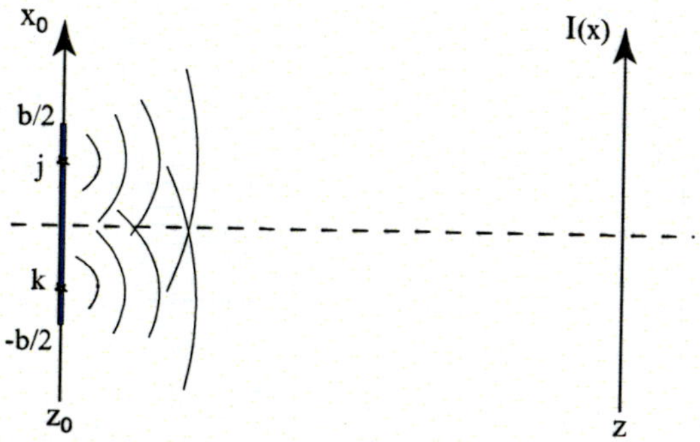

Fig. 14.12 A large number of independent point subsources, such as the jth and kth, are randomly distributed on the plane of a thermal source. These point subsources randomly radiate independent spherical waves to the object and image planes, respectively. Due to the chaotic nature of the source, these independent subintensities simply added together yielding a constant total intensity in space and in time on any transverse planes

which produce the point-to-point or point-to-"spot" image-forming function in that particular imaging system. In a classical imaging system, it is the first-order constructive–destructive interference that causes the point-to-point correspondence between the object and image planes, i.e., any radiation which is radiated (or reflected) from a point on the object plane will arrive at an unique point on the image plane. In the lensless ghost imaging system of thermal light, it is the two-photon interference that causes the nonlocal second-order correlation between the object and image planes. Analogous to the ghost image of entangled photon pairs, this natural, nonfactorizable, point-to-point correlation represents a nonlocal behavior of a measured pair of independent and randomly radiated photons: neither photon-one nor photon-two "knows" precisely where to go when they are created at each independent subsource; however, if one is observed at a point on the object plane, the other one has twice the probability of arriving at a unique corresponding point on the image plane.[5]

There is no doubt that random and chaotic radiation would propagate to any transverse plane in a random and chaotic manner. What is the physical cause of this natural, nonfactorizable, partial point-to-point image-forming function of $1 + \delta(\vec{\rho}_1 - \vec{\rho}_{obj})$? There seems no reason to have such a statistical correlation for thermal light. Figure 14.12 schematically illustrates this situation. To simplify the picture, we assume the source in 1-D with a large number of independent point subsources randomly distributed from $-b/2$ to $b/2$. Each point subsource, such as the jth and the kth subsource, randomly radiates independent spherical waves to the object and image planes, respectively. Due to the chaotic nature of the source, these independent and incoherent sub-intensities simply add together yielding a constant

[5]Similar to the far-field HBT correlation, the contrast of the "near-surface-field" point-to-point image-forming function is 50%, i.e., a two to one ratio between the maximum value and the constant background.

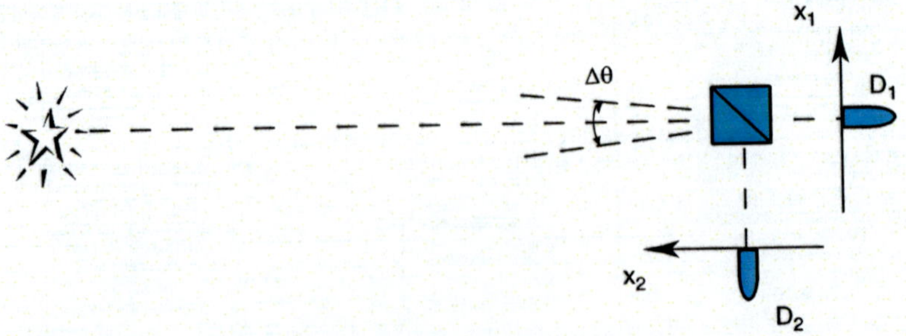

Fig. 14.13 Schematic of the historical Hanbury Brown and Twiss experiment which measures the transverse spatial correlation of far-field thermal radiation

total intensity spatially and temporally on any transverse plane. The more chaotic subfields that contribute to the intensity sum, the less value of $\Delta I/I$ is expected. For any two transverse planes, such as the object and the image planes in Fig. 14.9, each with independent and randomly distributed intensities, statistically, there is no reason to expect any spatial or temporal correlations. What is the physical cause that forces a twice large probability for the thermal radiation to jointly appear at $\vec{\rho}_1 = \vec{\rho}_{obj}$?

In fact, we have been facing this question since 1956, after the discovery of Hanbury Brown and Twiss (HBT). The lensless ghost imaging setup looks similar to that of the historical HBT spatial interferometer, which was used for measuring the angular size of distant stars. A significant difference is that the lensless ghost imaging measurement is in near-field[6] for imaging purposes [7]. The HBT experiment created quite a surprise in the physics community with an enduring debate about the classical or quantum nature of the phenomenon [13, 14].

Figure 14.13 is a schematic of the historical HBT experiment, which measures the transverse spatial correlation of far-field thermal radiation. Performing the measurement in 1-D by scanning photodetectors D_1 and/or D_2 along the axes x_1 and x_2, the second-order transverse spatial correlation function $G^{(2)}(x_1, x_2)$ was found to be

$$G^{(2)}(x_1, x_2) \sim I_0^2 \left\{1 + \mathrm{sinc}^2\left[\frac{\pi \Delta\theta (x_1 - x_2)}{\lambda}\right]\right\}, \tag{14.33}$$

where $\Delta\theta$ is the angular size of the star, λ the wavelength of the radiation. The far-field HBT correlation of (14.33) has been interpreted as the result of classical statistical correlation of the intensity fluctuations

$$\langle I_1 I_2 \rangle = \langle (\bar{I}_1 + \Delta I_1)(\bar{I}_2 + \Delta I_2) \rangle = \bar{I}_1 \bar{I}_2 + \langle \Delta I_1 \Delta I_2 \rangle,$$

---

[6]The concept of "near-field" was defined by Fresnel to be distinct from the Fraunhofer far-field. The Fresnel near-field is different from the "near-surface-field", that considers a distance of a few wavelengths from a surface.

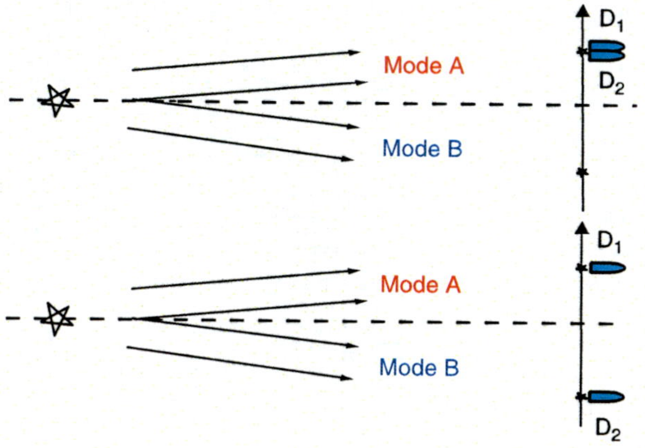

Fig. 14.14 A phenomenological interpretation of the historical HBT experiment. *Upper*: the two photodetectors receive identical modes of the far-field radiation and thus experience identical intensity fluctuations. The joint measurement of D_1 and D_2 gives a maximum value of $\langle \Delta I_1 \Delta I_2 \rangle$. *Lower*: the two photodetectors receive different modes of the far-field radiation. In this case, the joint measurement gives $\langle \Delta I_1 \Delta I_2 \rangle = 0$. Unfortunately, this hand-waving interpretation does not reflect the correct physics in the case of $\Delta\theta \neq 0$. For a finite angular sized source, there is no chance, at least realistically, for D_1 and D_2 to receive radiation from a single radiation mode only. Nevertheless, the above theory has convinced us to believe that the observation of the intensity fluctuation correlation only takes place in the far-field zone of the thermal source

where $\bar{I}_1$ and $\bar{I}_2$ are the mean intensities of the radiation measured by photodetectors D_1 and D_2, respectively. The second term in (14.33), $I_0^2 \,\mathrm{sinc}^2[\pi \Delta\theta (x_1 - x_2)/\lambda]$, is phenomenologically interpreted as the intensity fluctuation correlation $\langle \Delta I_1 \Delta I_2 \rangle$ in classical theory.

The well-accepted interpretation of the HBT phenomenon is the following: in HBT, the measurement is taken in the far-field zone of the radiation source, which is equivalent to the Fourier transform plane. Essentially, the HBT correlation measures the intensity fluctuations of the radiation field with the same **k** vector. When D_1 (D_2) is scanned in the neighborhood of $x_1 = x_2$, the two detectors measure the same mode (**k** vector) of the radiation field. The measured intensities have the same fluctuations and yield a maximum value of $\langle \Delta I_1 \Delta I_2 \rangle$. The two upper curves of $I(t)$ in Fig. 14.14 schematically illustrate this situation. When the two photodetectors move apart from $x_1 = x_2$, D_1 and D_2 measure different modes (**k** vector) of the radiation field. In this case, the measured two modes may have completely different fluctuations. The measurement yields $\langle \Delta I_1 \Delta I_2 \rangle = 0$ and gives $\langle I_1 I_2 \rangle = \bar{I}_1 \bar{I}_2$. This situation is illustrated in the two lower curves of $I(t)$ in Fig. 14.14. Unfortunately, this hand-waving interpretation does not reflect the correct physics in the case of $\Delta\theta \neq 0$. For a finite angular sized source, there is no chance, at least realistically, for D_1 and D_2 to receive radiation from a single radiation mode (**k** vector) only. Nevertheless, the above theory has convinced us to believe that the observation of the intensity fluctuation correlation only takes place in the far-field zone of the thermal source.

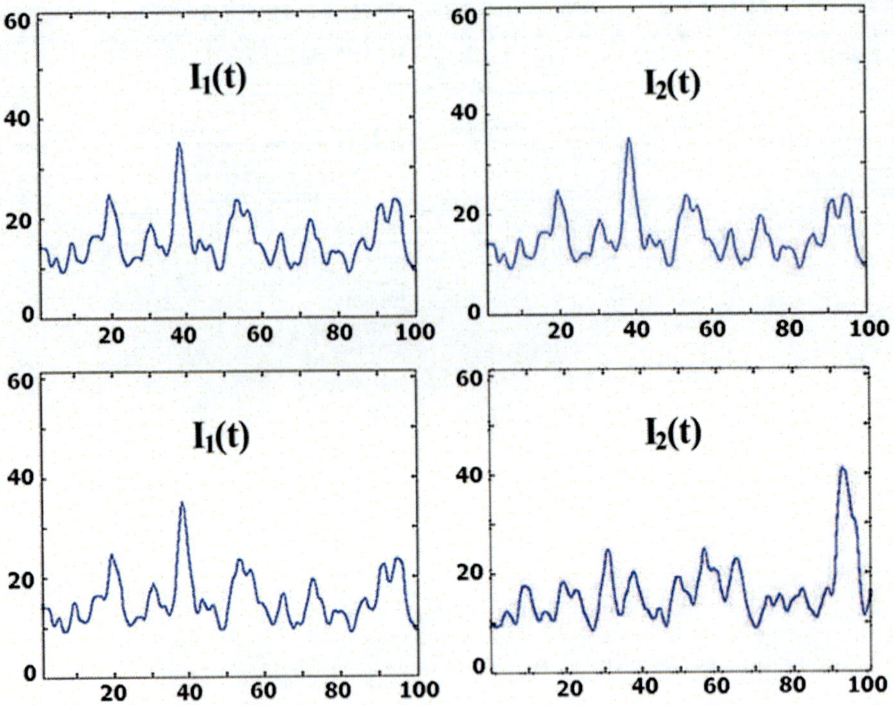

Fig. 14.15 Schematic illustration of the light intensities $I_1(t)$ at D_1 and $I_2(t)$ at D_2. The two upper (lower) curves of $I(t)$ correspond to the upper (lower) configuration in Fig. 14.14

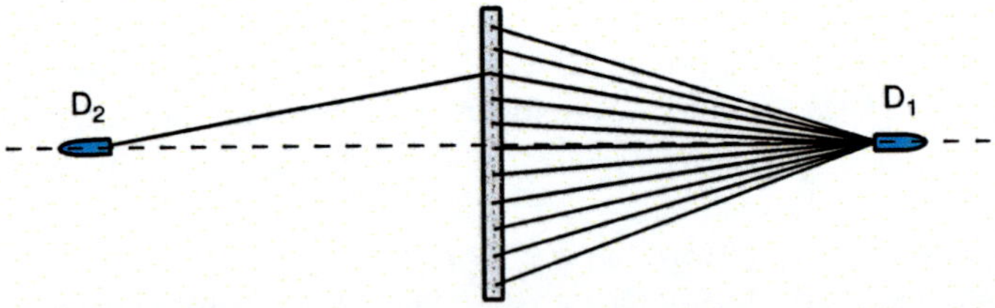

Fig. 14.16 Modified near-field HBT measurement – an unfolded Klyshko picture of the setup. Assuming a large sized disk-like near-field chaotic source, each point on the disk can be considered as an independent subsource. It is easy to see that (1) D_1 and D_2 are capable of receiving radiation from a large number of subsources; and (2) D_1 and D_2 have more chances to be triggered jointly by radiations from different subsources; (3) The ratio between the joint-detections triggered by radiation from a single subsource and from different subsources is roughly $N/N^2 = 1/N$ in any transverse position of D_1 and D_2

What will happen if we move the photodetectors D_1 and D_2 to the Fresnel "near-field" as shown in the unfolded schematic of Fig. 14.16? Does this hand-waving argument still predict the HBT correlation in this situation? We consider a disk-like thermal source with a large number of independent and randomly radiating

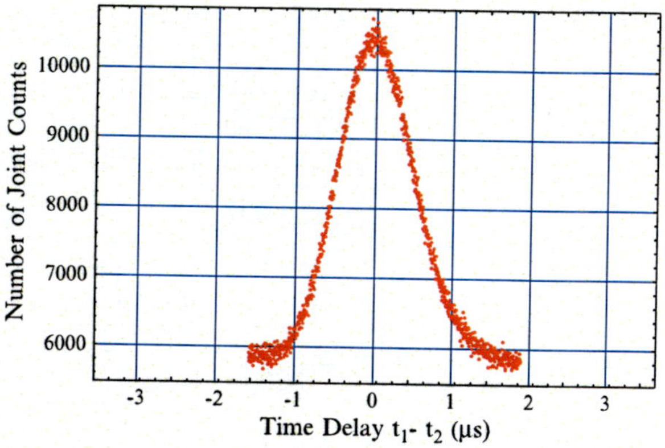

Fig. 14.17 $G^{(2)}(t_1 - t_2)$ of a chaotic-thermal sources. The temporal correlation width is measured $\sim 0.5\ \mu s$, which means that unless $t_1 - t_2 > 0.5\ \mu s$, the value of $G^{(2)}(t_1 - t_2)$ will stay at the neighborhood of its maximum

point subsources and assume the radiations coming from the same subsource have the same intensity fluctuation, and the radiations coming from different sub-sources have different intensity fluctuations. It is easy to see that in the near-field, (1) each photodetector, D_1 and D_2, is capable of receiving radiations from a large number of subsources; and (2) D_1 and D_2, have more chances to be triggered jointly by radiation from different subsources; (3) The ratio between the joint-detections triggered by radiation from a single subsource and from different sub-sources is roughly $N/N^2 = 1/N$ in any transverse position of D_1 and D_2. For a large value of N, the contribution of joint-detections triggered by radiation from a single subsource in any transverse position of D_1 and D_2 has the same negligible value $\langle \Delta I_1 \Delta I_2 \rangle / \bar{I}_1 \bar{I}_2 \sim 0$. Following the above philosophy, the near-field $G^{(2)}(\vec{\rho}_1, \vec{\rho}_2)$ should be a constant for any chosen transverse coordinates $\vec{\rho}_1$ and $\vec{\rho}_2$. The experimental observations, however, have shown a different story.

The nontrivial near-field point-to-point correlation was experimentally observed in a modified HBT experiment by Scarcelli et al.. in 2005 before the near-field lens-less ghost imaging demonstration. The modified HBT has a similar experimental setup as that of the historical HBT of Fig. 14.13, except replacing the distant star with a Fresnel near-field disk-like chaotic-thermal source. This light source has a considerably large angular diameter from the view of the photodetectors D_1 and D_2. The point photodetectors D_1 and D_2 are scannable along the axes of x_1 and x_2, respectively. The frequency bandwidth $\Delta\omega$ of this thermal source is chosen to be narrow enough to achieve $\sim\mu s$ correlation width of $G^{(2)}(t_1 - t_2)$, which is shown in Fig. 14.17. This means to change $G^{(2)}$ from its maximum (minimum) value to minimum (maximum) value requires a few hundred meters optical delay in the arm of either D_1 or D_2. The transverse intensity distributions were examined before the measurement of transverse correlation. The counting rate (weak light condition) or the output photocurrent (bright light condition) of each individual photodetector was found to be constant, i.e., $I(\vec{\rho}_1) \sim$ constant and $I(\vec{\rho}_2) \sim$ constant by scanning D_1 and D_2 in the transverse planes of $z_1 = z_0$ and $z_2 = z_0$, respectively. There is no surprise to have constant $I(\vec{\rho}_1)$ and $I(\vec{\rho}_2)$. The physics has been clearly illustrated in

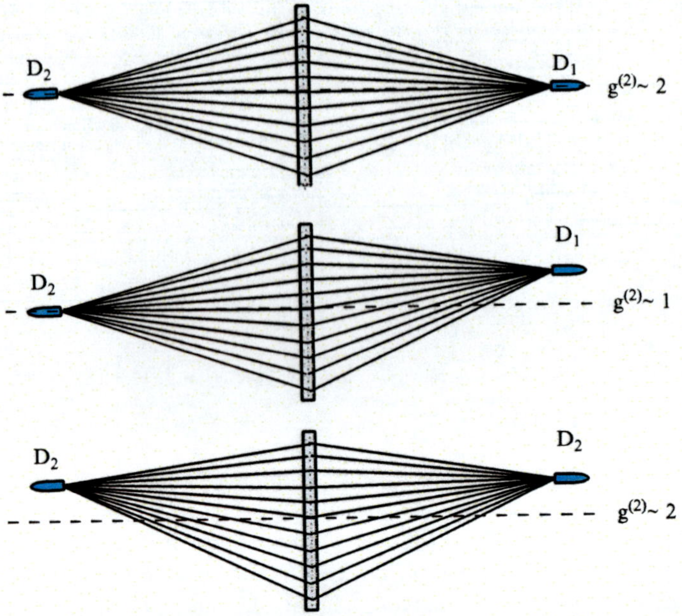

Fig. 14.18 Schematic of the Fresnel near-field spatial correlation measurement of Scarcelli et al.. *Upper*: $g^{(2)} \sim 2$, where D_1 and D_2 are placed at equal distances from the source and aligned symmetrically on the optical axis. In the spirit of the traditional interpretation of HBT, the intensities measured by D_1 and D_2 must have same fluctuations as shown in the figure. *Middle*: $g^{(2)} \sim 1$, where D_1 is moved up a few millimeters to an asymmetrical position. In the spirit of the traditional interpretation of HBT, the intensities measured by D_1 and D_2 must have different fluctuations. *Lower*: $g^{(2)} \sim 2$, where D_2 is moved up to a symmetrical position with respect to D_1, again. In the spirit of the traditional interpretation of HBT, the intensities measured by D_1 and D_2 must have same fluctuations again. What is the physical cause of the changes of the intensity fluctuations then? Remember the $G^{(2)}(t_1 - t_2)$ function has a width of $\sim 0.5\ \mu$, see Fig. 14.17

Fig. 14.12. By using this kind of chaotic-thermal source, Scarcelli et al.. measured the 1-D near-field normalized transverse spatial correlation of $g^{(2)}(x_1 - x_2)$ by scanning D_1 in the neighborhood of $x_1 = x_2$. The measurements confirmed the point-to-"spot" correlation of

$$g^{(2)}(x_1 - x_2) \sim 1 + \mathrm{sinc}^2\left[\frac{\pi\Delta\theta(x_1 - x_2)}{\lambda}\right], \tag{14.34}$$

where, again, $\Delta\theta$ is the angular diameter of the near-field disk-like chaotic source. It is worth emphasizing that $g^{(2)}(x_1 - x_2)$ dependents on $x_1 - x_2$ only. Taking $x_1 - x_2 =$ constant, $g^{(2)}(x_1 - x_2)$ is invariant under the displacements of transverse coordinates.

A simplified summary of the experimental observation is shown in Fig. 14.18: (1) In the upper figure, D_1 and D_2 are placed at equal distances from the source and aligned symmetrically on the optical axis. The normalized joint-detection, or the value of $g^{(2)}$ achieved its maximum of ~ 2. (2) In the middle figure, D_1 is moved up a few millimeters to a nonsymmetrical position, the normalized joint-detection, or the value of $g^{(2)}$ is measured to be ~ 1.

(3) In the lower figure, D_2 is moved a few millimeters up to a symmetrical position with respect to D_1. The normalized joint-detection, or the value of $g^{(2)}$ turned back to its maximum of ~ 2 again.

It is easy to see that the classical theory of statistical correlation of intensity fluctuations is facing difficulties in explaining the experimental results. In near-field, D_1 and D_2 receive the same large number of modes at any $\vec{\rho}_1$ and $\vec{\rho}_2$. In the spirit of the traditional interpretation of HBT, there seems no reason to have a different intensity fluctuation correlation between $\vec{\rho}_1 = \vec{\rho}_2$ and $\vec{\rho}_1 \neq \vec{\rho}_2$ for the $G^{(2)}$ function shown in Fig. 14.17. In the upper measurement, we have obtained the maximum value of $g^{(2)} \sim 2$ at $z_1 = z_2$ and $x_1 = x_2$, which indicates the achievement of a maximum intensity fluctuation correlation as shown in Fig. 14.17 with $|t_1 - t_2| \sim 0$. In the middle measurement, $g^{(2)} \sim 1$ indicates a minimum intensity fluctuation correlation by moving D_1 a few millimeters up, which means the intensities measured by D_1 and D_2 must have different fluctuations. In the lower measurement, D_2 is moved up a few millimeters to a new symmetrical position with respect to D_1, the measurements obtain $g^{(2)} \sim 2$ again. The intensities measured by D_1 and D_2 must have same fluctuations again. What is the physical cause of the changes of the intensity fluctuations then? Remember the $G^{(2)}(t_1 - t_2)$ function has a width of $\sim 0.5\mu$.

For half a century since 1956, we have believed that the HBT correlation is observable in the far-field only. It was quite a surprise that in 2005 Scarcelli et al.. successfully demonstrated a Fresnel near-field point-to-point transverse correlation of chaotic-thermal light, indicating that the nontrivial HBT spatial correlation is observable in the near-field and is useful for reproducing ghost images in a nonlocal manner.[7] The experiment of Scarcelli et al.. raised a question: "Can two-photon correlation of chaotic light be considered as correlation of intensity fluctuations?" [7]. At least, this experiment suggested we reexamine the relationship between the quantum mechanical concept of joint-detection probability with the classical concept of intensity fluctuation correlation. It seems that jointly observing a pair of photons at space-time point $(\mathbf{r}_1, t_1)$ and $(\mathbf{r}_2, t_2)$ is perhaps only phenomenologically connected but not physically caused by the classical statistical correlation of intensity fluctuations. The point-to-point image-forming correlation is more likely the result of an interference. In the view of two-photon interference, far-field is not a necessary condition for observing the partial point-to-point correlation of thermal light. Furthermore, it is quite common in two-photon interference-type experiments to observe constant counting rates or intensities in individual photodetectors D_1 and D_2, respectively, and simultaneously observe nontrivial space-time correlation

[7] We cannot help but stop to ask: What has been preventing this simple move from far-field to near-field for half a century? The hand-waving argument of intensity fluctuation correlation may have played a role.

in the joint-detection between D_1 and D_2. These observations are consistent with the quantum theory of two-photon interferometry [11].

14.4.2 Quantum Theory of Thermal Light Ghost Imaging

According to the quantum theory of light, the observed partial point-to-point image-forming correlation is the result of multi-photon interference. In Glauber's theory of photo-detection [23], an idealized point photodetector measures the probability of observing a photo-detection event at space-time point $(\mathbf{r}, t)$

$$G^{(1)}(\mathbf{r}, t) = tr\{\hat{\rho}\, E^{(-)}(\mathbf{r}, t) E^{(+)}(\mathbf{r}, t)\}, \tag{14.35}$$

where $\hat{\rho}$ is the density operator which characterizes the state of the quantized electromagnetic field, $E^{(-)}(\mathbf{r}, t)$ and $E^{(+)}(\mathbf{r}, t)$ the negative and positive field operators at space-time coordinate $(\mathbf{r}, t)$. The counting rate of a point photon counting detector, or the output current of a point analog photodetector, is proportional to $G^{(1)}(\mathbf{r}, t)$. A joint-detection of two independent point photodetectors measures the probability of observing a joint-detection event of two photons at space-time points $(\mathbf{r}_1, t_1)$ and $(\mathbf{r}_2, t_2)$

$$G^{(2)}(\mathbf{r}_1, t_1; \mathbf{r}_2, t_2) = tr\{\hat{\rho}\, E^{(-)}(\mathbf{r}_1, t_1) E^{(-)}(\mathbf{r}_2, t_2) E^{(+)}(\mathbf{r}_2, t_2) E^{(+)}(\mathbf{r}_1, t_1)\}, \tag{14.36}$$

where $(\mathbf{r}_j, t_j)$, $j = 1, 2$, is the space-time coordinate of the jth photo-detection event. The coincidence counting rate of two photon counting detectors, or the output reading of a linear multiplier (RF mixer) between two photodetectors, is proportional to $G^{(2)}(\mathbf{r}_1, t_1; \mathbf{r}_2, t_2)$. To calculate the partial point-to-point correlation between the object plane and the image plane, we need (1) to estimate the state, or the density matrix, of the thermal radiation; and (2) to propagate the field operators from the radiation source to the object and the image planes. We will first calculate the state of thermal radiation at the single-photon level for photon counting measurements to explore the physics behind ghost imaging as two-photon interference and then generalize the result to any intensity of thermal radiation.

We assume a large transverse sized chaotic source consisting of a large number of independent and randomly radiating point subsources. Each point subsource may also consist of a large number of independent atoms that are ready for two-level atomic transitions in a random manner. Most of the time, the atoms are in their ground state. There is, however, a small chance for each atom to be excited to a higher energy level E_2 ($\Delta E_2 \neq 0$) and later return to its ground state E_1. It is reasonable to assume that each atomic transition generates a field in the following single-photon state

$$|\,\Psi\,\rangle \simeq |\,0\,\rangle + \epsilon \sum_{\mathbf{k},s} f(\mathbf{k}, s)\, \hat{a}^{\dagger}_{\mathbf{k},s}\, |\,0\,\rangle, \tag{14.37}$$

where $|\epsilon| \ll 1$ is the probability amplitude for the atomic transition, $f(\mathbf{k}, s) = \langle \Psi_{\mathbf{k},s} \,|\, \Psi \rangle$ is the probability amplitude for the radiation field to be in the single-photon state of wave number $\mathbf{k}$ and polarization s: $|\,\Psi_{\mathbf{k},s}\rangle = |\,1_{\mathbf{k},s}\rangle = \hat{a}^{\dagger}_{\mathbf{k},s}\,|\,0\,\rangle$. For this simplified two-level system, the density matrix that characterizes the state of the radiation field excited by a large number of possible atomic transitions is thus

$$
\begin{aligned}
\hat{\rho} &= \prod_{t_{0j}} \left\{ |\,0\,\rangle + \epsilon \sum_{\mathbf{k},s} f(\mathbf{k}, s)\, \mathrm{e}^{-\mathrm{i}\omega t_{0j}}\, \hat{a}^{\dagger}_{\mathbf{k},s}\, |\,0\,\rangle \right\} \\
&\quad \times \prod_{t_{0k}} \left\{ \langle\,0\,| + \epsilon^{*} \sum_{\mathbf{k}',s'} f(\mathbf{k}', s')\, \mathrm{e}^{\mathrm{i}\omega' t_{0k}}\, \langle\,0\,|\hat{a}_{\mathbf{k}',s'} \right\} \\
&\simeq \left\{ |\,0\,\rangle + \epsilon \left[\sum_{t_{0j}} \sum_{\mathbf{k},s} f(\mathbf{k}, s)\, \mathrm{e}^{-\mathrm{i}\omega t_{0j}}\, \hat{a}^{\dagger}_{\mathbf{k},s}\, |\,0\,\rangle \right] + \epsilon^{2}[\ldots] \right\} \\
&\quad \times \left\{ \langle\,0\,| + \epsilon^{*} \left[\sum_{t_{0k}} \sum_{\mathbf{k}',s'} f(\mathbf{k}', s')\, \mathrm{e}^{\mathrm{i}\omega' t_{0k}}\, \langle\,0\,|\hat{a}_{\mathbf{k}',s'} \right] + \epsilon^{*2}[\ldots] \right\},
\end{aligned}
\tag{14.38}
$$

where $\mathrm{e}^{-\mathrm{i}\omega t_{0j}}$ is a random phase factor associated with the jth atomic transition. Since $|\epsilon| \ll 1$, it is a good approximation to keep the necessary lower-order terms of ϵ in (14.38). After summing over t_{0j} (t_{0k}) by taking into account all its possible values, we obtain

$$
\begin{aligned}
\hat{\rho} &\simeq |\,0\,\rangle\langle\,0\,| + |\epsilon|^{2} \sum_{\mathbf{k},\mathbf{s}} |f(\mathbf{k}, \mathbf{s})|^{2}\, |\,1_{\mathbf{k},\mathbf{s}}\,\rangle\langle\,1_{\mathbf{k},\mathbf{s}}\,| \\
&\quad + |\epsilon|^{4} \sum_{\mathbf{k},\mathbf{s}} \sum_{\mathbf{k}',s'} |f(\mathbf{k}, \mathbf{s})|^{2}\, \left|f(\mathbf{k}', s')\right|^{2}\, |\,1_{\mathbf{k},s} 1_{\mathbf{k}',s'}\,\rangle\langle\,1_{\mathbf{k},s} 1_{\mathbf{k}',s'}\,| + \ldots
\end{aligned}
\tag{14.39}
$$

Similar to our earlier discussion, we will focus our calculation on the transverse correlation by assuming a narrow enough frequency bandwidth in (14.39). In the experiments of Scarcelli et al.. the coherence time of the radiation was chosen $\sim\mu$s, the maximum achievable optical path differences $\sim$ps by the scanning of D_1 and D_2, and the response time of the photodetectors is much less than the coherence time. The transverse spatial correlation measurement is under the condition of achieving a maximum temporal coherence of $G^{(2)}(t_1 - t_2) \sim 2$ during the scanning of D_1 and D_2 at any $\vec{\rho}_1$ and $\vec{\rho}_2$. In the *photon counting* regime, under the above condition, it is reasonable to model the thermal light in the following mixed state

$$
\begin{aligned}
\hat{\rho} &\simeq |\,0\,\rangle\langle\,0\,| + |\epsilon|^{2} \sum_{\vec{\kappa}} \hat{a}^{\dagger}(\vec{\kappa})|\,0\,\rangle\langle\,0\,|\hat{a}(\vec{\kappa}) \\
&\quad + |\epsilon|^{4} \sum_{\vec{\kappa}} \sum_{\vec{\kappa}'} \hat{a}^{\dagger}(\vec{\kappa})\hat{a}^{\dagger}(\vec{\kappa}')\,|\,0\,\rangle\langle\,0\,|\,\hat{a}(\vec{\kappa}')\hat{a}(\vec{\kappa}).
\end{aligned}
\tag{14.40}
$$

Basically, we are modeling the light source as an incoherent statistical mixture of single-photon states and two-photon states with equal probability of having any transverse momentum. The spatial part of the second-order coherence function is thus calculated as

$$\begin{aligned}G^{(2)}(\vec{\rho}_1, z_1; \vec{\rho}_2, z_2) &= tr[\hat{\rho}\, E^{(-)}(\vec{\rho}_1, z_1) E^{(-)}(\vec{\rho}_2, z_2) E^{(+)}(\vec{\rho}_2, z_2) E^{(+)}(\vec{\rho}_1, z_1)] \\ &= \sum_{\vec{\kappa},\vec{\kappa}'} \langle 1_{\vec{\kappa}} 1_{\vec{\kappa}'} | E^{(-)}(\vec{\rho}_1, z_1) E^{(-)}(\vec{\rho}_2, z_2) \\ &\qquad E^{(+)}(\vec{\rho}_2, z_2) E^{(+)}(\vec{\rho}_1, z_1) | 1_{\vec{\kappa}} 1_{\vec{\kappa}'} \rangle \\ &\equiv \sum_{\vec{\kappa},\vec{\kappa}'} \left| \Psi_{\vec{\kappa},\vec{\kappa}'}(\vec{\rho}_1, z_1; \vec{\rho}_2, z_2) \right|^2, \end{aligned} \tag{14.41}$$

where we have defined an effective two-photon wavefunction in transverse spatial coordinates

$$\Psi_{\vec{\kappa},\vec{\kappa}'}(\vec{\rho}_1, z_1; \vec{\rho}_2, z_2) = \langle 0 | E^{(+)}(\vec{\rho}_2, z_2) E^{(+)}(\vec{\rho}_1, z_1) | 1_{\vec{\kappa}} 1_{\vec{\kappa}'} \rangle. \tag{14.42}$$

The transverse part of the electric field operator can be written as

$$E^{(+)}(\vec{\rho}_j, z_j) \propto \sum_{\vec{\kappa}} g_j(\vec{\rho}_j, z_j; \vec{\kappa})\, \hat{a}(\vec{\kappa}), \tag{14.43}$$

again, $g_j(\vec{\rho}_j, z_j; \vec{\kappa})$ is the Green's function. Substituting the field operators into (14.42), we have

$$\Psi_{\vec{\kappa},\vec{\kappa}'}(\vec{\rho}_1, z_1; \vec{\rho}_2, z_2) = \frac{1}{\sqrt{2}} [g_2(\vec{\rho}_2, z_2; \vec{\kappa}) g_1(\vec{\rho}_1, z_1; \vec{\kappa}') + g_2(\vec{\rho}_2, z_2; \vec{\kappa}') g_1(\vec{\rho}_1, z_1; \vec{\kappa})] \tag{14.44}$$

and

$$\begin{aligned}G^{(2)}(\vec{\rho}_1, z_1; \vec{\rho}_2, z_2) &= \textstyle\sum_{\vec{\kappa},\vec{\kappa}'} \Big| \frac{1}{\sqrt{2}} [g_2(\vec{\rho}_2, z_2; \vec{\kappa}) g_1(\vec{\rho}_1, z_1; \vec{\kappa}') \\ &\qquad + g_2(\vec{\rho}_2, z_2; \vec{\kappa}') g_1(\vec{\rho}_1, z_1; \vec{\kappa})] \Big|^2, \end{aligned} \tag{14.45}$$

representing the key result for our understanding of the phenomenon. Equations (14.44) and (14.45) indicates an interference between two alternatives, different yet indistinguishable, which leads to a joint photo-detection event. This interference phenomenon is not, as in classical optics, due to the superposition of electromagnetic fields at a local point of space-time. This interference is the result of the superposition between $g_2(\vec{\rho}_2, z_2; \vec{\kappa}) g_1(\vec{\rho}_1, z_1; \vec{\kappa}')$ and $g_2(\vec{\rho}_2, z_2; \vec{\kappa}') g_1(\vec{\rho}_1, z_1; \vec{\kappa})$, the so-called two-photon amplitudes, nonclassical entities that involve both arms of the optical setup as well as two distant photo-detection events at $(\vec{\rho}_1, z_1)$ and

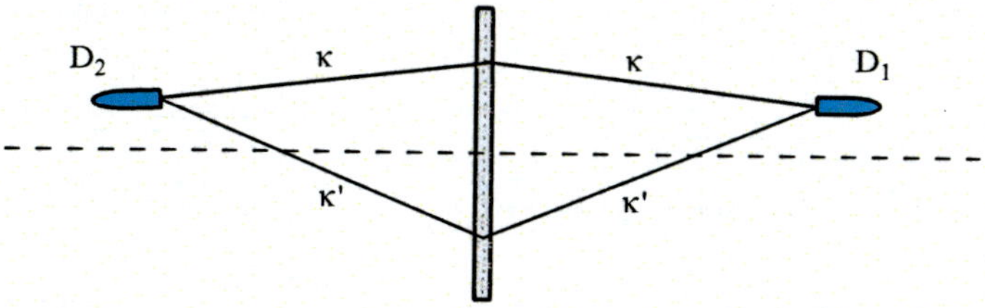

Fig. 14.19 Schematic illustration of two-photon interference: a superposition between two-photon amplitudes $g_2(\vec{\rho}_2, z_2\vec{\kappa})g_1(\vec{\rho}_1, z_1; \vec{\kappa}')$ and $g_2(\vec{\rho}_2, z_2; \vec{\kappa}')g_1(\vec{\rho}_1, z_1; \vec{\kappa})$. It is clear that the amplitudes $g_2(\vec{\rho}_2, z_2; \vec{\kappa})g_1(\vec{\rho}_1, z_1; \vec{\kappa}')$ and $g_2(\vec{\rho}_2, z_2; \vec{\kappa}')g_1(\vec{\rho}_1, z_1; \vec{\kappa})$ will experience equal optical path propagation and superpose constructively when D_1 and D_2 are located at $\vec{\rho}_1 \simeq \vec{\rho}_2$ and $z_1 \simeq z_2$. This nonlocal superposition has no classical correspondence

$(\vec{\rho}_2, z_2)$, respectively. Examining the effective wavefunction of (14.44), we find this symmeterized effective wavefunction plays the same role as that of the symmetrized wavefunction of identical particles in quantum mechanics. This peculiar nonlocal superposition has no classical correspondence, and makes the type-two ghost image turbulence-free, i.e., any phase disturbance in the optical path has no influence on the ghost image [27]. Figure 14.19 schematically illustrates the two alternatives for a pair of mode $\vec{\kappa}$ and $\vec{\kappa}'$ to produce a joint photo-detection event: $\vec{\kappa}1 \times \vec{\kappa}'2$ and $\vec{\kappa}2 \times \vec{\kappa}'1$. The superposition of each pair of these amplitudes produces an individual subinterference-pattern in the joint-detection space of $(\vec{\rho}_1, z_1, t_1; \vec{\rho}_2, z_2, t_2)$. A large number of these subinterference-patterns simply add together resulting in a nontrivial $G^{(2)}(\vec{\rho}_1, z_1; \vec{\rho}_2, z_2)$ function. It is easy to see that each pair of the two-photon amplitudes, illustrated in Fig. 14.19, will superpose constructively whenever D_1 and D_2 are placed in the positions satisfying $\vec{\rho}_1 \simeq \vec{\rho}_2$ and $z_1 \simeq z_2$; and consequently, $G^{(2)}(\vec{\rho}_1, z_1; \vec{\rho}_2, z_2)$ achieves its maximum value as the result of the sum of these individual constructive interferences. In other coordinates, however, the superposition of each individual pair of the two-photon amplitudes may yield different values between constructive maximum and destructive minimum due to unequal optical path propagation, resulting in an averaged sum.

Before calculating $G^{(2)}(\vec{\rho}_1, z_1; \vec{\rho}_2, z_2)$, we examine the single counting rate of the point photodetectors D_1 and D_2, which are placed at $(\vec{\rho}_1, z_1)$ and $(\vec{\rho}_2, z_2)$, respectively. With reference to the experimental setup of Fig. 14.9, the Green's function of free-propagation is derived in the Appendix

$$g_1(\vec{\rho}_1, z_1; \vec{\kappa}) = \int d\vec{\rho}_s \left\{ \frac{-i\omega}{2\pi c} \frac{e^{i\frac{\omega}{c}z_1}}{z_1} e^{i\frac{\omega}{2cz_1}|\vec{\rho}_1 - \vec{\rho}_s|^2} \right\} e^{-i\vec{\kappa}\cdot\vec{\rho}_s},$$

$$g_2(\vec{\rho}_2, z_2; \vec{\kappa}) = \int d\vec{\rho}_s' \left\{ \frac{-i\omega}{2\pi c} \frac{e^{i\frac{\omega}{c}z_2}}{z_2} e^{i\frac{\omega}{2cz_2}|\vec{\rho}_2 - \vec{\rho}_s'|^2} \right\} e^{-i\vec{\kappa}\cdot\vec{\rho}_s'},$$

where $\vec{\rho}_s$ is the transverse vector in the source plane, and the field has propagated from the source to the $\vec{\rho}_1$ plane and $\vec{\rho}_2$ plane in arms 1 and 2, respectively. The single detector counting rate or the output photocurrent is proportional to $G^{(1)}(\mathbf{r},t)$ as shown in (14.35),

$$G^{(1)}(\vec{\rho}_j,z_j) = tr\{\hat{\rho}\,E^{(-)}(\vec{\rho}_j,z_j)E^{(+)}(\vec{\rho}_j,z_j)\}$$
$$\propto \sum_{\vec{\kappa}} \langle 0\,|\,\hat{a}(\vec{\kappa})\,E^{(-)}(\vec{\rho}_j,z_j)E^{(+)}(\vec{\rho}_j,z_j)\,\hat{a}^{\dagger}(\vec{\kappa})|\,0\,\rangle$$
$$\sim \text{constant}, \tag{14.46}$$

where $j = 1,2$ indicating the jth photodetector.

Although $G^{(1)}(\vec{\rho}_1,z_1)$ and $G^{(1)}(\vec{\rho}_2,z_2)$ are both constants, $G^{(2)}(\vec{\rho}_1,z_1;\vec{\rho}_2,z_2)$ turns to be a nontrivial function of $(\vec{\rho}_1,z_1)$ and $(\vec{\rho}_2,z_2)$,

$$G^{(2)}(\vec{\rho}_1,z_1;\vec{\rho}_2,z_2) = \sum_{\vec{\kappa},\vec{\kappa}'} \left| \frac{1}{\sqrt{2}}[g_2(\vec{\rho}_2,z_2;\vec{\kappa})g_1(\vec{\rho}_1,z_1;\vec{\kappa}') + g_2(\vec{\rho}_2,z_2;\vec{\kappa}')g_1(\vec{\rho}_1,z_1;\vec{\kappa})] \right|^2$$
$$\equiv G^{(1)}(\vec{\rho}_1,z_1)G^{(1)}(\vec{\rho}_2,z_2) + \left|G^{(1)}(\vec{\rho}_1,z_1;\vec{\rho}_2,z_2)\right|^2, \tag{14.47}$$

where

$$\left|G^{(1)}(\vec{\rho}_1,z_1;\vec{\rho}_2,z_2)\right|^2 = \left| \int \mathrm{d}\vec{\kappa}\; g_1^{*}(\vec{\rho}_1,z_1;\vec{\kappa})\, g_2(\vec{\rho}_2,z_2;\vec{\kappa}) \right|^2$$
$$\propto \left| \int \mathrm{d}\vec{\rho}_s\, \mathrm{e}^{-\mathrm{i}\frac{\omega}{c}z_1}\, \mathrm{e}^{-\mathrm{i}\frac{\omega}{2cz_1}|\vec{\rho}_1-\vec{\rho}_s|^2}\, \mathrm{e}^{\mathrm{i}\frac{\omega}{c}z_2}\, \mathrm{e}^{\mathrm{i}\frac{\omega}{2cz_2}|\vec{\rho}_2-\vec{\rho}_s|^2} \right|^2.$$

If we choose the distances from the source to the two detectors to be equal ($z_1 = z_2 = d$), the above integral of $\mathrm{d}\vec{\rho}_s$ yields a point-to-point correlation between the transverse planes $z_1 = d$ and $z_2 = d$,

$$|\,G^{(1)}_{12}(\vec{\rho}_1;\vec{\rho}_2)\,|^2 \propto \left| \int \mathrm{d}\vec{\rho}_s\, \mathrm{e}^{\mathrm{i}\frac{\omega}{cd}(\vec{\rho}_1-\vec{\rho}_2)\cdot\vec{\rho}_s} \right|^2$$
$$= somb^2\left[\frac{R}{d}\frac{\omega}{c}|\vec{\rho}_1-\vec{\rho}_2|\right] \sim \delta(\vec{\rho}_1-\vec{\rho}_2), \tag{14.48}$$

where the δ-function is an approximation by assuming a large enough thermal source of angular size $\Delta\theta \sim R/d$ and high enough frequency ω, such as a visible light source. The nontrivial $G^{(2)}$ function is therefore

$$G^{(2)}(\vec{\rho}_1;\vec{\rho}_2) \sim 1 + \delta(\vec{\rho}_1-\vec{\rho}_2). \tag{14.49}$$

In the ghost imaging experiment, the joint-detection counting rate is thus

$$R_{12} \propto \int \mathrm{d}\vec{\rho}_2\, |A(\vec{\rho}_2)|^2\, G^{(2)}(\vec{\rho}_1;\vec{\rho}_2) \sim R_0 + |A(\vec{\rho}_1)|^2, \tag{14.50}$$

where R_0 is a constant and $A(\vec{\rho}_2)$ is the aperture function of the object.

So far, we have successfully derived an analytical solution for ghost imaging with thermal radiation at the single-photon level. We have shown that the partial point-to-point correlation of thermal radiation is the result of a constructive–destructive interference caused by the superposition of two two-photon amplitudes, corresponding to two alternative ways for a pair of jointly measured photons to produce a joint-detection event. In fact, the above analysis is not restricted to single-photon states. The partial point-to-point correlation of $G^{(2)}(\vec{\rho}_1;\vec{\rho}_2)$ is generally true for any order of quantized thermal radiation [28]. Now we generalize the calculation to an arbitrary quantized thermal field with occupation number from $n_{\mathbf{k},\mathbf{s}} = 0$ to $n_{\mathbf{k},\mathbf{s}} \gg 1$ by keeping all higher order terms in (14.38). After summing over t_{0j} and t_{0k}, the density matrix can be written as

$$\hat{\rho} = \sum_{\{n\}} p_{\{n\}} \, |\{n\}\rangle\langle\{n\}|, \tag{14.51}$$

where $p_{\{n\}}$ is the probability for the thermal field in the state

$$|\{n\}\rangle \equiv \prod_{\mathbf{k},s} |n_{\mathbf{k},s}\rangle = |n_{\mathbf{k},s}\rangle |n_{\mathbf{k}',s'}\rangle \dots |n_{\mathbf{k}''\dots',s''\dots'}\rangle.$$

The summation of (14.51) includes all possible modes $\mathbf{k}$, polarizations s, occupation numbers $n_{\mathbf{k},s}$ for the mode $(\mathbf{k}, s)$ and all possible combinations of occupation numbers for different modes in a set of $\{n\}$. Substituting the field operators and the density operator of (14.51) into (14.35), we obtain the constant $G^{(1)}(\vec{\rho}_j, z_j, t_j)$, $j = 1, 2$, which corresponds to the intensities $I(\vec{\rho}_1, z_1, t_1)$ and $I(\vec{\rho}_2, z_2, t_2)$,

$$\begin{aligned} &G^{(1)}(\vec{\rho}_j, z_j, t_j) \\ &= \sum_{\{n\}} \int d\vec{\kappa} \int d\vec{\kappa}' g_j^*(\vec{\rho}_j, z_1, t_j; \vec{\kappa}) g_j(\vec{\rho}_j, z_1, t_j; \vec{\kappa}') \, p_{\{n\}} \, \langle\{n\}|\, a(\vec{\kappa}) a^\dagger(\vec{\kappa}') \,|\{n\}\rangle \\ &\propto \sum_{\{n\}} n_{\vec{\kappa}} n_{\vec{\kappa}'} \int d\vec{\kappa} \, |\, g_j^*(\vec{\rho}_j, z_j, t_j; \vec{\kappa}) \,|^2 \\ &\simeq \text{constant.} \end{aligned} \tag{14.52}$$

Although $G^{(1)}(\vec{\rho}_1, z_1, t_1)$ and $G^{(1)}(\vec{\rho}_2, z_2, t_2)$ are both constants, substituting the field operators and the density operator of (14.51) into (14.36), we obtain a nontrivial point-to-point correlation function of $G^{(2)}(\vec{\rho}_1;\vec{\rho}_2)$ at the two transverse planes $z_1 = d$ and $z_2 = d$,

$$\begin{aligned} G^{(2)}(\vec{\rho}_1;\vec{\rho}_2) = \sum_{\{n\}} \int d\vec{\kappa} \int d\vec{\kappa}' \int d\vec{\kappa}'' \int d\vec{\kappa}''' g_1^*(\vec{\rho}_1, z_1; \vec{\kappa}) g_2^*(\vec{\rho}_2, z_2; \vec{\kappa}') \\ g_2(\vec{\rho}_2, z_2; \vec{\kappa}'') g_1(\vec{\rho}_1, z_1; \vec{\kappa}''') \times \, p_{\{n\}} \, \langle\{n\}|\, a(\vec{\kappa}) a(\vec{\kappa}') a^\dagger(\vec{\kappa}'') a^\dagger(\vec{\kappa}''') \,|\{n\}\rangle \end{aligned}$$

$$\propto \sum_{\{n\}} n_{\vec{\kappa}} n_{\vec{\kappa}'} \int d\vec{\kappa} \int d\vec{\kappa}' \int d\vec{\kappa}'' \int d\vec{\kappa}''' g_1^*(\vec{\rho}_1, z_1; \vec{\kappa})$$

$$g_2^*(\vec{\rho}_2, z_2; \vec{\kappa}') g_2(\vec{\rho}_2, z_2; \vec{\kappa}'') g_1(\vec{\rho}_1, z_1; \vec{\kappa}''') \times p_{\{n\}}(\delta_{\vec{\kappa}\vec{\kappa}'''}\delta_{\vec{\kappa}'\vec{\kappa}''} + \delta_{\vec{\kappa}\vec{\kappa}''}\delta_{\vec{\kappa}'\vec{\kappa}'''})$$

$$= \sum_{n_{\vec{\kappa}} n_{\vec{\kappa}'}} p_{\{\ldots n_{\vec{\kappa}} \ldots n_{\vec{\kappa}'} \ldots\}} n_{\vec{\kappa}} n_{\vec{\kappa}'} \times \left\{ \int d\vec{\kappa} \int d\vec{\kappa}' \left| \frac{1}{\sqrt{2}} [g_1(\vec{\rho}_1, z_1; \vec{\kappa}) g_2(\vec{\rho}_2, z_2; \vec{\kappa}') \right.\right.$$

$$\left.\left. + g_2(\vec{\rho}_2, z_2; \vec{\kappa}) g_1(\vec{\rho}_1, z_1; \vec{\kappa}')] \right|^2 \right\} \propto \left\{ 1 + somb^2 \left[\frac{R}{d} \frac{\omega}{c} (\vec{\rho}_1 - \vec{\rho}_2) \right] \right\}. \quad (14.53)$$

It is clear that in (14.53) the partial point-to-point correlation of thermal light is the result of a constructive–destructive interference between two quantum-mechanical amplitudes. We also note from (14.53) that the partial point-to-point correlation is independent of the occupation numbers, $\{n\}$, and the probability distribution, $p_{\{n\}}$, of the quantized thermal radiation.

It is interesting but not surprising to see that the effective two-photon wavefunction in bright light condition

$$\Psi_{\vec{\kappa},\vec{\kappa}'}(\vec{\rho}_1, z_1; \vec{\rho}_2, z_2) = \frac{1}{\sqrt{2}} [g_2(\vec{\rho}_2, z_2; \vec{\kappa}) g_1(\vec{\rho}_1, z_1; \vec{\kappa}') + g_2(\vec{\rho}_2, z_2; \vec{\kappa}') g_1(\vec{\rho}_1, z_1; \vec{\kappa})]$$

is the same as that of weak light at single-photon level. In fact, the above effective wavefunction does play the same role in specifying two different yet indistinguishable alternatives for the two annihilated photons contributing to a joint-detection event of D_1 and D_2, which implies that the partial point-to-point correlation is the result of two-photon interference in bright light condition. This nonlocal partial correlation indicates that a 50% contrast ghost image is observable at bright light condition provided registering no more than one coincidence event within the joint-detection time window. This requirement can be easily achieved by using adjustable ND-filters with D_1 and D_2.

Quantum theory predicts and calculates the probability of observing a certain physical event. The output photocurrent of an idealized point photodetector is proportional to the probability of observing a photo-detection event at space-time point $(\mathbf{r}, t)$. The joint-detection between two idealized point photodetectors is proportional to the probability of observing a joint photo-detection event at space-time points $(\mathbf{r}_1, t_1)$ and $(\mathbf{r}_2, t_2)$. In most of the experimental situations, there exists more than one possible alternative ways to produce a photo-detection event, or a joint photo-detection event. These probability amplitudes, which are defined as the single-photon amplitudes and the two-photon amplitudes, respectively, are superposed to contribute to the final measured probability, and consequently determine the probability of observing a photo-detection event or a joint photo-detection event. In the view of quantum theory, whenever the state of the quantum system and the alternative ways to produce a photo-detection event or a joint photo-detection event are determined, the result of a measurement is determined. We may consider this as a basic criterion of quantum measurement theory.

14.4.3 A Semiclassical Model of Nonlocal Interference

The multiphoton interference nature of type-two ghost imaging can be seen intuitively from the superposition of paired-sub-fields of chaotic radiation. Let us consider a similar experimental setup as that of the modified HBT experiment of Scarcelli et al. We assume a large angular sized disk-like chaotic source that contains a large number of randomly radiating independent point "subsources", such as trillions of independent atomic transitions randomly distributed spatially and temporally. It should be emphasized that a large number of independent or incoherent sub-sources is the only requirement for type-two ghost imaging. What we need is an ensemble of point-sub-sources with random relative phases so that the sub-fields coming from these sub-sources are able to take all possible values of relative phases in their superposition. It is *unnecessary* to require the radiation source to have either nature or artificial intensity fluctuations at all. In this model, each point sub-source contributes to the measurement an independent spherical wave as a sub-field of complex amplitude $E_j = a_j e^{i\varphi_j}$, where a_j is the real and positive amplitude of the jth sub-field and φ_j is a *random* phase associated with the jth sub-field. We have the following picture for the source: (1) a large number of independent point-sources distribute randomly on the transverse plane of the source (counted spatially); (2) each point-source contains a large number of independently and randomly radiating atoms (counted temporally); (3) a large number of sub-sources, either counted spatially or temporally, may contribute to each of the independent radiation mode $(\vec{\kappa}, \omega)$ at D_1 and D_2 (counted by mode). The instantaneous intensity at space-time $(\mathbf{r}_j, t_j)$, measured by the jth idealized point photodetector D_j, $j = 1, 2$, is calculated as

$$\begin{aligned} I(\mathbf{r}_j, t_j) &= E^*(\mathbf{r}_j, t_j)E(\mathbf{r}_j, t_j) = \sum_l E_l^*(\mathbf{r}_j, t_j) \sum_m E_m(\mathbf{r}_j, t_j) \\ &= \sum_{l=m} E_l^*(\mathbf{r}_j, t_j)E_l(\mathbf{r}_j, t_j) + \sum_{l \neq m} E_l^*(\mathbf{r}_j, t_j)E_m(\mathbf{r}_j, t_j), \end{aligned} \quad (14.54)$$

where the subfields are identified by the index l and m originated from the l and m subsources. The first term is a constant representing the sum of the subintensities, where the lth subintensity is originated from the lth subsource. The second term adds the "cross" terms corresponding to different subsources. When *taking into account all possible realizations of the fields*, it is easy to find that the only surviving terms in the sum are these terms in which the field and its conjugate come from the same subsource, i.e., the first term in (14.54). The second term in (14.54) vanishes if $\varphi_l - \varphi_m$ *takes all possible values*. We may write (14.54) into the following form

$$I(\mathbf{r}_j, t_j) = \langle I(\mathbf{r}, t)\rangle + \Delta I(\mathbf{r}, t), \tag{14.55}$$

where

$$\langle I(\mathbf{r}, t)\rangle \equiv \left\langle \sum_l E_l^*(\mathbf{r}_j, t_j) \sum_m E_m(\mathbf{r}_j, t_j) \right\rangle = \sum_l E_l^*(\mathbf{r}_j, t_j) E_l(\mathbf{r}_j, t_j). \tag{14.56}$$

The notation $\langle \ldots \rangle$ denotes the mathematical expectation, when *taking into account all possible realizations of the fields*, i.e., taking into account all possible complex amplitudes for the large number of subfields in the superposition. In the probability theory, the expectation value of a measurement equals the mean value of an ensemble. In a real measurement, the superposition may not take all possible realizations of the fields and consequently the measured instantaneous intensity $I(\mathbf{r}, t)$ may differ from its expectation value $\langle I(\mathbf{r}, t)\rangle$ from time to time. The variation $\delta I(\mathbf{r}, t)$ turns to be a random function of time. The measured $I(\mathbf{r}, t)$ fluctuate randomly in the neighborhood of $\langle I(\mathbf{r}, t)\rangle$ non-deterministically.

In the classical limit, a large number of independent and randomly radiated subsources contribute to the instantaneous intensity $I(\mathbf{r}_j, t_j)$. These large number of independent randomly distributed subfields may have *taken all possible realizations of their complex amplitudes* in the superposition. In this case, the sum of the cross terms vanishes,

$$\Delta I(\mathbf{r}, t) = \sum_{l \neq m} E_l^*(\mathbf{r}_j, t_j) E_m(\mathbf{r}_j, t_j) \simeq 0, \tag{14.57}$$

therefore,

$$I(\mathbf{r}_j, t_j) \simeq \sum_l E_l^*(\mathbf{r}_j, t_j) E_l(\mathbf{r}_j, t_j) = \langle I(\mathbf{r}, t)\rangle.$$

Now we calculate the second-order correlation function $G^{(2)}(\mathbf{r}_1, t_1; \mathbf{r}_2, t_2)$, which is defined as

$$G^{(2)}(\mathbf{r}_1, t_1; \mathbf{r}_2, t_2) \equiv \left\langle \sum_{j,k,l,m} E_j^*(\mathbf{r}_1, t_1)\, E_k(\mathbf{r}_1, t_1) E_l^*(\mathbf{r}_2, t_2)\, E_m(\mathbf{r}_2, t_2) \right\rangle, \tag{14.58}$$

where the notation $\langle \ldots \rangle$, again, denotes an expectation operation by *taking into account all possible realizations of the fields*, i.e., averaging all possible complex amplitudes for the subfields in the superposition. In the following calculation we only take into account the random phases of the subfields without considering the amplitude variations. Due to the chaotic nature of the independent subsources, after taking into account all possible realizations of the phases associated with the subfields, the only surviving terms in the summation are those with: (1) $j = k$,

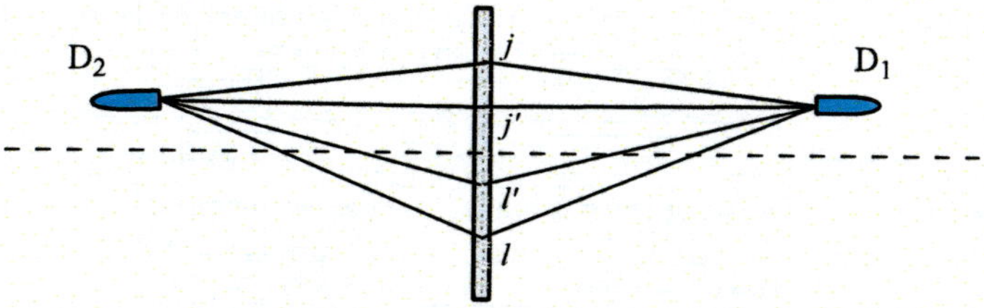

Fig. 14.20 Schematic illustration of $\sum_{j,l} |E_{j1}E_{l2} + E_{l1}E_{j2}|^2$. It is clear that the amplitude pairs $j1 \times l2$ with $l1 \times j2$, where j and l represent all point subsources, pair by pair, will experience equal optical path propagation and superpose constructively when D_1 and D_2 are located at $\vec{\rho}_1 \simeq \vec{\rho}_2$, $z_1 \simeq z_2$. This interference is similar to symmetrizing the wavefunction of identical particles in quantum mechanics

$l = m$, (2) $j = m, k = l$. Therefore, $G^{(2)}(\mathbf{r}_1, t_1; \mathbf{r}_2, t_2)$ reduces to the sum of the following two groups:

$$G^{(2)}(\mathbf{r}_1, t_1; \mathbf{r}_2, t_2) = \Big\langle \sum_j E_j^*(\mathbf{r}_1, t_1)\, E_j(\mathbf{r}_1, t_1) \sum_l E_l^*(\mathbf{r}_2, t_2)\, E_l(\mathbf{r}_2, t_2)$$
$$+ \sum_j E_j^*(\mathbf{r}_1, t_1)\, E_j(\mathbf{r}_2, t_2) \sum_l E_l^*(\mathbf{r}_2, t_2)\, E_l(\mathbf{r}_1, t_1) \Big\rangle$$
$$= \Big\langle \sum_j \sum_l \Big| \frac{1}{\sqrt{2}} [E_j(\mathbf{r}_1, t_1) E_l(\mathbf{r}_2, t_2) + E_l(\mathbf{r}_1, t_1) E_j(\mathbf{r}_2, t_2)] \Big|^2 \Big\rangle. \tag{14.59}$$

It is not difficult to see the nonlocal nature of the superposition shown in (14.59). In (14.59), $G^{(2)}(\mathbf{r}_1, t_1; \mathbf{r}_2, t_2)$ is written as a superposition between the paired sub-fields $E_j(\mathbf{r}_1, t_1)E_l(\mathbf{r}_2, t_2)$ and $E_l(\mathbf{r}_1, t_1)E_j(\mathbf{r}_2, t_2)$. The first term in the superposition corresponds to the situation in which the field at D_1 was generated by the jth sub-source, and the field at D_2 was generated by the lth sub-source. The second term in the superposition corresponds to a different yet indistinguishable situation in which the field at D_1 was generated by the lth sub-source, and the field at D_2 was generated by the jth sub-source. Therefore, an interference is concealed in the joint measurement of D_1 and D_2, which physically occurs at two space-time points $(\mathbf{r}_1, t_1)$ and $(\mathbf{r}_2, t_2)$. The interference corresponds to $|E_{j1}E_{l2} + E_{l1}E_{j2}|^2$. It is easy to see from Fig. 14.20, the amplitude pairs $j1 \times l2$ with $l1 \times j2$, $j'1 \times l'2$ with $l'1 \times j'2$, $j1 \times l'2$ with $l'1 \times j2$, and $j'1 \times l2$ with $l1 \times j'2$, etc., pair by pair, experience equal total optical path propagation, which involves two arms of D_1 and D_2, and thus superpose constructively when D_1 and D_2 are placed in the neighborhood of $\vec{\rho}_1 = \vec{\rho}_2$, $z_1 = z_2$. Consequently, the summation of these individual constructive interference terms will yield a maximum value. When $\vec{\rho}_1 \neq \vec{\rho}_2$, $z_1 = z_2$, however, each pair of the amplitudes may achieve different relative phase and contribute a different value to the summation, resulting in an averaged constant value.

It does not seem to make sense to claim a nonlocal interference between [(E_j goes to D_1) × (E_l goes to D_2)] and [(E_l goes to D_1) × (E_j goes to D_2)] in the framework of Maxwell's electromagnetic wave theory of light. This statement is more likely adapted from particle physics, similar to symmetrizing the wavefunction of identical particles, and is more suitable to describe the interference between quantum amplitudes: [(particle-j goes to D_1) × (particle-l goes to D_2)] and [(particle-l goes to D_1) × (particle-j goes to D_2)], rather than waves. Classical waves do not behave in such a way. In fact, in this model each subsource corresponds to an independent spontaneous atomic transition in nature, and consequently corresponds to the creation of a photon. Therefore, the above superposition corresponds to the superposition between two indistinguishable two-photon amplitudes, and is thus called *two-photon interference* [11]. In Dirac's theory, this interference is the result of a measured pair of photons interfering with itself.

In the following, we attempt a Fresnel near-field calculation to derive the point-to-point correlation of $G^{(2)}(\vec{\rho}_1, z_1; \vec{\rho}_2, z_2)$. We start from (14.59) and concentrate to the transverse spatial correlation

$$G^{(2)}(\vec{\rho}_1, z_1; \vec{\rho}_2, z_2) = \left\langle \sum_j \sum_l \left| \frac{1}{\sqrt{2}} \left[E_j(\vec{\rho}_1, z_1) E_l(\vec{\rho}_2, z_2) + E_l(\vec{\rho}_1, z_1) E_j(\vec{\rho}_2, z_2) \right] \right|^2 \right\rangle. \tag{14.60}$$

In the near-field, we apply the Fresnel approximation as usual to propagate the field from each subsource to the photodetectors. $G^{(2)}(\vec{\rho}_1, z_1; \vec{\rho}_2, z_2)$ can be formally written in terms of the Green's function,

$$\begin{aligned} G^{(2)}(\vec{\rho}_1, z_1; \vec{\rho}_2, z_2) &= \left\langle \int \mathrm{d}\vec{\kappa}\, \mathrm{d}\vec{\kappa}' \left| \frac{1}{\sqrt{2}} \left[g(\vec{\rho}_1, z_1, \vec{\kappa}) g(\vec{\rho}_2, z_2, \vec{\kappa}') + g(\vec{\rho}_2, z_2, \vec{\kappa}) g(\vec{\rho}_1, z_1, \vec{\kappa}') \right] \right|^2 \right\rangle \\ &= \left\langle \int \mathrm{d}\vec{\kappa} \left| g(\vec{\rho}_1, z_1, \vec{\kappa}) \right|^2 \int \mathrm{d}\vec{\kappa}' \left| g(\vec{\rho}_2, z_2, \vec{\kappa}') \right|^2 + \left| \int \mathrm{d}\vec{\kappa}\, g^*(\vec{\rho}_1, z_1, \vec{\kappa})\, g(\vec{\rho}_2, z_2, \vec{\kappa}) \right|^2 \right\rangle \\ &\equiv G^{(1)}(\vec{\rho}_1, z_1) G^{(1)}(\vec{\rho}_2, z_2) + \left| G^{(1)}(\vec{\rho}_1, z_1; \vec{\rho}_2, z_2) \right|^2. \end{aligned} \tag{14.61}$$

In (14.61), we have formally written $G^{(2)}$ in terms of the first-order correlation functions $G^{(1)}$, but keep in mind that the first-order correlation function $G^{(1)}$ and the second-order correlation function $G^{(2)}$ represent different physics based on different measurements. Substituting the Green's function derived in the Appendix for free propagation

$$g(\vec{\rho}_j, z_j, \vec{\kappa}) = \frac{-\mathrm{i}\omega}{2\pi c} \frac{\mathrm{e}^{\mathrm{i}\frac{\omega}{c} z_j}}{z_j} \int \mathrm{d}\vec{\rho}_0\, a(\vec{\rho}_0)\, \mathrm{e}^{\mathrm{i}\varphi(\vec{\rho}_0)}\, \mathrm{e}^{\mathrm{i}\frac{\omega}{2cz_j} |\vec{\rho}_j - \vec{\rho}_0|^2}$$

into (14.61), we obtain $G^{(1)}(\vec{\rho}_1, z_1)G^{(1)}(\vec{\rho}_2, z_2) \sim$ constant and

$$\left|G^{(1)}(\vec{\rho}_1, z_1; \vec{\rho}_2, z_2)\right|^2 \propto \left|\left\langle \frac{1}{z_1 z_2} \int d\vec{\rho}_0\, a^2(\vec{\rho}_0)\, e^{-i\frac{\omega}{c}z_1}\, e^{-i\frac{\omega}{2cz_1}|\vec{\rho}_1-\vec{\rho}_0|^2} e^{i\frac{\omega}{c}z_2}\, e^{i\frac{\omega}{2cz_2}|\vec{\rho}_2-\vec{\rho}_0|^2} \right\rangle\right|^2.$$

Assuming $a^2(\vec{\rho}_0) \sim$ constant, and taking $z_1 = z_2 = d$, we obtain

$$\begin{aligned}\left|G^{(1)}_{12}(\vec{\rho}_1; \vec{\rho}_2)\right|^2 &\propto \left| \int d\vec{\rho}_0\, a^2(\vec{\rho}_0)\, e^{-i\frac{\omega}{2cd}|\vec{\rho}_1-\vec{\rho}_0|^2}\, e^{i\frac{\omega}{2cd}|\vec{\rho}_2-\vec{\rho}_0|^2}\right|^2 \\ &\propto \left| e^{-i\frac{\omega}{2cd}(|\vec{\rho}_1|^2-|\vec{\rho}_2|^2)} \int d\vec{\rho}_0\, a^2(\vec{\rho}_0)\, e^{i\frac{\omega}{cd}(\vec{\rho}_1-\vec{\rho}_2)\cdot\vec{\rho}_0}\right|^2 \\ &\propto somb^2\left[\frac{R}{d}\frac{\omega}{c}|\vec{\rho}_1-\vec{\rho}_2|\right],\end{aligned} \tag{14.62}$$

where we have assumed a disk-like light source with a finite radius of R. The transverse spatial correlation function $G^{(2)}(\vec{\rho}_1; \vec{\rho}_2)$ is thus

$$G^{(2)}\left(|\vec{\rho}_1-\vec{\rho}_2|\right) = I_0^2\left[1 + somb^2\left(\frac{R}{d}\frac{\omega}{c}|\vec{\rho}_1-\vec{\rho}_2|\right)\right]. \tag{14.63}$$

Consequently, the degree of the second-order spatial coherence is

$$g^{(2)}(|\vec{\rho}_1-\vec{\rho}_2|) = 1 + somb^2\left(\frac{R}{d}\frac{\omega}{c}|\vec{\rho}_1-\vec{\rho}_2|\right). \tag{14.64}$$

For a large value of $2R/d \sim \Delta\theta$, where $\Delta\theta$ is the angular size of the radiation source viewed at the photodetectors, the point-spread *somb*-function can be approximated as a δ-function of $|\vec{\rho}_1-\vec{\rho}_2|$. We effectively have a "point-to-point" correlation between the transverse planes of $z_1 = d$ and $z_2 = d$. In 1-D (14.63) and (14.64) become

$$G^{(2)}(x_1 - x_2) = I_0^2\left[1 + \mathrm{sinc}^2\left(\frac{\pi\Delta\theta(x_1-x_2)}{\lambda}\right)\right] \tag{14.65}$$

and

$$g^{(2)}(x_1 - x_2) = 1 + \mathrm{sinc}^2\left(\frac{\pi\Delta\theta(x_1-x_2)}{\lambda}\right), \tag{14.66}$$

which has been experimentally demonstrated and reported in Fig. 14.18.

We have thus derived the same second-order correlation and coherence functions as that of the quantum theory. The nonfactorizable point-to-point correlation is expected at any intensity. The only requirement is a large number of point subsources with random relative phases participating to the measurement, such as trillions of independent atomic transitions. There is no surprise to derive the same result as that of the quantum theory from this simple model. Although the fields are not quantized and no quantum formula was used in the above calculation, this model has implied the same nonlocal two-photon interference mechanism as that of the quantum theory. Different from the phenomenological theory of intensity fluctuations, this semiclassical model explores the physical cause of the phenomenon.

14.5 Classical Simulation

Can the natural, nonfactorizeable, point-to-point correlation of thermal light or entangled photon pairs be simulated by classical intensity correlation or intensity fluctuation correlations to reproduce ghost images? The answer is positive provided the "light knows where to go" when it is generated. There have been quite a few classical approaches to simulate the point-to-point ghost image-forming correlation. Different from the randomly distributed and propagated chaotic-thermal radiation or entangled photon pairs in ghost imaging, in these classical simulations the radiations are prepared in such a way that each light beam or intensity "speckle" is propagated to a precise chosen "spot" of the object. For each detection, the coordinate $\vec{\rho}_{\rm obj}$, which is chosen by the light source, is recorded against the counting rate of the bucket detector at that coordinate, which is proportional to $A(\vec{\rho}_{\rm obj})$. The object function $A(\vec{\rho}_{\rm obj})$ is thus reproduced after a large number of such records. In fact, for this kind of light source, it is unnecessary to use two photo-detectors for joint-detection. One bucket detector is good enough to reproduce the image of the object. We briefly discuss three classical simulations in the following.

(I) Correlated laser beams.

In 2002, Bennink et al. simulated ghost imaging by two correlated laser beams [29, 30]. The authors intended to show that two correlated co-rotating laser beams can simulate similar physical effects of entangled photon pairs. Figure 14.21 is a schematic picture of the experiment of Bennink et al.. Different from ghost imaging, here the point-to-point correspondence between the object plane and the "image plane" is made artificially by two co-rotating laser beams "shot by shot". In each shot, the laser beams are propagated to a well-defined direction and focused on the object and image planes, respectively. If laser beam-1 is blocked by the object mask, there would be no joint-detection between D_1 and D_2 for that "shot", while if laser beam-1 is unblocked, a coincidence count will be recorded against that angular position of the co-rotating laser beams. A shadow of the object mask is then reconstructed in coincidences by the blocking-partial blocking-unblocking of laser beam-1.

The point-to-point correlation of Bennink et al.. is not only different from that of ghost imaging, but also different from the standard statistical intensity

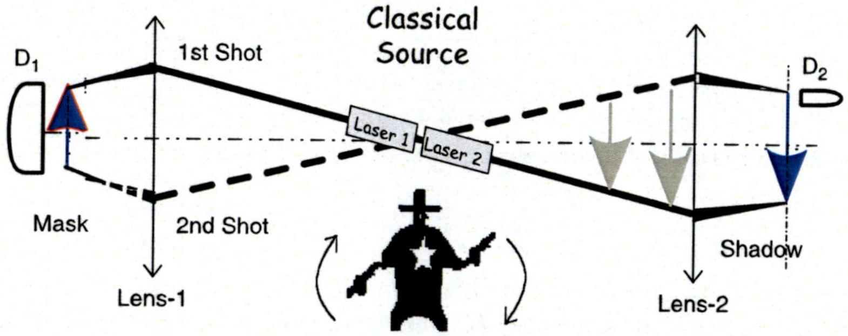

Fig. 14.21 The point-to-point correlation is made shot by shot by two co-rotating laser beams. A ghost shadow can be made in coincidences by "blocking-unblocking" of the correlated laser beams, or simply by "blocking-unblocking" two correlated gun shots

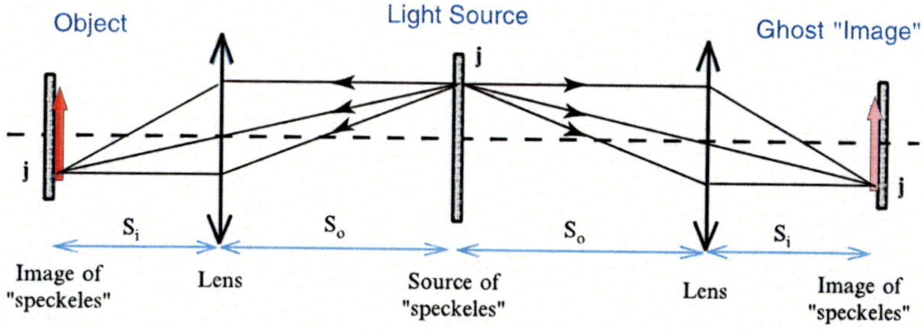

Fig. 14.22 A ghost image is made by a man-made correlation of "speckles". The two identical sets of speckles are the *classical images* of the speckles of the light source. The lens, which may be part of a CCD *camera* used for the joint measurement, reconstructs classical images of the speckles of the source onto the object plane and the image plane, respectively. s_o and s_i satisfy the Gaussian thin lens equation $1/s_o + 1/s_i = 1/f$

fluctuation correlations. Nevertheless, the experiment of Bennink et al.. obtained a ghost shadow, which may be useful for certain purposes. In fact, this experiment can be considered as a good example to distinguish a man-made factorizeable classical intensity–intensity correlation from a natural, nonfactorizeable second-order correlation that is caused by nonlocal two-photon interference.

(II) Correlated speckles.

Following a similar philosophy, Gatti et al.. proposed a classical correlation between "speckles" [16]. The experimental setup of Gatti et al.. is depicted in Fig. 14.22. Their experiments use either entangled photon pairs of spontaneous parametric down-conversion (SPDC) or chaotic-thermal light for simulating ghost images in coincidences. The "ghost image" observed in coincidences comes from a man-made classical speckle-to-speckle correlation. The speckles observed on the object and image planes are the classical images of the speckles of the radiation source, reconstructed by the lenses shown in the figure (the lens may be part of a CCD *camera* used for the joint measurement). Each speckle on the source, such as the jth speckle near the top of the source, has two identical images on the object plane and on the image plane. Mathematically, the speckle-to-speckle correlation is factorizeable into a product of two classical images,

$$\gamma(\vec{\rho}_1, z_1; \vec{\rho}_2, z_2) \simeq \delta(\vec{\rho}_0 - \vec{\rho}_1/m)\delta(\vec{\rho}_0 - \vec{\rho}_2/m),$$

where $\vec{\rho}_0$ is the transverse coordinate of the light source plane, $m = s_i/s_o$ is the classical imaging magnification factor,[8] z_1 and z_2 are defined as the optical distance between the plane of the light source and the planes of the object and the ghost image. The choices of z_1 and z_2 must satisfy the Gaussian thin lens equation, respectively, see Fig. 14.22. It is easy to see from Fig. 14.22 that D_1 and D_2 will have more chance to be triggered jointly when they are in the position within the two identical speckles, such as the two jth speckles near the bottom of the object plane and the image plane. It is also easy to see that the size of the identical speckles determines the spatial resolution of the ghost shadow. This observation has been confirmed by quite a few experimental demonstrations. There is no surprise that Gatti et al. consider ghost imaging classical [31, 32]. Their speckle–speckle correlation is a man-made classical correlation and their ghost imager is indeed classical. The classical simulation of Gatti et al. might be useful for certain applications. However, the man-made speckle-to-speckle correlation of Gatti et al. is fundamentally different from the natural, nonfactorizeable point-to-point image-forming correlation observed in the ghost imaging experiment of Pittman et al. with entangled photon pairs and the lensless ghost imaging experiment of Scarcelli et al. with chaotic light.

(III) "Computational" imaging.

Recently, Shapiro proposed a "computational" ghost imaging experiment, which consists of a controllable (deterministic) light source, an object for imaging, and a bucket photo-detector [33]. Different from the randomly distributed and propagated chaotic-thermal radiation or entangled photon pairs in ghost imaging, this controllable (deterministic) light source propagates light beams or "speckles" to a precise chosen "spot" of the object in each "shot" of its operation. From shot to shot, the light source illuminates different "spot" of the object. The bucket detector records the coordinate $\vec{\rho}_{obj}$, which is chosen by the light source for each shot, and its counting rate at that coordinate, which is proportional to the value of the aperture function at that chosen coordinate, for each detection of the bucket detector. The object function $A(\vec{\rho}_{obj})$ is thus reproduced after a large number of such records.

---

[8]The original publications of Gatti et al.. choose $m = 2f/2f = 1$ with $1/2f + 1/2f = 1/f$ to image the speckles of the source onto the object plane and the ghost image plane.

In fact, this kind of imaging technology has been studied in the filed of computer science. In these studies spatial modulators of light, statistical modeling and computation decoding are usually involved in the process of image reconstruction.

14.6 Local? Nonlocal?

We have discussed the physics of both type-one and type-two ghost imaging. Although different radiation sources are used for different cases, these two types of experiments demonstrated a similar nonfactorizable point-to-point image-forming correlation:

Type-one:

$$\delta(\vec{\rho}_1 - \vec{\rho}_2) \sim \left| \int d\vec{\kappa}_s \, d\vec{\kappa}_i \, \delta(\vec{\kappa}_s + \vec{\kappa}_i) \, g_1(\vec{\kappa}_s, \vec{\rho}_1) \, g_2(\vec{\kappa}_i, \vec{\rho}_2) \right|^2, \tag{14.67}$$

Type-two:

$$\begin{aligned} 1 + \delta(\vec{\rho}_1 - \vec{\rho}_2) &\propto \left\langle \sum_j \sum_l \left| \frac{1}{\sqrt{2}} [E_j(\vec{\rho}_1) E_l(\vec{\rho}_2) + E_l(\vec{\rho}_1) E_j(\vec{\rho}_2)] \right|^2 \right\rangle \\ &= \left\langle \int d\vec{\kappa} \, d\vec{\kappa}' \left| \frac{1}{\sqrt{2}} [g_1(\vec{\kappa}, \vec{\rho}_1) g_2(\vec{\kappa}', \vec{\rho}_2) + g_2(\vec{\kappa}, \vec{\rho}_2) g_1(\vec{\kappa}', \vec{\rho}_1)] \right|^2 \right\rangle. \end{aligned} \tag{14.68}$$

Equations (14.67) and (14.68) indicate that the point-to-point correlation of ghost imaging, either type-one or type-two, is the results of two-photon interference. Unfortunately, neither of them is in the form of $|\sum_j E_j|^2$ or $|E_1 + E_2|^2$, and neither is measured at a local space-time point. The interference shown in (14.67) and (14.68) occurs at different space-time points through the measurements of two spatially separated independent photodetectors.

In type-one ghost imaging, the δ-function in (14.67) means a typical EPR position–position correlation of an entangled photon pair. In EPR's language: when the pair is generated at the source the momentum and position of neither photon is determined, and neither photon-one nor photon-two "knows" where to go. However, if one of them is observed at a point at the object plane the other one must be found at a unique point in the image plane. In type-two ghost imaging, although the position–position determination in (14.68) is only partial, it generates more surprises because of the chaotic nature of the radiation source. Photon-one and photon-two, emitted from a thermal source, are completely random and independent, i.e., both propagate freely to any direction and may arrive at any position in the object and image planes. Analogous to EPR's language: when the measured two photons were emitted from the thermal source, neither the momentum nor the position of any photon is determined. However, if one of them is observed at a point on the object plane the other one must have twice large probability to be found at a unique point in the image plane. Where does this partial correlation come from? If one insists on the view point of intensity fluctuation correlation, then it is reasonable to ask why the intensities of the two light beams exhibit fluctuation correlations at $\vec{\rho}_1 = \vec{\rho}_2$ only? Recall that in the experiment of Sarcelli et al.. the ghost image is measured in the near-field. Regardless of position, D_1 and D_2 receive light from all (a large number) point subsources of the thermal source, and all subsources fluctuate randomly and independently. If $\Delta I_1 \Delta I_2 = 0$ for $\vec{\rho}_1 \neq \vec{\rho}_2$, what is the physics to cause $\Delta I_1 \Delta I_2 \neq 0$ at $\vec{\rho}_1 = \vec{\rho}_2$?

The classical superposition is considered "local". The Maxwell electromagnetic field theory requires the superposition of the electromagnetic fields, either $|\sum_j E_j|^2$ or $|E_1 + E_2|^2$, takes place at a local space-time point $(\mathbf{r}, t)$. However, the superposition shown in (14.67) and (14.68) happens at two different space-time points $(\mathbf{r}_1, t_1)$ and $(\mathbf{r}_2, t_2)$ and is measured by two independent photodetectors. Experimentally, it is not difficult to make the two photo-detection events space-like separated events. Following the definition given by EPR-Bell, we consider the superposition appearing in (14.67) and (14.68) *nonlocal*. Although the two-photon interference of thermal light can be written and calculated in terms of a semiclassical model, the nonlocal superposition appearing in (14.68) has no counterpart in the classical measurement theory of light, unless one forces a nonlocal classical theory by allowing the superposition to occur at a distance through the measurement of independent photodetectors, as we have done in (14.59). Perhaps, it would be more difficult to accept a nonlocal classical measurement theory of thermal light rather than to apply a quantum mechanical concept to "classical" thermal radiation.

14.7 Turbulence-Free Ghost Imaging

The multiphoton interference nature of ghost imaging determines its peculiar features. One of them is "turbulence-free", i.e., any index fluctuation type turbulence occur in the optical path would not affect the quality of the ghost image. Atmospheric turbulence is a serious problem for classical satellite and aircraft-ground-based distant imaging. Turbulence-free is thus attractive for these applications.

In the following, we analyze a recent ghost imaging experiment of Meyers et al., which demonstrated this interesting yet peculiar feature of ghost imaging [27]. The schematic setup of the experiment is shown in Fig. 14.23. It is a typical thermal light lensless ghost imaging setup, except for the addition of heating elements to produce laboratory atmospheric turbulence. In this experiment, turbulence is introduced to the optical paths by the adding heating elements underneath the optical paths. Figure 14.23 illustrates the most serious situation in which turbulence occurs in all optical paths of the setup. The heating elements can be isolated to produce turbulence for any individual optical path too. Heating of the air causes temporal and spatial fluctuations on its index of refraction that makes the classical image of the object to jitter about randomly on the image plane of a classical camera.

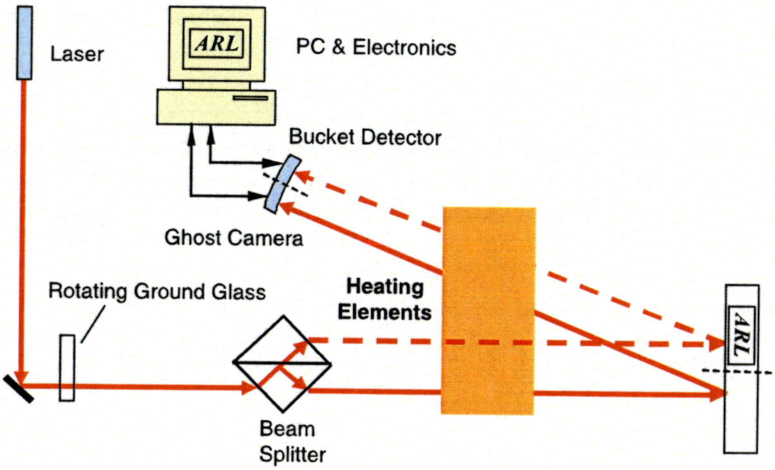

Fig. 14.23 Schematic setup of a typical thermal light lensless ghost imaging experiment in which significant turbulence is introduced in its optical paths. *Dashed line and arrows* indicate the optical path of the "bucket" detector. The *solid line and arrows* indicate the optical path of the ghost image arm. Here, we illustrate the most serious situation in which turbulence occurs in all optical paths of the setup. The heating elements can be isolated to produce turbulence for any individual optical path too

Similar to their earlier demonstration of ghost imaging, the chaotic-thermal light is generated from a fairly large angular sized pseudo-thermal source and is split into two by a 50%–50% beamsplitter. One of the beams illuminates an object located at z_1, such as the letters "ARL" as shown in Fig. 14.23. The scattered and reflected photons from the object are collected and counted by a bucket detector, which is simulated by the right-half of the CCDs in Fig. 14.23. The other beam propagates to the ghost image plane of $z_1 = z_2$. We have learned from early analysis of thermal light ghost imaging experiments, placing a CCD array on the ghost image plane, the CCD array will capture the ghost image of the object if its exposure is gated by the bucket detector. In this experiment, the CCD array is replaced by a piece of glossy white paper. The scattered and reflect light from the glossy white paper, which contains the information of the ghost image, is then captured by the left-half of the high resolution CCD camera, which is operated at the photon counting regime. The CCD camera is focused onto the ghost image plane and is gated by the bucket detector for the observation of the secondary ghost image. The secondary ghost image captured by the left-half CCD camera is the image of the primary ghost image located at $z_1 = z_2$. In this special setup the left-half and the right-half of the CCD camera may play the poles of two independent classical cameras in their "normal" ungated operation, and simultaneously capture the secondary ghost image in their gated joint-detection operation. The hardware circuit and the software program is designed to monitor the outputs of the left-half and the right-half of the CCDs, individually, as two independent classical cameras, and simultaneously to monitor the gated output of the left-half CCDs as a ghost camera. In the measurement, the classical image and the secondary ghost image of the object were captured and monitored simultaneously when the turbulence is introduced to each or to all optical paths.

The turbulence-free ghost imaging is the result of the turbulence-free nonfactorizable point-to-point image-forming correlation, which is caused by two-photon interference: superposition between paired two-photon amplitudes, corresponding to two different yet indistinguishable alternative ways of triggering a joint-detection event by two independent photons. We give a simple analysis in the following staring from (14.45)

$$G^{(2)}(\vec{\rho}_1, z_1; \vec{\rho}_2, z_2) = \int d\vec{\kappa}\, d\vec{\kappa}' \left| \frac{1}{\sqrt{2}} [g_2(\vec{\rho}_2, z_2; \vec{\kappa}) g_1(\vec{\rho}_1, z_1; \vec{\kappa}') + g_2(\vec{\rho}_2, z_2; \vec{\kappa}') g_1(\vec{\rho}_1, z_1; \vec{\kappa})] \right|^2 .$$

Equation (14.45) indicates an interference between two quantum amplitudes, corresponding to two alternatives, different yet indistinguishable, which leads to a joint photo-detection event. This interference involve both arms of the optical setup as well as two distant photo-detection events at $(\vec{\rho}_1, z_1)$ and $(\vec{\rho}_2, z_2)$, respectively.

Now, we introduce an arbitrary phase disturbance $e^{i\Delta\varphi_1(\vec{\rho}_1)}$ into the ghost image arm and another phase disturbance $e^{i\Delta\varphi_2(\vec{\rho}_2)}$ into the object-bucket-detector arm to simulate the turbulence, where $\Delta\varphi_1(\vec{\rho}_1)$ and $\Delta\varphi_2(\vec{\rho}_2)$ add random phase variations onto the radiation of transverse coordinate $\vec{\rho}_1$ and $\vec{\rho}_2$, respectively. The second-order correlation with turbulence turns to be

$$\begin{aligned} G^{(2)}_{\text{Turb}}(\vec{\rho}_1, z_1; \vec{\rho}_2, z_2) &= \int d\vec{\kappa}\, d\vec{\kappa}' \left| \frac{1}{\sqrt{2}} [g_2(\vec{\rho}_2, z_2; \vec{\kappa}) e^{i\Delta\varphi_2(\vec{\rho}_2)} g_1(\vec{\rho}_1, z_1; \vec{\kappa}') e^{i\Delta\varphi_1(\vec{\rho}_1)} \right. \\ &\quad \left. + g_2(\vec{\rho}_2, z_2; \vec{\kappa}') e^{i\Delta\varphi_2(\vec{\rho}_2)} g_1(\vec{\rho}_1, z_1; \vec{\kappa}) e^{i\Delta\varphi_1(\vec{\rho}_1)}] \right|^2 \\ &= G^{(2)}(\vec{\rho}_1, z_1; \vec{\rho}_2, z_2). \end{aligned} \quad (14.69)$$

It is easy to see that the phase turbulence has a null effect on the second-order correlation function $G^{(2)}(\vec{\rho}_1, z_1; \vec{\rho}_2, z_2)$ of (14.45). The normalized nonfactorizable point-to-point image-forming correlation $g^{(2)}(\vec{\rho}_1; \vec{\rho}_2)$ of thermal light is thus turbulence free. It is also true in the cases when turbulence occurs only in one arm of the optical paths, such as in the arm of D_2,

$$\begin{aligned} G^{(2)}_{\text{Turb}}(\vec{\rho}_1, z_1; \vec{\rho}_2, z_2) &= \int d\vec{\kappa}\, d\vec{\kappa}' \left| \frac{1}{\sqrt{2}} [g_2(\vec{\rho}_2, z_2; \vec{\kappa}) e^{i\Delta\varphi_2(\vec{\rho}_2)} g_1(\vec{\rho}_1, z_1; \vec{\kappa}') \right. \\ &\quad \left. + g_2(\vec{\rho}_2, z_2; \vec{\kappa}') e^{i\Delta\varphi_2(\vec{\rho}_2)} g_1(\vec{\rho}_1, z_1; \vec{\kappa})] \right|^2 \\ &= G^{(2)}(\vec{\rho}_1, z_1; \vec{\rho}_2, z_2). \end{aligned} \quad (14.70)$$

In either cases, the joint-detection counting rate between the bucket detector and the CCD array will therefore reproduce the aperture function as a turbulence-free ghost image

$$R_{12} \propto \int d\vec{\rho}_2 \, |A(\vec{\rho}_2)|^2 \, g^{(2)}(\vec{\rho}_1; \vec{\rho}_2) \sim R_0 + |A(\vec{\rho}_1)|^2,$$

where R_0 is a constant and $A(\vec{\rho}_2)$ is the aperture function of the object.

It should be emphasized that classical simulations of ghost imaging cannot obtain turbulence-free images. For instance, it is easy to see that the man-made factorizable speckle-to-speckle correlation

$$g^{(2)}(\vec{\rho}_o, \vec{\rho}_i) \simeq \delta(\vec{\rho}_s - \vec{\rho}_o/m)\delta(\vec{\rho}_s - \vec{\rho}_i/m), \tag{14.71}$$

which is made by classically imaging the speckles of the common light source onto the object and image planes (with magnification factor m), will be "blurred" under turbulence, because each classical image of the speckle will be "blurred".

14.8 Conclusion

In summary, we may conclude that ghost imaging is the result of quantum interference. Either type-one or type-two, ghost imaging is characterized by a non-factorizable point-to-point image-forming correlation, which is caused by constructive–destructive interferences involving the nonlocal superposition of two-photon amplitudes, a nonclassical entity corresponding to different yet indistinguishable alternative ways of producing a joint photo-detection event. The interference happens within a pair of photons and at two spatially separated coordinates. The multi-photon interference nature of ghost imaging determines its peculiar features: (1) it is nonlocal; (2) its imaging resolution differs from that of classical; and (3) the type-two ghost image is turbulence free. Taking advantage of its quantum interference nature, a ghost imaging system may turn a local "bucket" sensor into a nonlocal imaging camera with classically unachievable imaging resolution. For instance, using the Sun as light source for type-two ghost imaging, we may achieve an imaging spatial resolution equivalent to that of a classical imaging system with a lens of 92-meter diameter when taking pictures at 10 kilometers.[9] Furthermore, any phase disturbance in the optical path has no influence on the ghost image. To achieve these features, the realization of multi-photon interference is necessary.

[9] The angular size of Sun is about 0.53°. To achieve a compatible image spatial resolution, a traditional camera must have a lens of 92-meter diameter when taking pictures at 10 kilometers.

Acknowledgments The author thanks M. D'Angelo, G. Scarcelli, J.M. Wen, T.B. Pittman, M.H. Rubin, and L.A. Wu for helpful discussions. This work is partially supported by AFOSR and ARO-MURI program.

Appendix: Fresnel Free-Propagation

We are interested in knowing how a known field $E(\mathbf{r}_0, t_0)$ on the plane $z_0 = 0$ propagates or diffracts into $E(\mathbf{r}, t)$ on another plane $z =$ constant. We assume the field $E(\mathbf{r}_0, t_0)$ is excited by an arbitrary source, either point-like or spatially extended. The observation plane of $z =$ constant is located at an arbitrary distance from plane $z_0 = 0$, either far-field or near-field. Our goal is to find out a general solution $E(\mathbf{r}, t)$, or $I(\mathbf{r}, t)$, on the observation plane, based on our knowledge of $E(\mathbf{r}_0, t_0)$ and the laws of the Maxwell electromagnetic wave theory. It is not easy to find such a general solution. However, the use of the Green's function or the field transfer function, which describes the propagation of each mode from the plane of $z_0 = 0$ to the observation plane of $z =$ constant, makes this goal formally achievable.

Unless $E(\mathbf{r}_0, t_0)$ is a nonanalytic function in the space-time region of interest, there must exist a Fourier integral representation for $E(\mathbf{r}_0, t_0)$

$$E(\mathbf{r}_0, t_0) = \int \mathrm{d}\mathbf{k}\, E(\mathbf{k})\, w_{\mathbf{k}}(\mathbf{r}_0, t_0)\, \mathrm{e}^{-\mathrm{i}\omega t_0}, \tag{A.1}$$

where $w_{\mathbf{k}}(\mathbf{r}_0, t_0)$ is a solution of the Helmholtz wave equation under appropriate boundary conditions. The solution of the Maxwell wave equation $w_{\mathbf{k}}(\mathbf{r}_0, t_0)\, \mathrm{e}^{-\mathrm{i}\omega t_0}$, namely the Fourier mode, can be a set of plane-waves or spherical-waves depending on the chosen boundary condition. In (A.1), $E(\mathbf{k}) = a(\mathbf{k})\mathrm{e}^{\mathrm{i}\varphi(\mathbf{k})}$ is the complex amplitude of the Fourier mode $\mathbf{k}$. In principle, we should be able to find an appropriate Green's function which propagates each mode under the Fourier integral point by point from the plane of $z_0 = 0$ to the plane of observation,

$$\begin{aligned} E(\mathbf{r}, t) &= \int \mathrm{d}\mathbf{k}\, E(\mathbf{k})\, g(\mathbf{k}, \mathbf{r} - \mathbf{r}_0, t - t_0)\, w_{\mathbf{k}}(\mathbf{r}_0, t_0)\, \mathrm{e}^{-\mathrm{i}\omega t_0} \\ &= \int \mathrm{d}\mathbf{k}\, g(\mathbf{k}, \mathbf{r} - \mathbf{r}_0, t - t_0)\, E(\mathbf{k}, \mathbf{r}_0, t_0), \end{aligned} \tag{A.2}$$

where $E(\mathbf{k}, \mathbf{r}_0, t_0) = E(\mathbf{k})\, w_{\mathbf{k}}(\mathbf{r}_0, t_0)\, \mathrm{e}^{-\mathrm{i}\omega t_0}$. The secondary wavelets that originated from each point on the plane of $z_0 = 0$ are then superposed *coherently* on each point on the observation plane with their after-propagation amplitudes and phases. It is convenient to write (A.2) in the following form

$$E(\vec{\rho}, z, t) = \int \mathrm{d}\omega\, \mathrm{d}\vec{\kappa}\; g(\vec{\kappa}, \omega; \vec{\rho} - \vec{\rho}_0, z - z_0, t - t_0)\, E(\vec{\kappa}, \omega; \vec{\rho}_0, z_0, t_0), \tag{A.3}$$

where we have used the transverse-longitudinal coordinates in space-time ($\vec{\rho}$ and z) and in momentum ($\vec{\kappa}, \omega$).

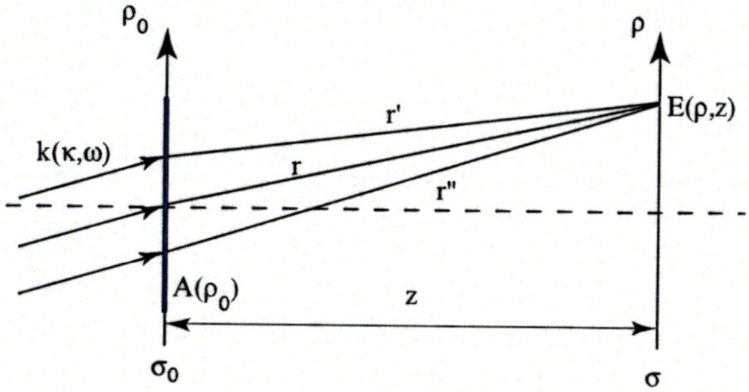

Fig. *A*.1 Schematic of free-space Fresnel propagation. The complex amplitude $\tilde{A}(\vec{\rho}_0)$ is composed of a real function $A(\vec{\rho}_0)$ and a phase $e^{-i\vec{\kappa}\cdot\vec{\rho}_0}$ associated with each of the transverse wavevectors $\vec{\kappa}$ in the plane of σ_0. Notice: only one mode of wavevector $\mathbf{k}(\vec{\kappa},\omega)$ is shown in the figure

Fig. *A*.1 is a simple example in which the field propagates freely from an aperture A of finite size on the plane σ_0 to the observation plane σ. Based on Fig. *A*.1, we evaluate $g(\vec{\kappa},\omega;\vec{\rho},z)$, namely the Green's function for free-space Fresnel propagation-diffraction.

According to the Huygens–Fresnel principle the field at a given space-time point $(\vec{\rho},z,t)$ is the result of a superposition of the spherical secondary wavelets that originated from each point on the σ_0 plane (see Fig. *A*.1),

$$E(\vec{\rho},z,t) = \int d\omega\, d\vec{\kappa}\; E(\vec{\kappa},\omega;0,0) \int_{\sigma_0} d\vec{\rho}_0\, \frac{\tilde{A}(\vec{\rho}_0)}{r}\, e^{-i(\omega t - kr)}, \tag{A.4}$$

where we have set $z_0 = 0$ and $t_0 = 0$ at plane σ_0, and defined $r = \sqrt{z^2 + |\vec{\rho}-\vec{\rho}_0|^2}$. In (*A*.4), $\tilde{A}(\vec{\rho}_0)$ is the complex amplitude or relative distribution of the field on the plane of σ_0, which may be written as a simple aperture function in terms of the transverse coordinate $\vec{\rho}_0$, as we have done in the earlier discussions.

In the near-field Fresnel paraxial approximation, when $|\vec{\rho}-\vec{\rho}_0|^2 \ll z^2$ we take the first-order expansion of r in terms of z and $\vec{\rho}$,

$$r = \sqrt{z^2 + |\vec{\rho}-\vec{\rho}_0|^2} \simeq z\left(1 + \frac{|\vec{\rho}-\vec{\rho}_0|^2}{2z^2}\right), \tag{A.5}$$

so that $E(\vec{\rho},z,t)$ can be approximated as

$$E(\vec{\rho},z,t) \simeq \int d\omega\, d\vec{\kappa}\; E(\vec{\kappa},\omega,0,0) \int d\vec{\rho}_0\, \frac{\tilde{A}(\vec{\rho}_0)}{z}\, e^{i\frac{\omega}{c}z}\, e^{i\frac{\omega}{2cz}|\vec{\rho}-\vec{\rho}_0|^2}\, e^{-i\omega t},$$

where $e^{i\frac{\omega}{2cz}|\vec{\rho}-\vec{\rho}_0|^2}$ is named the Fresnel phase factor.

Assuming that the complex amplitude $\tilde{A}(\vec{\rho}_0)$ is composed of a real function $A(\vec{\rho}_0)$ and a phase $e^{-i\vec{\kappa}\cdot\vec{\rho}_0}$, associated with the transverse wavevector and the transverse coordinate on the plane of σ_0, as is reasonable for the setup of Fig. *A*.1, we can then write $E(\vec{\rho},z,t)$ in the form

$$E(\vec{\rho},z,t) = \int d\omega\, d\vec{\kappa}\; E(\vec{\kappa},\omega;0,0)\, e^{-i\omega t}\, \frac{e^{i\frac{\omega}{c}z}}{z} \int d\vec{\rho}_0\, A(\vec{\rho}_0)\, e^{i\vec{\kappa}\cdot\vec{\rho}_0}\, e^{i\frac{\omega}{2cz}|\vec{\rho}-\vec{\rho}_0|^2}.$$

The Green's function $g(\vec{\kappa}, \omega; \vec{\rho}, z)$ for free-space Fresnel propagation is thus

$$g(\vec{\kappa}, \omega; \vec{\rho}, z) = \frac{e^{i\frac{\omega}{c}z}}{z} \int_{\sigma_0} d\vec{\rho}_0\, A(\vec{\rho}_0)\, e^{i\vec{\kappa}\cdot\vec{\rho}_0}\, G\left(|\vec{\rho} - \vec{\rho}_0|, \frac{\omega}{cz}\right). \tag{A.6}$$

In (A.6), we have defined a Gaussian function $G(|\vec{\alpha}|, \beta) = e^{i(\beta/2)|\alpha|^2}$, namely the Fresnel phase factor. It is straightforward to find that the Gaussian function $G(|\vec{\alpha}|, \beta)$ has the following properties:

$$\begin{aligned}
G^*(|\vec{\alpha}|, \beta) &= G(|\vec{\alpha}|, -\beta), \\
G(|\vec{\alpha}|, \beta_1 + \beta_2) &= G(|\vec{\alpha}|, \beta_1)\, G(|\vec{\alpha}|, \beta_2), \\
G(|\vec{\alpha}_1 + \vec{\alpha}_2|, \beta) &= G(|\vec{\alpha}_1|, \beta)\, G(|\vec{\alpha}_2|, \beta)\, e^{i\beta\vec{\alpha}_1\cdot\vec{\alpha}_2}, \\
\int d\vec{\alpha}\; G(|\vec{\alpha}|, \beta)\, e^{i\vec{\gamma}\cdot\vec{\alpha}} &= i\frac{2\pi}{\beta}\, G\left(|\vec{\gamma}|, -\frac{1}{\beta}\right).
\end{aligned} \tag{A.7}$$

Notice that the last equation in (A.7) is the Fourier transform of the $G(|\vec{\alpha}|, \beta)$ function. As we shall see in the following, these properties are very useful in simplifying the calculations of the Green's functions $g(\vec{\kappa}, \omega; \vec{\rho}, z)$.

Next, we consider inserting an imaginary plane σ' between σ_0 and σ. This is equivalent to having two consecutive Fresnel propagations with a diffraction-free σ' plane of infinity. Thus, the calculation of these consecutive Fresnel propagations should yield the same Green's function as that of the above direct Fresnel propagation shown in (A.6):

$$\begin{aligned}
g(\omega, \vec{\kappa}; \vec{\rho}, z) &= C^2 \frac{e^{i\frac{\omega}{c}(d_1+d_2)}}{d_1 d_2} \int_{\sigma'} d\vec{\rho}' \int_{\sigma_0} d\vec{\rho}_0\, \tilde{A}(\vec{\rho}_0) G\left(|\vec{\rho}' - \vec{\rho}_0|, \frac{\omega}{cd_1}\right) \\
&\quad \times G\left(|\vec{\rho} - \vec{\rho}'|, \frac{\omega}{cd_2}\right) \\
&= C\, \frac{e^{i\frac{\omega}{c}z}}{z} \int_{\sigma_0} d\vec{\rho}_0\, \tilde{A}(\vec{\rho}_0)\, G\left(|\vec{\rho} - \vec{\rho}_0|, \frac{\omega}{cz}\right),
\end{aligned} \tag{A.8}$$

where C is a necessary normalization constant for a valid (A.8), and $z = d_1 + d_2$. The double integral of $d\vec{\rho}_0$ and $d\vec{\rho}'$ in (A.8) can be evaluated as

$$\begin{aligned}
&\int_{\sigma'} d\vec{\rho}' \int_{\sigma_0} d\vec{\rho}_0\, \tilde{A}(\vec{\rho}_0)\, G\left(|\vec{\rho}' - \vec{\rho}_0|, \frac{\omega}{cd_1}\right) G\left(|\vec{\rho} - \vec{\rho}'|, \frac{\omega}{cd_2}\right) \\
&= \int_{\sigma_0} d\vec{\rho}_0\, \tilde{A}(\vec{\rho}_0)\, G\left(\vec{\rho}_0, \frac{\omega}{cd_1}\right) G\left(\vec{\rho}, \frac{\omega}{cd_2}\right) \\
&\quad \times \int_{\sigma'} d\vec{\rho}'\, G\left(\vec{\rho}', \frac{\omega}{c}\left(\frac{1}{d_1} + \frac{1}{d_2}\right)\right) e^{-i\frac{\omega}{c}\left(\frac{\vec{\rho}_0}{d_1} + \frac{\vec{\rho}}{d_2}\right)\cdot\vec{\rho}'}
\end{aligned}$$

$$= \frac{i2\pi c}{\omega}\frac{d_1 d_2}{d_1+d_2}\int_{\sigma_0} d\vec{\rho}_0\, \tilde{A}(\vec{\rho}_0)\, G\left(\vec{\rho}_0, \frac{\omega}{cd_1}\right) G\left(\vec{\rho}, \frac{\omega}{cd_2}\right)$$
$$\times G\left(\left|\frac{\vec{\rho}_0}{d_1}+\frac{\vec{\rho}}{d_2}\right|, \frac{\omega}{c}\left(\frac{d_1 d_2}{d_1+d_2}\right)\right)$$
$$= \frac{i2\pi c}{\omega}\frac{d_1 d_2}{d_1+d_2}\int_{\sigma_0} d\vec{\rho}_0\, \tilde{A}(\vec{\rho}_0)\, G\left(|\vec{\rho}-\vec{\rho}_0|, \frac{\omega}{c(d_1+d_2)}\right),$$

where we have applied ($A.7$), and the integral of $d\vec{\rho}'$ has been taken to infinity. Substituting this result into ($A.8$), we obtain

$$g(\vec{\kappa},\omega;\vec{\rho},z) = C^2\,\frac{i2\pi c}{\omega}\frac{e^{i\frac{\omega}{c}(d_1+d_2)}}{d_1+d_2}\int_{\sigma_0} d\vec{\rho}_0\, \tilde{A}(\vec{\rho}_0)\, G\left(|\vec{\rho}-\vec{\rho}_0|, \frac{\omega}{c(d_1+d_2)}\right)$$
$$= C\,\frac{e^{i\frac{\omega}{c}z}}{z}\int_{\sigma_0} d\vec{\rho}_0\, \tilde{A}(\vec{\rho}_0)\, G\left(|\vec{\rho}-\vec{\rho}_0|, \frac{\omega}{cz}\right).$$

Therefore, the normalization constant C must take the value of $C = -i\omega/2\pi c$. The normalized Green's function for free-space Fresnel propagation is thus

$$g(\vec{\kappa},\omega;\vec{\rho},z) = \frac{-i\omega}{2\pi c}\,\frac{e^{i\frac{\omega}{c}z}}{z}\int_{\sigma_0} d\vec{\rho}_0\, \tilde{A}(\vec{\rho}_0)\, G\left(|\vec{\rho}-\vec{\rho}_0|, \frac{\omega}{cz}\right). \qquad (A.9)$$

References

1. Pittman TB, Shih YH, Strekalov DV, Sergienko AV (1995) Phys Rev A 52:R3429
2. Klyshko DN (1988) Usp Fiz Nauk 154:133
3. Klyshko DN (1988) Sov Phys Usp 31:74
4. Klyshko DN (1988) Phys Lett A 132:299
5. Einstein A, Podolsky B, Rosen N (1935) Phys Rev 35:777
6. Strekalov DV, Sergienko AV, Klyshko DN, Shih YH (1995) Phys Rev Lett 74:3600. Due to its nonlocal behavior, this experiment was named "ghost" interference by the physics community
7. Scarcelli G, Berardi V, Shih YH (2006) Phys Rev Lett 96:063602
8. Valencia A, Scarcelli G, D'Angelo M, Shih YH (2005) Phys Rev Lett 94:063601
9. Scarcelli G, Valencia A, Shih YH (2004) Europhys Lett 68:618
10. Meyers R, Deacon KS, Shih YH (2008) Phys Rev A 77:041801
11. Shih YH (2003) IEEE J Sel Top Quant Electron 9:1455
12. Hanbury-Brown R, Twiss RQ (1956) Nature 177:27; 178:1046 (1956); 178:1447 (1956)
13. Hanbury-Brown R (1974) Intensity interferometer. Taylor and Francis, London
14. Scully MO, Zubairy MS (1997) Quantum optics. Cambridge University Press, Cambridge
15. Shih YH (2011) An Introduction of Quantum Optics-Photon and Biphoton physics. CRC Press, Taylor & Francis, London
16. Gatti A, Brambilla E, Bache M, Lugiato LA (2004) Phys Rev A 70:013802; Phys Rev Lett 93:093602 (2004)
17. Wang K, Cao D, quant-ph/0404078; Cao D, Xiong J, Wang K, quant-ph/0407065
18. Cai YJ, Zhu SY (2005) quant-ph/0407240, Phys Rev E 71:056607

19. Erkmen BI, Shapiro JH (2008) Phys Rev A 77:043809
20. Rubin MH (1996) Phys Rev A 54:5349
21. Goodman JW (1968) Introduction to Fourier optics. McGraw-Hill Publishing, New York, NY
22. Klyshko DN (1988) Photon and nonlinear optics. Gordon and Breach Science, New York
23. Glauber RJ (1963) Phys Rev 130:2529; Phys Rev 131:2766 (1963)
24. D'Angelo M, Valencia A, Rubin MH, Shih YH (2005) Phys Rev A 72:013810
25. Howell JC et al (2004) Phys Rev Lett 92:210403
26. Martienssen W, Spiller E (1964) Am J Phys 32:919
27. Meyers, RE, Deacon, KS, Shih, YH (2011) Applied Phys Lett 98:111115
28. Liu JB, Shih YH (2009) Phys Rev A 79:023818
29. Bennink RS, Bentley SJ, Boyd RW (2002) Phys Rev Lett 89:113601
30. Bennink RS et al (2004) Phys Rev Lett 92:033601
31. Gatti A et al (2007) Phys Rev Lett 98:039301 (comment)
32. Scarcelli G, Berardi V, Shih YH (2007) Phys Rev Lett 98:039302 (reply)
33. Shapiro JH (2008) Phys Rev A 78:061802(R)

Chapter 15
Milestones in the History of Probability

Michael F. Shlesinger

15.1 Introduction

David Middleton is the master of employing probability and statistics in communications theory. His book "Introduction to Statistical Communications Theory" [1] is the classic reference in this field of determining the effects of all types of noise on all aspects of communications. Many of the probabilistic concepts used in the book have their roots going back hundreds of years. We present here several of the developments in probability theory that created the foundation for modern works. Isaac Todhunter [2] produced the definitive review of probability in 1865 going up to the work of Pierre–Simon Laplace. Florence Nightingale David [3] collected much material and colorful stories up to the work of Abraham DeMoivre. Paradoxes in probability and statistics have been discussed by Gabor Szekely [4] and Nathan Rabinovitch [5] discussed probability in ancient and medieval Jewish literature. Elliott Montroll and the author [6] reviewed the history of random walks for applications in physics. We mostly rely on these sources, but supply our own views on the contributions of the pioneers in this field.

15.2 Bones and Sticks

The ankle bones of certain animal fit in the hand and when thrown land on different faces. Although somewhat cubical in nature, these bones only have four stable sides and will roll over to one of these if they land on the rounded unstable sides. The dreidel, a spinable four-sided object, is a modern descendant of the bones. The

M.F. Shlesinger (✉)
United States Naval Academy, Physics Dept., Annapolis MD 21402–5026, USA

Office of Naval Research, code 30, 875 N. Randolph St., Arlington VA 22203–1995, USA
e-mail: mike.shlesinger@navy.mil

L. Cohen et al. (eds.), *Classical, Semi-classical and Quantum Noise*,
DOI 10.1007/978-1-4419-6624-7_15,

Greeks played games of chance with the bones and labeled the four stable sides as 1,3,4, and 6. They played with five bones and the Roman latter played with four. Each culture labeled the outcomes. For the Greeks 1,3,3,4,4 was the Zeus throw, and for the Romans 1,3,4,6 was the Venus throw. Today, a pair of ones is called snake eyes. For the Romans, four ones was a going to dogs throw. Playing with the bones did not lead to a theory of probability, possibly because on a given bone landing on each side had different probabilities and there were variations from bone to bone. The permanence of statistical ratios was apparently not evident. That only came later when bones were turned into six-sided dice.

The Chinese did design a selection process that produced equally likely outcomes, the I Ching [7]. Sticks were chosen randomly that could be long (L) or short (S). Three sticks, called a trigram, led to eight possible outcomes, LLS, LSL, SSS, etc. Each of these outcomes was assigned a set of attributes. Long, long, long could mean heaven, strong, head, father, horse, northwest, creative, force. Short, short, short could be interpreted as earth, belly, cow, southwest, yielding, receptive. Two trigrams were selected and the resulting sixty-four outcomes were used as an oracle, rather than as an experiment in probabilities. The heaven and earth set of trigrams was a fortuitous draw.

The fact that the outcomes in games of bones, dice, and sticks were ascribed to luck or the intervention of gods should not seem that strange to today's gambler on a winning streak who wearing a lucky shirt or socks.

We note that the Arabic word for dice is al-zar, which became our word for hazard. Gambling certainly is hazardous because the odds are usually in favor of the house and against the player. Calculating the odds in a variety of dice and card games was the initial impetus for developing the theory of probability.

Dice games eventually became standardized. One game making the rounds in Italy was throwing three dice until their sum was a nine or ten. The house would bet that the ten throw would appear before a nine throw. There are six combinations that give a ten and also six combinations that produce a nine so the bet appears to be fair.

Add up to 10	# of ways	Add up to 9	# of ways
[6,2,2]	3	[6,2,1]	6
[5,2,3]	6	[5,1,3]	6
[4,2,4]	3	[4,2,3]	6
[6,3,1]	6	[5,2,2]	3
[4,3,3]	3	[4,4,1]	3
[5,4,1]	6	[3,3,3]	1

A gambler brought this puzzle to Galileo who correctly pointed out that there are $6 \times 6 \times 6 = 216$ possible outcomes. The [6, 2, 2] outcome occurs three different ways with the 6 on the first, second, or third die. Each combination needs to be counted in terms of permutations. There are 27 ways to get a ten and only 25 ways to get a nine out of the 216 possibilities, so the ten bet is a slightly bet. In Newton's time, this same problem was brought to his attention and he also provided the correct analysis. This points out the fact that the intelligent men of the day did understand probability. But knowledge is power and gamblers did not want the gambling public to be able to calculate odds. We should point out that just because a bet is fair or in one's favor that does not imply the bet should be made. Actual betting becomes a topic of risk-benefit analysis.

15.3 Pascal, Fermat, Huygens, Bernoulli and Montmort

Probability, as a mathematical topic, got its recognized start with the Pascal-Fermat letters. A gentleman and gambler Chevalier de Mere presented the following problem to Pascal. Throw one die four times. Are you more likely to see a six than to see a pair of sixes in twenty-four throws of a pair of dice? A first glance the probabilities might look the same. The probability of a six is 1/6, so perhaps the probability of seeing at least one six in four throws is 4/6. Similarly, one might suspect the probability of seeing at least a pair of six in twenty-four throws of a pair of dice to be $24/36 = 4/6$. Gamblers knew that the single die game won more often than the two-dice game. The 4/6 result cannot be correct because after seven throws the probability would not be 7/6, that number is the expected number of sixes, not the probability of seeing a six. The probability of not getting a six in four throws is $(5/6)^4 \approx 0.4823$ and 1–0.4823 ≈ 0.5177 is the probability of getting at least one six. In a similar manner, the probability of getting at least one pair of sixes in twenty-four throws of a pair of dice is $1 - (35/36)^{24} \approx 0.4914$. Equal expectations (averages) do not imply equal probabilities and the single-die game is a better bet. Pascal, in 1654, arrived at the correct answer, but wrote to Fermat to get his opinion.

Fermat agreed with Pascal's analysis and then suggested some problems of his own and the correspondence took off. As one example, they consider a game with three players A, B, and C. Any player can win in each round of play. Suppose A needs one more win, and B and C each needs two more wins and the players need to quit the game. How should they divide the ante? At most three more rounds are needed with $3^3 = 27$ possible outcomes. The winning sequences for B are: BBB, BBA, BBC, ABB, and CBB. There will also be five winning sequences for C, leaving seventeen winning sequences for A, so the ante should be divided as $A : B : C = 17 : 5 : 5$.

It might seem strange to consider the sequence BBB as B would have already won with the sequence BB. The need for the BBB sequence can be made clear with a simpler example. What is the probability of getting a head (H) in two throws of a coin. The four possible outcomes are HH, HT, TH, TT each with probability 1/4. Thus, the probability to get a head is 3/4. Note that the sequence HH is counted, even though the H on the first try satisfies the condition of getting an H. One might be tempted to only consider as equally likely outcomes, H, TH, and TT each with probability 1/3 and thus conclude the probability for seeing a head would be 2/3.

D'Alembert proposed this answer, and obviously he was not an experimentalist since a few rounds of coin flipping would provide results that converge to the correct answer. Try it your self.

Pascal and Fermat eventually lost interest, but their letters made the rounds of intellectual circles and influenced Christian Huygens to write a treatise on probability "De Ratiociniis in Aleae Ludo" in 1657. Huygens' (in Dutch the "g" is pronounced more like an "h") [8] book included a number of problems for choosing from a set with or without replacement. One of his questions had the solution 24414065/282429536481 showing the preference for vulgar fractions to decimal approximations.

Huygens' book was reprinted and discussed as the first part of Jacob Bernoulli's masterpiece "Ars Conjectandi." There were three other parts addressing combinations and permutations, games of chance, and applications to social phenomena. The long awaited "Ars Conjectandi" was published in 1713 eight years after Bernoulli's death in 1705. Among many topics, Bernoulli introduced the Bernoulli process

$$P_m(n) = [n!/(n-m)!m!]\, p^m(1-p)^{n-m} \tag{15.1}$$

for the probability $P_m(n)$ of m successes in n trials where the probability of success is p. For a fair game between two players with R coins between them and each wagering a coin on each trial, Bernoulli calculated that the mean number N of trials before one player is out of coins is proportional to R^2. This is a first passage time problem and most likely the *first* first passage time calculation. In a mathematical notation, $<N(R)>\sim R^2$. The tally of the coins is moving as a random walk between the players and as Einstein calculated in 1905 the mean square displacement of a random walker after N steps is $<R^2(N)>\sim N$, the equivalent of Bernoulli's result with the averaging over distance rather than number of trials.

Jacob Bernoulli's book, published posthumously, was edited by his nephew Nickolas who excused the long eight-year delay to his youth, inexperience and travels. Nickolas' travels brought him to spend time with the French mathematician Pierre Montmort whose own book on probability "Essai d'analyse sur les jeux de hazard" published in 1708 benefited from their discussions. One problem, in the second edition, ascribed to Nickolas, has become known as the St. Petersburg Paradox because Daniel Bernoulli latter wrote about it in the Commentaries of the St. Petersburg Academy of Sciences. The game is simple. Flip a coin until a head appears. If this occurs on the first flip, win one coin and this has probability of 1/2. If it occurs on the second flip (sequence TH), win two coins and this has a probability of 1/4. If the first head only appears on the Nth flip, then win 2^{N-1} coins with probability 2^{-N}. The expected winnings are $1 \times (1/2) + 2 \times (1/4) + 4 \times (1/8) + \ldots$ which is summing 1/2 an infinite number of times. The paradox arises when trying to determine the fair ante. The banker wants the player to ante an infinite number of coins because this is his expected loss. The player only wins one coin on the average half of the time, two or less coins 3/4 of the time, etc., and would need to flip a coin an infinite number of times to win the infinite number of coins. The problem is that one is trying to determine the characteristic size of the winnings when no characteristic size exists. This is an example of scaling; i.e., no scale size is characteristic. In this example, where probability theory encountered an infinite number of outcomes, the result was considered paradoxical. This casts a pall treating probability theory as a legitimate mathematics beyond its use of combinatorics when applied to discrete finite systems.

15.4 English, French, and Russian Schools

Abraham de Moivre, a contemporary and correspondent with Jacob Bernoulli and friend of Isaac Newton, produced the next great work in probability, "The Doctrine of Chances". A Huguenot refugee from France, at the age of eighteen, De Moivre settled in London. It is truly amazing how an eighteen-year-old refugee without an academic degree could become a world leading mathematician. Despite becoming Newton's friend and corresponding as an equal with European mathematicians included the Bernoullis and Montmort he was never able to gain an academic position in England or elsewhere. He earned a living by tutoring and writing books. De Moivre set himself up at nights in Slaughters coffee house where he calculated odds for gamblers. The coffee house, in St. Martin's Lane, London was actually an intellectual center and was frequented by patrons who founded the Royal Academy of Sciences. His "Annuities Upon Lives" was the first book on actuarial science. His "Doctrine of Chances" came out in three additions, the last in 1756. In addition to providing the odds for various dice and card games, De Moivre showed that the Bernoulli process (a type of random walk) in the many trials continuum limit approached the Gaussian distribution for the distribution of l winnings in n trials when n is large. He wrote, *"I conclude that if m or n/2 be a quantity infinitely great, then the hyperbolic logarithm of the ratio, which is a term distant from the middle by an interval l, has to the middle term, is −2 l l/n"* Of course, this was long before Gauss, so he did not call it the Gaussian, and DeMoivre's result was only for the Bernoulli process. Today, the result would be written as $P_l(n) = \exp-(2l^2/n)$.

For the Bernoulli probabilities, De Moivre needed to expand log (n!) for large n as

$$\log(n-1)! = \left(n - \frac{1}{2}\right)\log n - n + \log B + \sum_{r=1}^{\infty} \frac{(-1)^{r-1} B_r}{2r(2r-1)n^{2r-1}}$$

$$B = 1 - \frac{1}{12} + \frac{1}{360} - \cdots$$

He needed some help from his friend James Stirling to determine that the constant term B in the expansion was $\sqrt{2\pi}$, and today this is called the Sterling approximation. The B_r are the Bernoulli numbers. Years later, Poisson had a different approach to the Bernoulli process. He let the number of trials, n, go to infinity, but the probability of success, p, go to zero, in such a manner that np was a constant denoted by λ. His analysis found

$$\begin{aligned} P_m(n) &= \text{probability of m successes in n trials} \\ &= \frac{n!}{(n-m)!\,m!} p^m (1-p)^{n-m} = \frac{n(n-1)\cdots(n-m+1)}{m!} p^m (1-p)^{n-m} \\ &\approx \frac{n^m p^m}{m!} e^{-np} \\ &= \frac{\lambda^m}{m!} e^{-\lambda} \end{aligned} \tag{15.2}$$

which of course today is known as the Poisson distribution. The derivation was for rare events (small probability of success p), but once we go to the limit of np $= \lambda$ the value of λ need not be small. The Poisson distribution can describe radioactive decay where λ can be quite large. Poisson's work in probability was not well received, perhaps because it appeared in his book on social applications, "Recherches sur la probabilité des jugements en matière criminelle et en matière civile" i.e. application of probability to criminal and civil matters. The first application of Poisson's law was an 1894 study, of twenty years of data, of cavalry riders who were kicked to death by their horses, a rare event.

A memorable use of Possion's law was William Gossett's application to the brewing of beer. Gossett worked for Guinness and tested whether the correct amount of yeast was added to the brewing process. Too little and the beer was flat and too much and the beer was bitter. Counting yeast cells in drops of beer produced variations from sample to sample. Gossett calculated the Poisson λ from his data. He did not want to publish his work as an employee of Guinness where any information about their processes was probably a trade secret. Gossett publish this and many other works under the pseudonym "Student." His many works including the Student distribution are well known in statistics.

Returning to De Moivre, he also introduced generating functions to solve probability difference equations. De Moivre's well-known equation $(\cos\theta + i\sin\theta)^n = \cos n\theta + i\sin n\theta$ arose from such considerations. Somewhat later the generating function theme, in the hands of Laplace, became the Laplace transform. While Todhunter[2] devoted the bulk of his review of probability to the works of Laplace, we will just mention that Laplace's book, "Theorie analytique des probabilities" captured and improved upon all that went before him. First published in 1815, it was the masterpiece of probability mathematics in the 1800s. He introduced the integral into probability with problems with an infinite number of outcomes (usually in terms of selecting from urns with fixed ratios of black and white tickets). Laplace's analysis led to the calculation of integrals many of which now appear as standards in calculus books, e.g.

$$\int_0^\infty \frac{\cos(ax)}{1+x^2} dx = \frac{\pi}{2} e^{-a}$$

and

$$\int_{-\infty}^{\infty} \exp(-x^2)dx = \sqrt{\pi}$$

For approximations of integrals, Laplace introduced an early version of steepest descent analysis.

While Poisson's book had little impact in France, a student in Russia found a copy and did his thesis on probability. His name was Pafnuty Chebyshev. He is most famous in probability for an inequality and for two of his students, Lyapunov and Markov. They are best known for Lyapunov exponents and Markov chains. Chebyshev's inequality is remarkably simple to derive

$$\begin{aligned} P(|X| \geq t) &= \int_{|x|\geq t} p(x)dx \\ &\leq \int_{|x|\geq t} \frac{x^2}{t^2} p(x)dx \\ &= \frac{1}{t^2} \int_{|x|\geq t} x^2 p(x)dx \\ &\leq \frac{1}{t^2} \langle x^2 \rangle \end{aligned} \tag{15.3}$$

Probability fell somewhat out of favor in France and waited until J. Bertrand's book "Calcul de probabilites" (1879) and H. Poincare book "Calcul de probabilites" (1896). Over in Russia, Markov wrote a book, Ischislenie Veroyatnostei (1900) whose English translation is Calculus of Probabilites. This was followed by Paul Levy's "Calcul de probabilites" (1925). So there was not much originality in the names of probability books in this era.

15.5 Bayes' Inverse Probability

Thomas Bayes, an English minister, derived a formula for inverse probability, which was published posthumously in his book "Essay Towards Solving a Problem in the Doctrine of Chances" in 1764. It appears that publishing was slow and life spans were short as a number of works in the 1700s were posthumous, including DeMoive's 3rd edition of the Doctrine of Chances.

Consider two circles A and B that overlap. The probability that a randomly chosen point is in the overlap is P(A|B)P(B). This formula for being in the intersection is the conditional probability P(A|B) of being in A knowing that you are in B, times the probability of being in B. One could also start with P(B|A) and multiply by P(A) and equate the two expressions and arrive at the equation

$$P(A|B) = P(B|A)\frac{P(A)}{P(B)}. \tag{15.4}$$

This equation can be generalized to many sets or events. One application is to a discussion in the Talmud. There are three draws in a dresser. One contains two gold coins, one two silver coins, and one a gold and a silver coin. Without looking, a drawer and a coin from that draw are chosen at random and the coin is a gold one. What is the probability that the other coin in that drawer is also gold. One could argue correctly that having a gold coin means that the drawer with two gold coins or the drawer with one gold and one silver coin was chosen. With this knowledge, one might surmise equal probabilities of 1/2 for having selected either of these two drawers. An experiment would show that this is the wrong answer, but the discussion was only theoretical. Bayes' theorem will give the correct answer. Let us label the draws GG, GS, SS and they are chosen with equal probability of 1/3, e.g. P(GG) = P(GS) = P(SS) = $1/3$. In Bayes formula

$$\begin{aligned} P(GG|G) &= \frac{P(G|GG)\,P(GG)}{P(G|GG)P(GG) + P(G|GS)P(GS)} \\ &= \frac{1 \bullet 1/3}{1 \bullet 1/3 + 1/2 \bullet 1/3} = 2/3 \end{aligned} \tag{15.5}$$

Selecting gold gave a higher probability of being in GG than being in GS. This is easier to see when the three draws had ten gold coins, ten silver coins, and one drawer with nine silver coins and one gold coin. Selecting a gold coin would most likely be by picking the drawer with all gold coins.

Troubles arise with Bayes theorem when one is attempting to employing it to test hypotheses. Suppose A is the hypothesis that quantum mechanics has hidden variables and B is the result of an experiment. Then one calculates P(A|B) in terms of P(B|A), which is the conditional probability of result B given quantum mechanics has hidden variables. Besides having to know beforehand P(A) and P(B) and needing to knowing how to calculate P(B|A), Bayes theorem can give a positive probability for a false hypothesis being correct.

15.6 Bertrand's Paradox

We next encounter a paradox about uniform probability and measurement. Inscribe an equilateral triangle in a circle and then inscribe a circle inside of the triangle. Pick a chord at random and calculate the probability that it is of length longer than

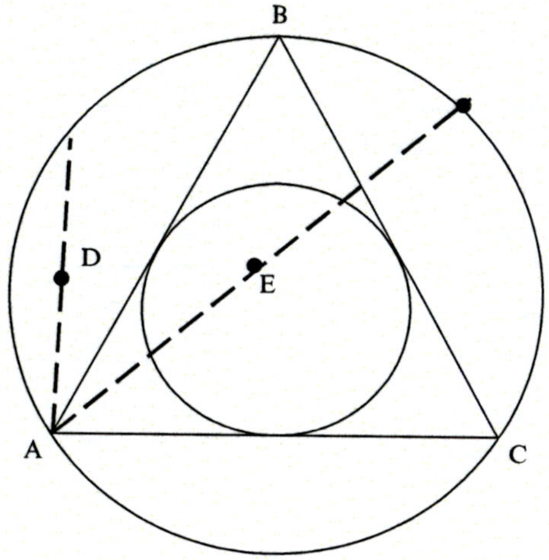

Fig. 15.1 Method 1. Choose a point on the circumference and call it A. Choose the next point randomly, also on the circumference. If that point lies between B and C, which comprises 1/3 of the circumference, then the chord drawn from A to that point will be larger than the side of the inscribed triangle. This method produces a probability of 1/3 to generate a chord longer than the side of the triangle. Method 2. Choosing a point randomly in the large circle and let that be the midpoint of a chord in the large circle. If that point (e.g., point E) lies inside the small circle, the chord will be larger than the side of the triangle, and if that point (e.g., point D) lies outside the small circle, the chord will be shorter than the side of the triangle, The small circle has area $1/4$ the area of the large circle so this method of selecting chords yields a probability of $1/4$ for a randomly drawn chord to be longer than the side of the triangle.

the side of the equilateral triangle. It turns out that this is not enough information to determine the probability. One must also decide what is uniformly distributed in the measurement process (see Fig. 15.1).

Choose one point on the large circle and consider this point to be at the vertex of the triangle, then any point on the arc of 120 degrees between the other two vertices creates a chord longer than the side of the triangle. Choose a second point on the circle and draw the chord connecting the two points. Assuming points on the circumference have an equal probability of being chosen this process gives an answer of probability 1/3.

But a chord can be chosen in a different way. Every chord has a midpoint. Choose an interior point of the circle as a midpoint of the chord. If the midpoint lies within the inscribed circle, then the chord will be longer than the side of the triangle. The area of the inscribed circle is $1/4$ the area of the outer circle. If the midpoints of chords are uniformly distributed, then the answer to the problem is $1/4$. Other answers are possible, all depending on how the points are chosen that will define the chord.

15.7 The Central Limit Theorem and Beyond

Gauss invented the method of least squares fitting to calculate celestial orbits whose observations had data errors. The method fits the errors to a Gaussian function, hence named after its creator. This was a form of the central limit theorem. Robert Adrain, an Irishman, who immigrated to New Jersey, and in 1808 introduced the "Gaussian" for the result of the distribution of errors in surveying measurements. This was a year before Gauss' publication of his method. Adrain's mathematics is somewhat verbal, in the notation of fluxions, and certain assumptions enter. More or less his mathematics seems to be that two consecutive errors in measurement are represented through a convolution integral and the result should be the same as the error distribution of a single error up to a scale factor. He found that $\exp(-x^2)$ provided a solution to this problem. This is equivalent to having a random walk with Gaussian single jump lengths have a Gaussian distribution after any number of jumps. Adrain published in his own journal, "The Analyst," which only survived for one issue, and the work of Gauss' is better remembered. Over the years, these results were refined are now known as the Central Limit Theorem.

This question of scaling was investigated in detail by Paul Levy in Paris. Levy again asked the question of, for summing identically distributed random variables, when does the sum have the same distribution as the single variables? If the random variables have a probability density in the form of $\exp(-x^4)$, their sum will eventually look Gaussian $\exp(-x^2)$ so the probability distribution for the sum will be different than for the single variables. Loosely speaking, the central limit theorem determines that any identically distributed random variables with a probability distribution with a finite second moment will converge to the Gaussian. A sum of random variables with Gaussian distributions produces a sum that is Gaussian. In terms of a random walk after performing N jumps chosen from a Gaussian distribution the position of the random walker will have a Gaussian distribution with N times the variance of the single jumps. Levy was interested if probability distributions, other than the Gaussian, had this property that the sum of these variables had the same distribution as the single variables. Levy called distributions satisfying this property, "stable distributions". The only way around the Gaussian, Levy concluded, was to examine probability distributions with infinite moments. We present a simplified version of the derivation of these Levy distributions.

For an n-step random walk,

$$P_n(x) = \int_{x'} P_{n-m}(x - x')P_m(x')\,dx'$$

$$\tilde{P}_n(k) = \tilde{P}_{n-m}(k)\tilde{P}_m(k)$$

$$\tilde{P}_n(k) = e^{-nD|k|^{\beta}}, \tag{15.6}$$

where the Fourier transform has been applied to both sides of the convolution integral equation and β so far is a free parameter. Levy's analysis provided a more complicated solution, but the result above is a good starting point. An understanding of the necessary restriction on β considers the following random walk [6], with $b^2 > n$,

$$p(x) = \frac{n-1}{2n}\sum_{j=0}^{\infty} n^{-j}\left(\delta_{x,b^j} + \delta_{x,b^{-j}}\right)$$

$$\tilde{p}(k) = \int_{-\infty}^{\infty} e^{ikx} p(x)\,dx = \frac{n-1}{n}\sum_{j=0}^{\infty} n^{-j}\cos\left(b^j k\right)$$

$$\left\langle x^2\right\rangle = \sum_{j=0}^{\infty} x^2 p(x) = \infty \text{ if } b^2 > n \tag{15.7}$$

Expanding the cosine term for small k values, in a Taylor series, would produce an infinite coefficient for the k^2 term. This can be handled in a clever way by representing the cosine in terms of its own Mellin transform,

$$\cos\left(b^j k\right) = \frac{1}{2\pi i}\oint \Gamma(s)\cos\left(\frac{\pi s}{2}\right) b^{-js}\,|k|^{-s}\,ds. \tag{15.8}$$

The sum, in eqn (15.7), becomes

$$\tilde{p}(k) = \frac{1}{2\pi i}\left(\frac{n-1}{n}\right)\sum_{j=0}^{\infty}\oint \Gamma(s)\cos\left(\frac{\pi s}{2}\right) b^{-js}\,|k|^{-s}\,n^{-j}\,ds. \tag{15.9}$$

After switching the contour integral and performing the sum, one has

$$\tilde{p}(k) = \frac{1}{2\pi i}\oint \frac{|k|^{-s}\,\Gamma(s)\cos(\pi s/2)}{1-n^{-1}b^{-s}}\mathrm{ds} \tag{15.10}$$

and this contour integral is dominated by the simple poles at $s = 0$ and $s = -\ln(n)/\ln(b) + 2\pi\, mi/\ln(b)$ (with m an integer) and $s = -2, -4, -6 \ldots$.. The small k expansion of $\tilde{p}(k)$ for the $m = 0$ term (other m terms cause log periodic oscillations) has the Levy form in Fourier space,

$$\lim_{k\to 0}\tilde{p}(k) = 1 - |k|^{\beta} + O\left(k^2\right)$$
$$\approx \exp\left(-|k|^{\beta}\right) \tag{15.11}$$

with $\beta = \ln(n)/\ln(b)$. The expansion holds when $\beta < 2$, otherwise the k^2 term dominates and Gaussian central limit theorem behavior returns as, for small k, $1 - k^2 \sim \exp(-k^2)$ and this transforms back into a Gaussian in real space. The St. Petersberg Paradox is a case of a Levy distribution of $\beta = 1$. Random walks with $\beta < 2$ were called Levy flights by Benoit Mandelbrot. The trajectory is a fractal where the set of points visited is self-similar and has a fractal dimension of β. This is bringing us past historical times so we will stop here.

Acknowledgement The author thanks the Kinnear Professorship at the United States Naval Academy for support.

References

1. Middleton D (1960) Introduction to statistical communications theory. McGraw-Hill, New York
2. Todhunter I (1865) History of the theory of probability. Chelsea, New York
3. David FN (1962) Games, gods, and gambling. Griffin & Co., London
4. Szekely GJ (1986) Paradoxes in probability theory and mathematical statistics. D. Reidel, Dordrecht
5. Rabinovitch NL (1973) Probability and statistical inference in ancient and medieval Jewish literature. University of Toronto Press, Toronto
6. Montroll EW, Shlesinger MF (1984) The wonderful world of random walks. In: JL Lebowitz and EW Montroll (eds), Studies in statistical mechanics, vol 11. North-Holland, Amsterdam, pp 1–121
7. http://en.wikipedia.org/wiki/I_Ching
8. http://frank.harvard.edu/$\sim$paulh/misc/pics/huygens_96.mp3

Chapter 16
Fluctuations in Two Component Interacting Bose–Einstein Condensate

Andrii S. Sizhuk, Anatoly A. Svidzinsky, and Marlan O. Scully

16.1 Introduction

The problem of particle number fluctuations in Bose–Einstein condensate (BEC) in the canonical ensemble is a subject of long-standing interest [1–14]. Recently, such a problem received growing attention due to possibility of measuring fluctuations in experiments with BEC of trapped atoms [15]. In the canonical ensemble, the total number of particles N is fixed, but the system can exchange energy with an external thermal reservoir at temperature T. Due to the constraint on particle number an analytical treatment of fluctuations is highly nontrivial.

For an ideal gas, a master equation analysis in the spirit of the quantum theory of the laser provided an analytical description of condensate fluctuations accurate at all temperatures [7, 8]. For a weakly interacting Bose gas a hybrid phenomenological approach of [13] is successful. Such a hybrid approach combines the master equation analysis [7, 8] and canonical ensemble quasiparticle formalism [9, 10].

Along with fluctuations in one-component BECs, fluctuations in a two-component condensate in which particles of different species interact with each other have also experimental interest. Systems of multicomponent condensates were first realized by the JILA group in a magnetic trap in ^{87}Rb [16, 17] and subsequently in an optical trap in ^{23}Na spinor condensate [18, 19]. Theoretical treatment of such systems began in the context of superfluid helium mixtures [20, 21] and spin-polarized

A.S. Sizhuk (✉) • A.A. Svidzinsky • M.O. Scully
Institute for Quantum Studies and Department of Physics, Texas A&M University, College Station, TX, USA
e-mail: cannabiss@ukr.net; asvid@physics.tamu.edu

M.O. Scully
Applied Physics and Materials Science Group, Engineering Quad, Princeton University, Princeton, NJ, USA
e-mail: scully@tamu.edu

L. Cohen et al. (eds.), *Classical, Semi-classical and Quantum Noise*,
DOI 10.1007/978-1-4419-6624-7_16, © Springer Science+Business Media, LLC 2012

hydrogen [22], and now has been extended to BEC in trapped gases [23–31]. In particular, the quasiparticle excitation spectrum and stability of the two-component BEC of atoms in different hyperfine states was studied using the canonical ensemble quasiparticle formalism in [32, 33].

Under a change of the interaction strength, the two-component BECs can undergo different types of macroscopic quantum phase transitions. For instance, an increase in interspecies interactions a_{12} results in a transition from a binary mixture to a phase-separated state. Such transition is possible to observe, for example, varying a_{12} by means of Feschbach resonances.

Here, we study particle number fluctuations in the ground (condensate) state of a weakly interacting two-component BEC and, in particular, analyze their behavior in the vicinity of the phase separation transition. Our analysis is based on the Bogoliubov theory extended for a two component system. Total numbers of particles of each species N_1 and N_2 are assumed to be fixed. That is, we are working in the canonical ensemble and must take into account the canonical ensemble constraint

$$\hat{N}_{0i} + \hat{N}_i^{\text{out}} = N_i, \quad i = 1, 2, \tag{16.1}$$

where $\hat{N}_{0i}$ and $\hat{N}_i^{\text{out}}$ are the particle number operators for the ground and excited states of the component i.

One can find an accurate expression for fluctuations in the ground state particle number using the following physical argument. For temperatures not too close to BEC transition T_c, one can treat particles in the excited states as being in contact with a large reservoir of condensate particles. Thus, fluctuations of particles in the excited states can be accurately described in the grand canonical ensemble which allows a simple analytical treatment. Then using the constraint (16.1) one can express fluctuations of N_{0i} in terms of fluctuations of the excited state particles and treat the later in the grand canonical picture. For example, (16.1) yields the following relation between central moments for condensate and noncondensate atoms

$$\langle (N_{0i} - \bar{N}_{0i})^m \rangle = (-1)^m \langle (N_i^{\text{out}} - \bar{N}_i^{\text{out}})^m \rangle. \tag{16.2}$$

Equation (16.2) is useful when fluctuations of particles in the excited states are statistically independent, that is, $\langle N_{i,k} N_{i,l} \rangle = \langle N_{i,k} \rangle \langle N_{i,l} \rangle$ for $k \neq l$. Here, $N_{i,k}$ stands for the particle number of the species i in the state k. We will assume this in the present paper. Such assumption becomes invalid near the critical temperature T_c when condensate particles no longer form a large particle reservoir. In this region, more sophisticated treatment (for example, similar to analysis of [8, 13] for one component BEC) is required. Thus, vicinity near T_c is beyond the scope of the present paper. Present analysis for the two component BEC is essentially similar to those of [6] for a one-component system.

The paper is organized as follows. We start from a model Hamiltonian, which describes a weakly interacting two-component Bose gas. Then we present Bogoliubov transformations which diagonalize the Hamiltonian in terms of the quasiparticle operators of each components. Next, we obtain analytical expressions for the average condensate particle number and fluctuations and analyze them in various limits. Finally, we show the results of numerical simulations.

16.1.1 Basic Formalism

Here, we study a weakly interacting two-component Bose gas confined in a cubic box of volume $V = L^3$ in the absence of interactions that inter-convert atoms of different internal states. The system's Hamiltonian in the second-quantized form is given by [32]

$$\hat{H} = \sum_{\mathbf{k}} \sum_{i=1}^{2} \varepsilon_k \hat{a}^+_{i,\mathbf{k}} \hat{a}_{i,\mathbf{k}} + \sum_{\mathbf{k}_1,\mathbf{k}_2,\mathbf{q}} \sum_{i=1}^{2} \frac{\lambda_i}{2V} \hat{a}^+_{i,\mathbf{k}_1+\mathbf{q}} \hat{a}^+_{i,\mathbf{k}_2-\mathbf{q}} \hat{a}_{i,\mathbf{k}_1} \hat{a}_{i,\mathbf{k}_2}$$
$$+ \frac{\lambda_{12}}{V} \sum_{\mathbf{k}_1,\mathbf{k}_2,\mathbf{q}} \hat{a}^+_{1,\mathbf{k}_1+\mathbf{q}} \hat{a}^+_{2,\mathbf{k}_2-\mathbf{q}} \hat{a}_{1,\mathbf{k}_1} \hat{a}_{2,\mathbf{k}_2}, \tag{16.3}$$

where $\varepsilon_k = \hbar^2 k^2/2m$ is the kinetic energy of a particle with momentum $\mathbf{k}$ (for simplicity, we assume that particles of different species have equal masses m), $\hat{a}_{i,\mathbf{k}}$ and $\hat{a}^+_{i,\mathbf{k}}$ are the annihilation and creation operators for particles of the species i ($i = 1,2$), λ_i are interaction parameters which are proportional to the s-wave scattering lengths a_{ii}: $\lambda_i = 4\pi a_{ii}\hbar^2/m$ for like-boson interactions and $\lambda_{12} = 4\pi a_{12}\hbar^2/m$ for unlike-boson interactions between atoms. We assume that the gas is dilute, that is the gas parameters are small $Na_{ii}^3/V \ll 1$, $Na_{12}^3/V \ll 1$, here N is the total number of particles.

Assuming that many atoms are condensed in the ground state, we replace $\hat{a}_{i,0}$ and $\hat{a}^+_{i,0}$ by c-numbers $\hat{a}_{i,0} \to \sqrt{Vn_i}$ and keep only the leading terms in $\sqrt{n_i}$ in the Hamiltonian (16.3). Here, n_i is the condensate density for the component i ($i = 1,2$). Then the Hamiltonian can be written as

$$\hat{H} = \hat{H}_0 + \hat{H}_q, \tag{16.4}$$

where $\hat{H}_q$ is a term quadratic in the particle operators with $\mathbf{k} \neq \mathbf{0}$

$$\hat{H}_q = \sum_{\mathbf{k}} \sum_{i=1}^{2} \varepsilon_k \hat{a}^+_{i,\mathbf{k}} \hat{a}_{i,\mathbf{k}} + \sum_{\mathbf{k}} \sum_{i=1}^{2} \frac{\lambda_i n_i}{2} (\hat{a}^+_{i,\mathbf{k}} + \hat{a}_{i,-\mathbf{k}})(\hat{a}^+_{i,-\mathbf{k}} + \hat{a}_{i,\mathbf{k}})$$
$$+ \lambda_{12}\sqrt{n_1 n_2} \sum_{\mathbf{k}} (\hat{a}^+_{1,\mathbf{k}} + \hat{a}_{1,-\mathbf{k}})(\hat{a}^+_{2,-\mathbf{k}} + \hat{a}_{2,\mathbf{k}}). \tag{16.5}$$

Hamiltonian (16.5) can be diagonalized by making the following Bogoliubov transformation [32]:

$$\hat{a}_{1,\mathbf{k}} = (u_{1,k}\hat{b}_{1,\mathbf{k}} + v_{1,k}\hat{b}^{+}_{1,-\mathbf{k}})\cos\gamma - (u_{2,k}\hat{b}_{2,\mathbf{k}} + v_{2,k}\hat{b}^{+}_{2,-\mathbf{k}})\sin\gamma,$$

$$\hat{a}_{2,\mathbf{k}} = (u_{1,k}\hat{b}_{1,\mathbf{k}} + v_{1,k}\hat{b}^{+}_{1,-\mathbf{k}})\sin\gamma + (u_{2,k}\hat{b}_{2,\mathbf{k}} + v_{2,k}\hat{b}^{+}_{2,-\mathbf{k}})\cos\gamma,$$

$$\hat{a}^{+}_{1,-\mathbf{k}} = (v_{1,k}\hat{b}_{1,\mathbf{k}} + u_{1,k}\hat{b}^{+}_{1,-\mathbf{k}})\cos\gamma - (v_{2,k}\hat{b}_{2,\mathbf{k}} + u_{2,k}\hat{b}^{+}_{2,-\mathbf{k}})\sin\gamma,$$

$$\hat{a}^{+}_{2,-\mathbf{k}} = (v_{1,k}\hat{b}_{1,\mathbf{k}} + u_{1,k}\hat{b}^{+}_{1,-\mathbf{k}})\sin\gamma + (v_{2,k}\hat{b}_{2,\mathbf{k}} + u_{2,k}\hat{b}^{+}_{2,-\mathbf{k}})\cos\gamma, \quad (16.6)$$

where

$$\sin\gamma = \frac{1}{\sqrt{2}}\sqrt{1 - \frac{\lambda_1 n_1 - \lambda_2 n_2}{\sqrt{(\lambda_1 n_1 - \lambda_2 n_2)^2 + 4\lambda_{12}^2 n_1 n_2}}}, \quad (16.7)$$

$$\cos\gamma = \frac{1}{\sqrt{2}}\sqrt{1 + \frac{\lambda_1 n_1 - \lambda_2 n_2}{\sqrt{(\lambda_1 n_1 - \lambda_2 n_2)^2 + 4\lambda_{12}^2 n_1 n_2}}}, \quad (16.8)$$

n_1 and n_2 are condensate densities for each component, respectively, and the Bogoliubov amplitudes $u_{1,k}$, $u_{2,k}$ and $v_{1,k}$, $v_{2,k}$ are expressed in terms of the quasiparticle energy

$$E_{1,k} = \sqrt{\varepsilon_k^2 + \varepsilon_k c_1}, \quad E_{2,k} = \sqrt{\varepsilon_k^2 + \varepsilon_k c_2}, \quad (16.9)$$

as

$$u_{1,k} = \frac{1}{2}\left(\sqrt{\frac{\varepsilon_k}{E_{1,k}}} + \sqrt{\frac{E_{1,k}}{\varepsilon_k}}\right), \quad v_{1,k} = \frac{1}{2}\left(\sqrt{\frac{\varepsilon_k}{E_{1,k}}} - \sqrt{\frac{E_{1,k}}{\varepsilon_k}}\right), \quad (16.10)$$

$$u_{2,k} = \frac{1}{2}\left(\sqrt{\frac{\varepsilon_k}{E_{2,k}}} + \sqrt{\frac{E_{2,k}}{\varepsilon_k}}\right), \quad v_{2,k} = \frac{1}{2}\left(\sqrt{\frac{\varepsilon_k}{E_{2,k}}} - \sqrt{\frac{E_{2,k}}{\varepsilon_k}}\right), \quad (16.11)$$

where

$$c_1 = \lambda_1 n_1 + \lambda_2 n_2 + \sqrt{(\lambda_1 n_1 - \lambda_2 n_2)^2 + 4\lambda_{12}^2 n_1 n_2}, \quad (16.12)$$

$$c_2 = \lambda_1 n_1 + \lambda_2 n_2 - \sqrt{(\lambda_1 n_1 - \lambda_2 n_2)^2 + 4\lambda_{12}^2 n_1 n_2} \quad (16.13)$$

are characteristic interaction energies.

The transformation (16.6) is canonical and the Bogoliubov quasiparticle operators $\hat{b}_{\mathbf{k}}$ and $\hat{b}^{\dagger}_{\mathbf{k}}$ also obey bosonic commutation relations $[\hat{b}_{i,\mathbf{k}}, \hat{b}^{+}_{i,\mathbf{k}'}] = \delta_{\mathbf{k},\mathbf{k}'}$ and $[\hat{b}_{i,\mathbf{k}}, \hat{b}_{i,\mathbf{k}'}] = [\hat{b}^{+}_{i,\mathbf{k}}, \hat{b}^{+}_{i,\mathbf{k}'}] = 0$. After the Bogoliubov transformation, one can write the $\hat{H}_q$ term as

$$\hat{H}_q = \sum_{\mathbf{k}} [E_{1,k}\, \hat{b}^{+}_{1,\mathbf{k}}\hat{b}_{1,\mathbf{k}} + E_{2,k}\, \hat{b}^{+}_{2,\mathbf{k}}\hat{b}_{2,\mathbf{k}}], \quad (16.14)$$

which describes ideal gas of the Bogoliubov quasiparticles. In thermal equilibrium

$$f_{1,k} \equiv \left\langle \hat{b}^+_{1,\mathbf{k}} \hat{b}_{1,\mathbf{k}} \right\rangle = \frac{1}{\exp(E_{1,k}/k_B T) - 1}, \tag{16.15}$$

$$f_{2,k} \equiv \left\langle \hat{b}^+_{2,\mathbf{k}} \hat{b}_{2,\mathbf{k}} \right\rangle = \frac{1}{\exp(E_{2,k}/k_B T) - 1}. \tag{16.16}$$

16.2 Average Condensate Density

As we mentioned in the Introduction, the grand canonical description is accurate for particles in the excited levels. Using the particle number constraint one can express the condensate particle operator for the component i, $\hat{N}_{0i}$, in terms of operators for noncondensate particles

$$\hat{N}_{0i} = N_i - \sum_{\mathbf{k}\neq 0} \hat{a}^+_{i,\mathbf{k}} \hat{a}_{i,\mathbf{k}}, \tag{16.17}$$

where N_i is the total number of particles of the species i which is assumed to be fixed. Using (16.17), we obtain for the average condensate particle number of component i the following expression:

$$\langle \hat{N}_{0i} \rangle = N_i - \sum_{\mathbf{k}\neq 0} \left\langle \hat{a}^+_{i,\mathbf{k}} \hat{a}_{i,\mathbf{k}} \right\rangle. \tag{16.18}$$

Bogoliubov transformation (16.6), and (16.15) and (16.16) yield

$$\left\langle \hat{a}^+_{1,\mathbf{k}} \hat{a}_{1,\mathbf{k}} \right\rangle = \left[(v^2_{1,k} + u^2_{1,k}) f_{1,k} + v^2_{1,k} \right] \cos^2\gamma + \left[(v^2_{2,k} + u^2_{2,k}) f_{2,k} + v^2_{2,k} \right] \sin^2\gamma, \tag{16.19}$$

$$\left\langle \hat{a}^+_{2,\mathbf{k}} \hat{a}_{2,\mathbf{k}} \right\rangle = \left[(v^2_{1,k} + u^2_{1,k}) f_{1,k} + v^2_{1,k} \right] \sin^2\gamma + \left[(v^2_{2,k} + u^2_{2,k}) f_{2,k} + v^2_{2,k} \right] \cos^2\gamma. \tag{16.20}$$

Bogoliubov amplitudes and parameter γ depend on $\langle \hat{N}_{0i} \rangle$ and thus (16.18), (16.19), and (16.20) must be solved self-consistently.

At $T = 0$, the factors $f_{1,k}$ and $f_{2,k}$ are equal to zero and replacing summation over $\mathbf{k}$ by integration we obtain the following expressions for the average condensate density of each component:

$$\begin{aligned} n_1 &= \frac{N_1}{V} - \frac{1}{V} \sum_{\mathbf{k}\neq 0} \left(v^2_{1,k} \cos^2\gamma + v^2_{2,k} \sin^2\gamma \right) \\ &= \frac{N_1}{V} - \frac{m^{3/2}}{6\sqrt{2}\pi^2\hbar^3} \left(c_1^{3/2} \cos^2\gamma + c_2^{3/2} \sin^2\gamma \right), \end{aligned} \tag{16.21}$$

$$n_2 = \frac{N_2}{V} - \frac{1}{V}\sum_{\mathbf{k}\neq 0}\left(v_{2,k}^2\cos^2\gamma + v_{1,k}^2\sin^2\gamma\right)$$

$$= \frac{N_2}{V} - \frac{m^{3/2}}{6\sqrt{2}\pi^2\hbar^3}\left(c_2^{3/2}\cos^2\gamma + c_1^{3/2}\sin^2\gamma\right). \tag{16.22}$$

The total condensate density at $T = 0$ is given by

$$n_1 + n_2 = \frac{N}{V} - \frac{m^{3/2}}{6\sqrt{2}\pi^2\hbar^3}\left(c_1^{3/2} + c_2^{3/2}\right). \tag{16.23}$$

For weakly interacting gas, depletion of the condensate at $T = 0$ is small. Therefore, in formulas for c_1, c_2, and γ one can take $n_1 \approx N_1/V$ and $n_2 \approx N_2/V$. Then (16.21), (16.22), and (16.23) give a simple analytical answer for n_1 and n_2 for a two-component BEC at zero temperature.

16.3 Fluctuations

For a two-component BEC fluctuations in the condensate particle number can be calculated in a way similar to those used in [6] for a one component system. Namely, we use (16.17) and express the square variance $\Delta N_{0i}^2 \equiv \langle(\hat{N}_{0i} - \langle\hat{N}_{0i}\rangle)^2\rangle = \langle\hat{N}_{0i}^2\rangle - \langle\hat{N}_{0i}\rangle^2$ in terms of the noncondensate particle operators which we treat in the grand canonical ensemble

$$\Delta N_{0i}^2 = \sum_{\mathbf{k},\mathbf{k}'\neq 0}\left\langle\hat{N}_{i,\mathbf{k}}\hat{N}_{i,\mathbf{k}'}\right\rangle - \left(\sum_{\mathbf{k}\neq 0}\langle\hat{N}_{i,\mathbf{k}}\rangle\right)^2, \tag{16.24}$$

where $\hat{N}_{i,\mathbf{k}} \equiv \hat{a}_{i,\mathbf{k}}^{+}\hat{a}_{i,\mathbf{k}}$. We assume that fluctuations of particles with different $\mathbf{k}$ are statistically independent, that is $\langle\hat{N}_{i,\mathbf{k}}\hat{N}_{i,\mathbf{k}'}\rangle = \langle\hat{N}_{i,\mathbf{k}}\rangle\langle\hat{N}_{i,\mathbf{k}'}\rangle$ for $\mathbf{k} \neq \mathbf{k}'$. Such assumption is valid when there are many particles in the condensate and fails near the BEC transition temperature. Then we use (16.6) and express the particle operators $\hat{a}_{i,\mathbf{k}}$ via operators of the Bogoliubov quasiparticles $\hat{b}_{i,\mathbf{k}}$. Average of a pair of quasiparticle operators is given by (16.15) and (16.16). For $\mathbf{k}' = \mathbf{k}$ (16.24) contains averages of four quasiparticle operators. Such averages can be calculated using Wick's theorem.

Finally, we obtain the following expressions for square variance of condensate particle number of each components:

$$\Delta N_{01}^2 = \sum_{\mathbf{k}\neq 0}\Big\{\left[2u_{1,k}^2v_{1,k}^2 + f_{1,k}(1+f_{1,k})\left\{6u_{1,k}^2v_{1,k}^2 + v_{1,k}^4 + u_{1,k}^4\right\}\right]\cos^4\gamma$$

$$+\left[2u_{2,k}^2v_{2,k}^2 + f_{2,k}(1+f_{2,k})\left\{6u_{2,k}^2v_{2,k}^2 + v_{2,k}^4 + u_{2,k}^4\right\}\right]\sin^4\gamma$$

$$
+ \left[\{f_{2,k}(f_{1,k}+1) + f_{1,k}(f_{2,k}+1)\} \left\{(u_{2,k}v_{1,k}+u_{1,k}v_{2,k})^2 + (v_{1,k}v_{2,k}+u_{1,k}u_{2,k})^2\right\} + (u_{2,k}v_{1,k}+u_{1,k}v_{2,k})^2 \right] \sin^2\gamma\cos^2\gamma \Big\}, \tag{16.25}
$$

$$
\begin{aligned}
\Delta N_{02}^2 = \sum_{\mathbf{k}\neq 0} \Big\{ & \left[2u_{2,k}^2v_{2,k}^2 + f_{2,k}(1+f_{2,k})\left\{6u_{2,k}^2v_{2,k}^2 + v_{2,k}^4 + u_{2,k}^4\right\}\right]\cos^4\gamma \\
& + \left[2u_{1,k}^2v_{1,k}^2 + f_{1,k}(1+f_{1,k})\left\{6u_{1,k}^2v_{1,k}^2 + v_{1,k}^4 + u_{1,k}^4\right\}\right]\sin^4\gamma \\
& + \left[\{f_{2,k}(f_{1,k}+1) + f_{1,k}(f_{2,k}+1)\} \left\{(u_{2,k}v_{1,k}+u_{1,k}v_{2,k})^2 \right.\right. \\
& \left.\left. + (v_{1,k}v_{2,k}+u_{1,k}u_{2,k})^2\right\} + (u_{2,k}v_{1,k}+u_{1,k}v_{2,k})^2 \right] \sin^2\gamma\cos^2\gamma \Big\}.
\end{aligned} \tag{16.26}
$$

Fluctuations in the total condensate particle number $N_0 = N_{01} + N_{02}$ are given by equation:

$$
\begin{aligned}
\Delta N_0^2 &\equiv \langle (\hat{N}_{01} + \hat{N}_{02} - \langle \hat{N}_{01} + \hat{N}_{02}\rangle)^2\rangle \\
&= \Delta N_{01}^2 + \Delta N_{02}^2 + 2\sum_{\mathbf{k},\mathbf{k}'\neq 0} \Big(\langle \hat{N}_{1,\mathbf{k}}\hat{N}_{2,\mathbf{k}'}\rangle - \langle \hat{N}_{1,\mathbf{k}}\rangle\langle \hat{N}_{2,\mathbf{k}'}\rangle \Big),
\end{aligned} \tag{16.27}
$$

which yields

$$
\begin{aligned}
\Delta N_0^2 = \sum_{\mathbf{k}\neq 0} \Big\{ & 2u_{1,k}^2v_{1,k}^2 + f_{1,k}(1+f_{1,k})\left[6u_{1,k}^2v_{1,k}^2 + v_{1,k}^4 + u_{1,k}^4\right] \\
& + 2u_{2,k}^2v_{2,k}^2 + f_{2,k}(1+f_{2,k})\left[6u_{2,k}^2v_{2,k}^2 + v_{2,k}^4 + u_{2,k}^4\right] \Big\}.
\end{aligned} \tag{16.28}
$$

Equations (16.25), (16.26), and (16.28) generalize the result of [6] for a two-component BEC. Next, we consider several limiting cases.

16.3.1 Zero Temperature

At $T = 0$ replacing summation over $\mathbf{k}$ by integration we find the following expressions for square variances:

$$
\begin{aligned}
\Delta N_{01}^2 = \frac{1}{4\pi}\left(\sqrt{\frac{m}{2}\frac{L}{\hbar}}\right)^3 \Big[& c_1^{3/2}\cos^4\gamma + c_2^{3/2}\sin^4\gamma + \frac{4}{3\pi}\sqrt{c_1}\Big\{c_2\, F\left(\frac{\pi}{2},\kappa\right) \\
& + (c_1+c_2)\, E\left(\frac{\pi}{2},\kappa\right)\Big\}\sin^2\gamma\cos^2\gamma\Big],
\end{aligned} \tag{16.29}
$$

$$\Delta N_{02}^2 = \frac{1}{4\pi}\left(\sqrt{\frac{m}{2}\frac{L}{\hbar}}\right)^3 \left[c_2^{3/2}\cos^4\gamma + c_1^{3/2}\sin^4\gamma + \frac{4}{3\pi}\sqrt{c_1}\left\{c_2 F\left(\frac{\pi}{2},\kappa\right)\right.\right.$$

$$\left.\left. + (c_1 + c_2)E\left(\frac{\pi}{2},\kappa\right)\right\}\sin^2\gamma\cos^2\gamma\right], \qquad (16.30)$$

where $F\left(\frac{\pi}{2},\kappa\right)$ and $E\left(\frac{\pi}{2},\kappa\right)$ are full elliptical integrals of type 1 and 2, respectively

$$F\left(\frac{\pi}{2},\kappa\right) = \int_0^{\pi/2} \frac{d\varphi}{\sqrt{1-\kappa^2\sin^2\varphi}}, \quad E\left(\frac{\pi}{2},\kappa\right) = \int_0^{\pi/2} \sqrt{1-\kappa^2\sin^2\varphi}\, d\varphi \quad (16.31)$$

and $\kappa = \sqrt{1 - c_2/c_1}$.

For the square variance of the total condensate particle number at $T = 0$, we find

$$\Delta N_0^2 = \frac{1}{4\pi}\left(\sqrt{\frac{m}{2}\frac{L}{\hbar}}\right)^3 \left(c_1^{3/2} + c_2^{3/2}\right). \qquad (16.32)$$

Thus, the square variance of the total condensate particle number is proportional to the total number of noncondensate particles at $T = 0$: $\Delta N_0^2 \propto N - N_{01} - N_{02}$.

16.3.2 Thomas–Fermi Limit

Now we consider the Thomas–Fermi regime in which the interaction energy is much larger than the kinetic energy of an atom in the condensate $\lambda_i n_i \gg \varepsilon_1$, $i = 1, 2$ and $\lambda_{12}\sqrt{n_1 n_2} \gg \varepsilon_1$, where $\varepsilon_1 = (2\pi\hbar/L)^2(1/2m)$ is the kinetic energy of a single atom in the box of size L in the ground state. These conditions can be written as

$$c_i \gg \varepsilon_1, \quad i = 1, 2. \qquad (16.33)$$

In the terms of E_i $(i = 1, 2)$ (16.25) and (16.26) have the form

$$\Delta N_{01}^2 = \sum_{\mathbf{k}\neq 0}\left\{\frac{1}{8}\left[\left(\frac{\varepsilon_k^2}{E_{1,k}^2} + \frac{E_{1,k}^2}{\varepsilon_k^2} - 2\right)\cos^4\gamma + \left(\frac{\varepsilon_k^2}{E_{2,k}^2} + \frac{E_{2,k}^2}{\varepsilon_k^2} - 2\right)\sin^4\gamma\right.\right.$$

$$\left. +2\left(\frac{\varepsilon_k^2}{E_{1,k}E_{2,k}} + \frac{E_{1,k}E_{2,k}}{\varepsilon_k^2} - 2\right)\cos^2\gamma\sin^2\gamma\right]$$

$$+\frac{1}{2}f_{1,k}(1 + f_{1,k})\left(\frac{\varepsilon_k^2}{E_{1,k}^2} + \frac{E_{1,k}^2}{\varepsilon_k^2}\right)\cos^4\gamma$$

$$+\frac{1}{2}f_{2,k}(1+f_{2,k})\left(\frac{\varepsilon_k^2}{E_{2,k}^2}+\frac{E_{2,k}^2}{\varepsilon_k^2}\right)\sin^4\gamma$$
$$+\frac{1}{2}\{f_{2,k}(f_{1,k}+1)+f_{1,k}(f_{2,k}+1)\}\left(\frac{\varepsilon_k^2}{E_{1,k}E_{2,k}}+\frac{E_{1,k}E_{2,k}}{\varepsilon_k^2}\right)$$
$$\times\sin^2\gamma\cos^2\gamma\Bigg\}, \tag{16.34}$$

$$\Delta N_{02}^2=\sum_{\mathbf{k}\neq 0}\Bigg\{\frac{1}{8}\Bigg[\left(\frac{\varepsilon_k^2}{E_{1,k}^2}+\frac{E_{1,k}^2}{\varepsilon_k^2}-2\right)\sin^4\gamma+\left(\frac{\varepsilon_k^2}{E_{2,k}^2}+\frac{E_{2,k}^2}{\varepsilon_k^2}-2\right)\cos^4\gamma$$
$$+2\left(\frac{\varepsilon_k^2}{E_{1,k}E_{2,k}}+\frac{E_{1,k}E_{2,k}}{\varepsilon_k^2}-2\right)\cos^2\gamma\sin^2\gamma\Bigg]$$
$$+\frac{1}{2}f_{1,k}(1+f_{1,k})\left(\frac{\varepsilon_k^2}{E_{1,k}^2}+\frac{E_{1,k}^2}{\varepsilon_k^2}\right)\sin^4\gamma$$
$$+\frac{1}{2}f_{2,k}(1+f_{2,k})\left(\frac{\varepsilon_k^2}{E_{2,k}^2}+\frac{E_{2,k}^2}{\varepsilon_k^2}\right)\cos^4\gamma$$
$$+\frac{1}{2}\{f_{2,k}(f_{1,k}+1)+f_{1,k}(f_{2,k}+1)\}\left(\frac{\varepsilon_k^2}{E_{1,k}E_{2,k}}+\frac{E_{1,k}E_{2,k}}{\varepsilon_k^2}\right)$$
$$\times\sin^2\gamma\cos^2\gamma\Bigg\}, \tag{16.35}$$

where $E_{i,k}=\varepsilon_1\sqrt{\mathbf{I}^4+(c_i/\varepsilon_1)\mathbf{I}^2}$, $i=1,2$, $\mathbf{I}=(I_1,\ I_2,\ I_3)$, and $\varepsilon_k=\mathbf{I}^2\varepsilon_1$.

In the Thomas–Fermi limit $E_{i,k}\approx\sqrt{\varepsilon_k c_i}=\sqrt{\mathbf{I}^2\varepsilon_1 c_i}$. Because terms with $E_{i,k}<k_\mathrm{B}T$ contribute to the variance we keep such terms only. As a result, in the Thomas–Fermi regime at temperature $k_\mathrm{B}T\gg c_i$, $i=1,2$, we obtain the following answer for fluctuations in the condensate particle numbers for each component:

$$\Delta N_{01}^2\approx\Delta N_{02}^2\approx\frac{1}{2}\frac{(k_\mathrm{B}T)^2}{\varepsilon_1^2}s_4=\frac{N_i^{4/3}\,s_4}{2\pi^2\,(\zeta(3/2))^{4/3}}\left(\frac{T}{T_{ic}}\right)^2, \tag{16.36}$$

where

$$s_4=\sum_{\mathbf{I}\neq 0}\frac{1}{\mathbf{I}^4}=16.53,\quad \mathbf{I}=(I_1,I_2,I_3), \tag{16.37}$$

T_{ic} and N_i are the critical temperature and the total number of particles of the component $i = 1, 2$. Equation (16.36) shows that under such conditions the fluctuations for each components behave as if the components are independent. Formula (16.36) does not depend on interaction between particles; however, the square variance given by (16.36) is smaller by a factor of two than those for an ideal gas [9, 10].

16.3.3 Fluctuations in the Vicinity of Phase Separation Transition

Under certain conditions, the two-component BEC undergoes a phase transition from a uniform mixture of components into a phase when components become spatially separated. At zero temperature and large condensate volume, the transition occurs at the point $\lambda_1\lambda_2 = \lambda_{12}^2$. If $\lambda_1\lambda_2 > \lambda_{12}^2$ the system is in a mixed state, while for $\lambda_1\lambda_2 < \lambda_{12}^2$ the two components are separated. Here, we consider fluctuations in the mixed state near the phase separation transition $\lambda_1\lambda_2 \to \lambda_{12}$ at $T = 0$ and apply (16.29) and (16.30).

Near the phase separation transition (16.12) and (16.13) yield

$$c_1 \approx 2(\lambda_1 n_1 + \lambda_2 n_2), \tag{16.38}$$

$$c_2 \approx 2n_1 n_2 \frac{\lambda_1\lambda_2 - \lambda_{12}^2}{\lambda_1 n_1 + \lambda_2 n_2} \ll c_1, \tag{16.39}$$

and therefore we can approximate

$$F\left(\frac{\pi}{2}, \sqrt{1 - \frac{c_2}{c_1}}\right) \approx -\frac{1}{2}\ln\left(\frac{1}{2}\frac{c_2}{c_1}\right). \tag{16.40}$$

As a result, we obtain the following expressions for fluctuations:

$$\Delta N_{01}^2 \approx \frac{1}{4\pi}\left(\frac{\sqrt{m}L}{\hbar}\right)^3 \frac{(\lambda_1 n_1)^2}{\sqrt{\lambda_1 n_1 + \lambda_2 n_2}} \left[1 + \frac{4}{3\pi} \times \left(1 - \frac{1}{2}\frac{(\lambda_1\lambda_2 - \lambda_{12}^2)n_1 n_2}{(\lambda_1 n_1 + \lambda_2 n_2)^2} \ln\left\{\frac{1}{2}\frac{(\lambda_1\lambda_2 - \lambda_{12}^2)n_1 n_2}{(\lambda_1 n_1 + \lambda_2 n_2)^2}\right\}\right)\frac{\lambda_2 n_2}{\lambda_1 n_1}\right], \tag{16.41}$$

$$\Delta N_{02}^2 \approx \frac{1}{4\pi}\left(\frac{\sqrt{m}L}{\hbar}\right)^3 \frac{(\lambda_2 n_2)^2}{\sqrt{\lambda_1 n_1 + \lambda_2 n_2}} \left[1 + \frac{4}{3\pi} \times \left(1 - \frac{1}{2}\frac{(\lambda_1\lambda_2 - \lambda_{12}^2)n_1 n_2}{(\lambda_1 n_1 + \lambda_2 n_2)^2} \ln\left\{\frac{1}{2}\frac{(\lambda_1\lambda_2 - \lambda_{12}^2)n_1 n_2}{(\lambda_1 n_1 + \lambda_2 n_2)^2}\right\}\right)\frac{\lambda_1 n_1}{\lambda_2 n_2}\right]. \tag{16.42}$$

Equations (16.41) and (16.42) show that at the point of phase separation transition the variance of the particle number has a cusp (infinite derivative).

16.4 Numerical Results

We solved numerically (16.18), (16.19), and (16.20) for the number of condensate particle for each species. In simulations, we assume that particles are trapped in a box with volume V. Equations are solved self-consistently. In calculations we fix the total numbers of particles for each components $N_1 = 300$ and $N_2 = 200$, respectively, and fix the gas parameter for the first component $a_{11}(N_1/V)^{1/3} = 0.2$. Also, we fix the ratio between the s-wave scattering lengths a_{11} and a_{22}, namely we take $a_{11}/a_{22} = 2$.

Figure 16.1 shows the average condensate particle number of each species as a function of temperature for different values of the interspecies scattering length a_{12}, $a_{12}/a_{22} = 0, 1.2, 1.5$. The temperature is normalized by the critical temperature T_{c1} of an ideal gas in the box with total particle number N_1. As we mentioned before, (16.18), (16.19), and (16.20) become invalid near the critical temperature and yield a kink in $n_1(T)$ and $n_2(T)$ at the BEC transition points for each components. In a proper description, instead of a kink, $n_1(T)$ and $n_2(T)$ undergo a smooth crossover near BEC transition for mesoscopic number of particles.

For $a_{12} = 0$, the two components do not interact with each other. Figure 16.1 shows that increase in a_{12} (at fixed other parameters) yields depletion of the condensate for both components at all temperatures. This is qualitatively different from a one-component system in which increase in the interaction strength results

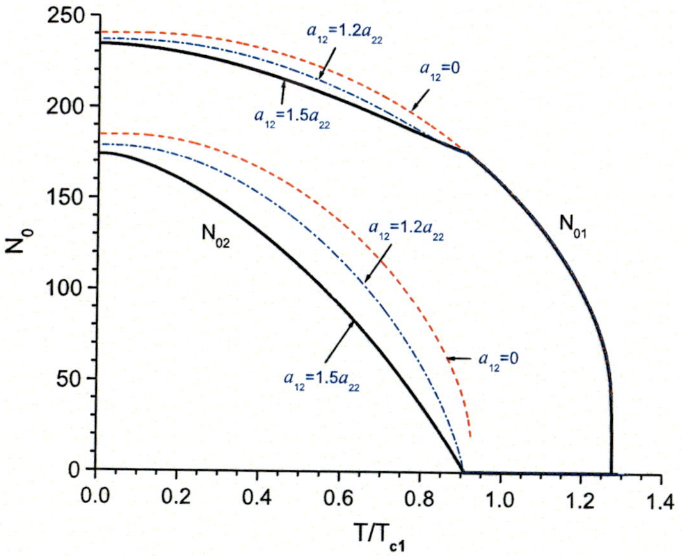

Fig. 16.1 Average number of condensate particles for each components as a function of temperature for $N_1 = 300$, $N_2 = 200$, $a_{11}(N_1/V)^{1/3} = 0.2$, $a_{11}/a_{22} = 2$, and $a_{12}/a_{22} = 0$, 1.2, 1.5

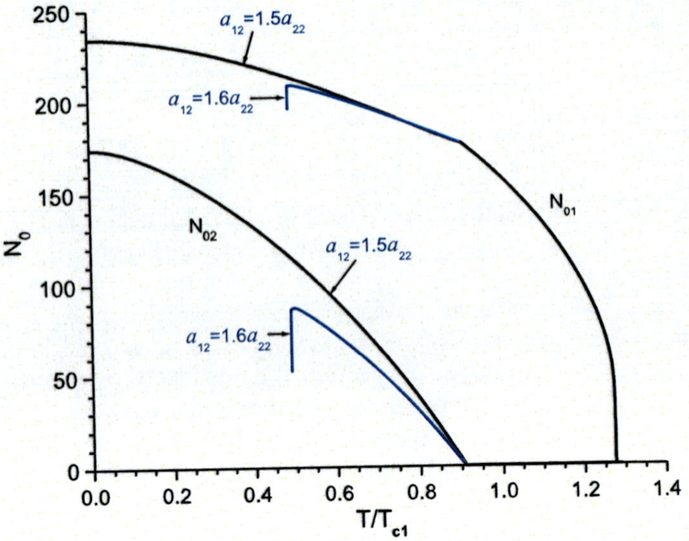

Fig. 16.2 Average number of condensate particles for each components as a function of temperature for parameters of Fig. 16.1, but with $a_{12}/a_{22} = 1.5$ and 1.6. For $a_{12}/a_{22} = 1.6$, the system undergoes a phase separation transition at a certain temperature

in BEC stimulation at intermediate temperatures due to the effect of attraction in momentum space [34]. However, such effect ceases to exist if atoms are distinguishable.

In Fig. 16.2, we compare $n_1(T)$ and $n_2(T)$ for $a_{12}/a_{22} = 1.5$ and 1.6. For $a_{12}/a_{22} = 1.6$, the two-component system undergoes a phase separation transition at a certain temperature as one can see from the figure. When temperature approaches the phase separation transition point, the average condensate particle number of each components suddenly drops. This can be explained by "softening" of normal mode spectrum. For example, at $T = 0$ the transition occurs when c_2 in (16.39) vanishes and therefore levels in the quasiparticle spectrum $E_{2,k} = \sqrt{\varepsilon_k^2 + \varepsilon_k c_2}$ become close to each other. As a result, particles are redistributed between several lowest energy levels which leads to a sudden decrease in the ground state population.

In Fig. 16.3, we plot the variance of the particle number for each components as a function of temperature obtained from (16.25) and (16.26) for $a_{12}/a_{22} = 1.5, 1.6$. For $a_{12}/a_{22} = 1.6$, the system undergoes a phase separation transition at certain temperature. Near the transition temperature the fluctuations suddenly drop. Such drop in fluctuations occur due to decrease in $n_1(T)$ and $n_2(T)$ near the point of phase separation transition.

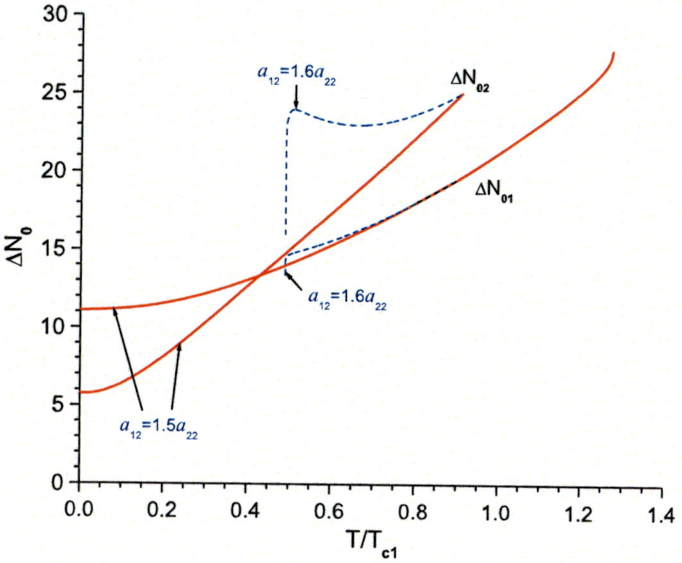

Fig. 16.3 Variance of the condensate particle number for each components as a function of temperature for $N_1 = 300$, $N_2 = 200$, $a_{11}(N_1/V)^{1/3} = 0.2$, $a_{11}/a_{22} = 2$, and $a_{12}/a_{22} = 1.5, 1.6$

16.5 Conclusions

In this paper, we study fluctuations in the condensate particle number in a two-component weakly interacting Bose gas. We assume that the total number of particles of each species is fixed, that is the system is described by the canonical ensemble. Using the particle number constraint, one can express fluctuations of the condensate particles in terms of fluctuations in the noncondensate particles and treat the later in the grand canonical description. Such approach is accurate at all temperatures apart from a vicinity of the BEC transition point. Using this method, we obtain general expressions for the condensate particle number of each components and their fluctuations. In the particular cases of zero temperature and the Thomas–Fermi limit, we find answers in a closed analytical form. We also analyze in detail fluctuations in the vicinity of a quantum phase separation transition at $T = 0$ and found that variance of the particle number has a cusp at the transition point. In addition, the phase separation transition can occur under variation of temperature. We study fluctuations near such transition numerically and found that both the average particle number and fluctuations undergo a sudden drop in the vicinity of the phase separation transition.

Acknowledgments It is a pleasure to dedicate this chapter to David Middleton whose interests run from theoretical physics (e.g., his Ph.D. background) to subtle statistical analysis (e.g., his famous books). We gratefully acknowledge support of the National Science Foundation Grant EEC-0540832 (MIRTHE ERC), the Office of Naval Research and the Robert A. Welch Foundation (Award A-1261).

References

1. Ziff RM, Uhlenbeck GE, Kac M (1977) Phys Rep 32:169
2. ter Haar D (1977) Lectures on selected topics in statistical mechanics. Elsevier, Amsterdam (Chap. 4)
3. Politzer HD (1996) Phys Rev A 54:5048
4. Wilkens M, Weiss C (1997) J Mod Opt 44:1801
5. Holthaus M, Kalinowski E, Kirsten K (1998) Ann Phys (NY) 270:198
6. Giorgini S, Pitaevskii LP, Stringari S (1998) Phys Rev Lett 80:5040
7. Scully MO (1999) Phys Rev Lett 82:3927
8. Kocharovsky VV, Scully MO, Zhu SY, Zubairy MS (2000) Phys Rev A 61:023609
9. Kocharovsky VV, Kocharovsky VlV, Scully MO (2000) Phys Rev Lett 84:2306
10. Kocharovsky VV, Kocharovsky VlV, Scully MO (2000) Phys Rev A 61:053606
11. Idziaszek Z, Rzazewski K, Lewenstein M (2000) Phys Rev A 61:053608
12. For a recent review see Kocharovsky VV, Kocharovsky VlV, Holthaus M, Ooi CHR, Svidzinsky AA, Ketterle W, Scully MO (2006) Adv Atom Mol Opt Phys 53:291
13. Svidzinsky AA, Scully MO (2006) Phys Rev Lett 97:190402
14. Scully MO, Svidzinsky AA (2007) J Mod Opt 53:2399
15. Chuu CS, Schreck F, Meyrath TP, Hanssen JL, Price GN, Raizen MG (2005) Phys Rev Lett 95:260403
16. Myatt CJ, Burt EA, Ghrist RW, Cornell EA, Wieman CE (1997) Phys Rev Lett 78:586
17. Hall DS, Matthews MR, Ensher JR, Wieman CE, Cornell EA (1998), Phys Rev Lett 81:1539
18. Stamper-Kurn DM, Andrews MR, Chikkatur AP, Inouye S, Miesner H-J, Stenger J, Ketterle W (1998) Phys Rev Lett 80:2027
19. Stenger J, Inouye S, Stamper-Kurn DM, Miesner H-J, Chikkatur AP, Ketterle W (1998) Nature (London) 396:345
20. Khalatnikov IM (1957) Zh Eksp Teor Fiz 32:653
21. Khalatnikov IM (1957) Sov Phys JETP 5:542
22. Siggia ED, Ruckenstein AE (1980) Phys Rev Lett 44:1423
23. Ho T-L, Shenoy VB (1996) Phys Rev Lett 77:3276
24. Law CK, Pu H, Bigelow NP, Eberly JH (1997) Phys Rev Lett 79:3105
25. Esry BD, Greene CH, Burke JP, Bohn JL (1997) Phys Rev Lett 78:3594
26. Pu H, Bigelow NP (1998) Phys Rev Lett 80:1130
27. Ao P, Chui ST (1998), Phys Rev A 58:4836
28. Chui ST, Ao P (1999) Phys Rev A 59:1473
29. Chui ST, Ao P, Tanatar B (1999) J Phys Soc Jpn 15:142
30. Svidzinsky AA, Chui ST (2003) Phys Rev A 67:053608
31. Svidzinsky AA, Chui ST (2003) Phys Rev A 68:013612
32. Tommasini P, de Passos EJV, de Toledo Piza AFR, Hussein MS, Timmermans E (2003) Phys Rev A, 67:023606
33. Search CP, Rojo AG, Berman PR (2001) Phys Rev A, 64:013615
34. Leggett A (2001) Rev Mod Phys 73:307

Chapter 17
Entanglement Criteria for Continuous-Variable Systems

Qingqing Sun and M. Suhail Zubairy

17.1 Introduction

In 1935, Einstein, Podolsky, and Rosen (EPR) questioned the completeness of quantum theory in their seminal work [1]. The argument is based on two spatially separated systems, which are prepared into a bipartite wavefunction and then cease to interact with each other. According to quantum mechanics, a measurement on system I will reduce system II into certain wavefunction. A different measurement setting will reduce system II into another wavefunction. These two wavefunctions could be eigenfunctions of noncommuting operators. On the one hand, *localism* states that the choice of measurement in system I should not change system II. On the other hand, without disturbance, the eigenfunction of an operator provides the value of the physical quantity with certain, which corresponds to *an element of physical reality*. So by choosing different measurement settings, two noncommuting physical quantities could have simultaneous reality, which obviously contradicts with the uncertainty relation. Therefore, EPR argued that the wave function description in quantum theory cannot be complete.

EPR paradox has since triggered a lot of theoretical and experimental interests due to its fundamental importance. The original example in EPR's paper [1] used the continuous variables position and momentum. Bohm [2] presented a discrete version of this paradox. Bell [3] established his famous inequalities to test EPR's theory using measurable physical quantities, which are subject to experimental verifications. Since then, experiments in various systems have been conducted [4–11]. Almost all of them violate the Bell-CHSH inequalities [3, 12] and support the quantum theory. Therefore, the assumptions in EPR's argument cannot be

Q. Sun • M.S. Zubairy (✉)
Department of Physics and Institute of Quantum Studies, Texas A&M University, College Station, TX 77843, USA
e-mail: qsun@physics.tamu.edu; zubairy@tamu.edu

L. Cohen et al. (eds.), *Classical, Semi-classical and Quantum Noise*,
DOI 10.1007/978-1-4419-6624-7_17, © Springer Science+Business Media, LLC 2012

both true. We have to give up either the locality or the reality. The nonlocal correlation between space-like separated particles was termed *entanglement* by Schödinger [13].

Entanglement enables many effects that are classically impossible. It is essential in quantum information processes such as quantum teleportation [14], quantum dense coding [15], quantum cryptography [16], and quantum computing [17]. Naturally, it is important to know whether a state is entangled or not before any application. The entanglement in a pure bipartite state can be easily determined from the Schmidt decomposition [18]. The situation for a mixed state is much more complicated since we have to deal with density matrix, which has numerous ways of decomposition. The partial transposition method proposed by Peres [19] has proven to be quite universal and efficient in detecting bipartite entanglement. The negative density matrix after partial transposition is a sufficient condition for entanglement. It is both necessary and sufficient for 2×2 and 2×3 systems [20]. In continuous-variable systems, several necessary and sufficient conditions for bipartite Gaussian states entanglement have been found [21–23]. Non-Gaussian states entanglement is more complicated and in general require higher order moments. Series of inequalities have been established as sufficient conditions for entanglement [24–28]. There are some good reviews on this topic [29, 30].

In Sect. 17.2, we give the general definition of entanglement. In Sect. 17.3, we review some important criteria in continuous-variable systems, and apply them to detect some entangled states. Section 17.4 is the summary.

17.2 Definition of Entanglement

The operational definition for entanglement is inseparable. For a bipartite pure state, separable means it can be written as a product of two states of the subsystems, which can be easily tested by the Schmidt decomposition [18].

Assuming the Hilbert spaces of both subsystems A and B have finite dimensions, a general pure state can be written as

$$|\Psi_{\mathrm{AB}}\rangle = \sum_{i,j} m_{ij} |a_i\rangle \otimes |b_j\rangle, \tag{17.1}$$

where $\{|a_i\rangle\}$ and $\{|b_j\rangle\}$ are the orthonormal basis vectors for their corresponding Hilbert spaces. According to Schmidt's theorem, the above double sum can always be converted into a single sum

$$|\Psi_{\mathrm{AB}}\rangle = \sum_{\nu} \sqrt{p_\nu} |\psi_\nu\rangle \otimes |\varphi_\nu\rangle. \tag{17.2}$$

Clearly, if there is only one nonzero coefficient p_ν, the state is separable; if there are more nonzero coefficients, the state is entangled.

In real world, we always find mixed states, due to the interaction with the environment. Instead of wavefunction, we can use density matrix to describe a mixed state. The definition of separability is [31]

$$\hat{\rho} = \sum_i p_i \hat{\rho}_i^{(A)} \otimes \hat{\rho}_i^{(B)}, \tag{17.3}$$

where p_i is the probability of being in the ith product state, $\hat{\rho}_i^{(A)}$ and $\hat{\rho}_i^{(B)}$ are the density matrices of the subsystems, respectively. Any state that cannot be written in the above form is an entangled state. However, the density matrix of a mixed state can be decomposed in numerous different ways. It is practically impossible to test all these decomposition. Instead, people explore various properties that should be satisfied by a separable state and use them as criteria, as we shall elaborate in the next section.

17.3 Entanglement Criteria

17.3.1 Peres–Horodecki Criterion

The positive partial transposition (PPT) proposed by Peres [19] and Horodecki [20] is a powerful and fundamental criterion. It leads to several other criteria. Their argument is based on the point that any legitimate density matrix (corresponding to a physical state) should only have nonnegative eigenvalues. In the two Hilbert spaces, they apply two positive operations, in particular, they choose the identity map I and the transposition T. An initial separable state like (17.3) then become

$$\hat{\rho}^{\mathrm{PT}} = \sum_i p_i \hat{\rho}_i^{(A)} \otimes \left(\hat{\rho}_i^{(B)}\right)^{\mathrm{T}}. \tag{17.4}$$

The transposed matrices $(\hat{\rho}_i^{(B)})^{\mathrm{T}}$ are still legitimate since they are non-negative and have unit trace. Therefore, the density matrix for the whole system $\hat{\rho}^{\mathrm{PT}}$ is still legitimate and should only have nonnegative eigenvalues. This is a necessary condition for separability. For an initial entangled state, the above argument may no longer be true since the operation in one subsystem will affect the other subsystem as well. If after the partial transposition the matrix became negative, it is a sufficient condition for entanglement. This is the so-called NPT criterion. Peres applied it to the example of Werner state [31]. After partial transposition, the minimum eigenvalue becomes negative if the parameter $x > 1/3$. This entanglement condition is much stronger than $x > 1/\sqrt{2}$ from Bell's inequality and $x > 1/\sqrt{3}$ from the α-entropic inequality [32].

Horodecki et al. showed that NPT criterion is both necessary and sufficient for entanglement in 2×2 and 2×3 systems [20]. But for higher dimensional systems, there exist states that remain positive after partial transposition but nevertheless entangled [33]. Since the PPT criterion is not violated, these states cannot be distilled into singlet forms and hence the name *bound entangled states* [34].

17.3.2 Simon criterion

Simon [21] applied the NPT criterion to continuous variable bipartite states. First, arrange the phase space variables and the Hermitian canonical operators into four-dimensional column vectors

$$\xi = (q_1, p_1, q_2, p_2), \quad \hat{\xi} = (\hat{q}_1, \hat{p}_1, \hat{q}_2, \hat{p}_2). \tag{17.5}$$

The commutation relations take the compact form [35]

$$[\hat{\xi}_\alpha, \hat{\xi}_\beta] = \mathrm{i}\Omega_{\alpha\beta}, \quad \alpha, \beta = 1, 2, 3, 4, \tag{17.6}$$

$$\Omega = \begin{pmatrix} J & 0 \\ 0 & J \end{pmatrix}, \quad J = \begin{pmatrix} 0 & 1 \\ -1 & 0 \end{pmatrix}. \tag{17.7}$$

For a given bipartite density operator $\hat{\rho}$, we define $\Delta\hat{\xi}_i = \hat{\xi}_i - \langle\hat{\xi}_i\rangle$, and the Hermitian operators $\{\Delta\hat{\xi}_\alpha, \Delta\hat{\xi}_\beta\} = (\Delta\hat{\xi}_\alpha\Delta\hat{\xi}_\beta + \Delta\hat{\xi}_\beta\Delta\hat{\xi}_\alpha)/2$. We can arrange their expectation values into a 4×4 real variance matrix V, defined through $V_{\alpha\beta} = \langle\{\Delta\hat{\xi}_\alpha, \Delta\hat{\xi}_\beta\}\rangle$. Then we have the uncertainty principle in the following compact form:

$$V + \frac{\mathrm{i}}{2}\Omega \geq 0. \tag{17.8}$$

This is an inequality which should be satisfied by all physical states.

Simon showed that under partial transposition, the Wigner distribution undergoes mirror reflection and the variance matrix is changed to $\tilde{V} = \Lambda V \Lambda$, where $\Lambda = diag(1, 1, 1, -1)$. If the initial $\hat{\rho}$ is separable, after partial transposition the density matrix is still legitimate. So $\tilde{V}$ should satisfy

$$\tilde{V} + \frac{\mathrm{i}}{2}\Omega \geq 0. \tag{17.9}$$

This restriction, to be obeyed by all separable states, is generically stronger than the usual uncertainty principle equation (17.8).

We can write the variance matrix V in the block form consisting of 2×2 real submatrices

$$V = \begin{pmatrix} A & C \\ C^{\mathrm{T}} & B \end{pmatrix}. \tag{17.10}$$

Then the condition (17.9) can be simplified into

$$\det A \det B + \left(\tfrac{1}{4} - |\det C|\right)^2 - \mathrm{Tr}(AJCJBJC^{\mathrm{T}}J) \geq \tfrac{1}{4}(\det A + \det B). \quad (17.11)$$

It is a necessary condition for separability. Furthermore, it is both necessary and sufficient for separability, for all bipartite Gaussian states.

17.3.3 Duan–Giedke–Cirac–Zoller Criterion

Based on the uncertainty relation and the Cauchy–Schwarz inequality, Duan et al. derive equivalent conditions following a different approach [22]. We define the EPR-like operators

$$\hat{u} = |a|\hat{x}_1 + \frac{1}{c}\hat{x}_2, \quad (17.12)$$

$$\hat{v} = |b|\hat{p}_1 - \frac{1}{d}\hat{p}_2 \quad (17.13)$$

For a separable state, we have $\langle \hat{x}_1\hat{x}_2\rangle_i = \langle \hat{x}_1\rangle_i \langle \hat{x}_2\rangle_i$ and similarly for $\langle \hat{p}_1\hat{p}_2\rangle_i$. Therefore, the summation of invariances

$$\begin{aligned}
\langle(\Delta\hat{u})^2\rangle + \langle(\Delta\hat{v})^2\rangle &= \sum_i p_i \left(a^2\langle\hat{x}_1^2\rangle_i + \frac{1}{c^2}\langle\hat{x}_2^2\rangle_i + b^2\langle\hat{p}_1^2\rangle_i + \frac{1}{d^2}\langle\hat{p}_2^2\rangle_i \right. \\
&\quad \left. + 2\frac{|a|}{c}\langle\hat{x}_1\rangle_i\langle\hat{x}_2\rangle_i - 2\frac{|b|}{d}\langle\hat{p}_1\rangle_i\langle\hat{p}_2\rangle_i \right) - \langle\hat{u}\rangle^2 - \langle\hat{v}\rangle^2 \\
&= \sum_i p_i \left[a^2\langle(\Delta\hat{x}_1)^2\rangle_i + \frac{1}{c^2}\langle(\Delta\hat{x}_2)^2\rangle_i + b^2\langle(\Delta\hat{p}_1)^2\rangle_i + \frac{1}{d^2}\langle(\Delta\hat{p}_2)^2\rangle_i \right] \\
&\quad + \sum_i p_i\langle\hat{u}\rangle_i^2 - \left(\sum_i p_i\langle\hat{u}\rangle_i\right)^2 + \sum_i p_i\langle\hat{v}\rangle_i^2 - \left(\sum_i p_i\langle\hat{v}\rangle_i\right)^2. \quad (17.14)
\end{aligned}$$

From uncertainty relation we have $a^2\langle(\Delta\hat{x}_1)^2\rangle_i + b^2\langle(\Delta\hat{p}_1)^2\rangle_i \geq |ab|$ and $1/c^2\langle(\Delta\hat{x}_2)^2\rangle_i + 1/d^2\langle(\Delta\hat{p}_2)^2\rangle_i \geq 1/|cd|$. From the Cauchy–Schwarz inequality, we know the last line of the above equation is nonnegative. Therefore, we have the necessary condition for separability

$$\langle(\Delta\hat{u})^2\rangle + \langle(\Delta\hat{v})^2\rangle \geq |ab| + \frac{1}{|cd|}. \quad (17.15)$$

Using local linear unitary the Bogoliubov operations on the Wigner characteristic function, Duan et al. proved that the above inequality is both sufficient and necessary for any Gaussian two-mode state.

17.3.4 Hillery–Zubairy criterion

After the establishment of entanglement criteria for bipartite Gaussian states, people started to explore the criteria for non-Gaussian states. In general, they are expressed in inequalities of higher order moments.

Hillery and Zubairy [24] considered two modes of the electromagnetic field, with a and $a^\dagger$ be the annihilation and creation operators of the first mode, and b and $b^\dagger$ of the second. For the quadratic operators $L_1 = ab^\dagger + a^\dagger b$, $L_2 = \mathrm{i}(ab^\dagger - a^\dagger b)$, their variances add up to

$$\langle(\Delta L_1)^2\rangle + \langle(\Delta L_2)^2\rangle = 2\Big[\langle(N_a+1)N_b\rangle + \langle N_a(N_b+1)\rangle \\ - 2\left|\langle ab^\dagger\rangle\right|^2\Big]. \tag{17.16}$$

For a product state, all the terms in the right-hand side can be factorized. The Cauchy–Schwarz inequality gives $|\langle a\rangle|^2 \le \langle N_a\rangle$ and $|\langle b\rangle|^2 \le \langle N_b\rangle$. So for a product state

$$\langle(\Delta L_1)^2\rangle + \langle(\Delta L_2)^2\rangle \ge 2(\langle N_a\rangle + \langle N_b\rangle). \tag{17.17}$$

The above inequality is shown to hold for all separable states. We can compare it to the uncertainty relation which holds for all states:

$$\langle(\Delta L_1)^2\rangle + \langle(\Delta L_2)^2\rangle \ge 2\left|\langle N_a\rangle - \langle N_b\rangle\right|. \tag{17.18}$$

Since the right-hand side of (17.17) is no less than that of (17.18), there could exist some states which violate (17.17). A good example is the Bell state $|\psi_{01}\rangle = (|0\rangle_a|1\rangle_b + |1\rangle_a|0\rangle_b)/\sqrt{2}$.

For another pair of quadratic operators $K_1 = ab + a^\dagger b^\dagger$, $K_2 = -\mathrm{i}(a^\dagger b^\dagger - ab)$, adding up the variances and assuming separability provides an inequality which coincides with the uncertainty relation. So it cannot be used to determine entanglement. However, we can employ the Schwarz inequality to find the variance lower bound for a general operator $K(\phi) = \mathrm{e}^{\mathrm{i}\phi}a^\dagger b^\dagger + \mathrm{e}^{-\mathrm{i}\phi}ab$. Any separable state should satisfy $\langle(\Delta K(\phi))^2\rangle \ge 1$.

Hillery and Zubairy [24] also showed a more general form of their approach. For two operators $A(a)$ and $B(b)$, using the Schwarz inequality and separability assumption we have

$$\begin{aligned}\left|\langle AB^\dagger\rangle\right| &\le \sum_k p_k \left|\langle AB^\dagger\rangle_k\right| \le \sum_k p_k(\langle A^\dagger AB^\dagger B\rangle_k)^{1/2} \\ &\le \left(\sum_k p_k\right)^{1/2}\left(\sum_k p_k\langle A^\dagger AB^\dagger B\rangle_k\right)^{1/2} \\ &= (\langle A^\dagger AB^\dagger B\rangle)^{1/2}.\end{aligned} \tag{17.19}$$

For different choices of A and B, some necessary conditions for separability

$$|\langle a^m (b^\dagger)^n \rangle|^2 \text{ and } |\langle a^m b^n \rangle|^2 \leq \langle (a^\dagger)^m a^m (b^\dagger)^n b^n \rangle, \tag{17.20}$$

where m and n are positive integers. As an example, the two-mode squeezed vacuum state

$$|\psi\rangle = (1 - x^2)^{1/2} \sum_{n=0}^{\infty} x^n |n\rangle_a |n\rangle_b, \quad (0 < x \leq 1) \tag{17.21}$$

violates the condition $|\langle ab \rangle| \leq [\langle N_a + 1 \rangle \langle N_b \rangle]^{1/2}$, so it is entangled.

17.3.5 Agarwal–Biswas Criterion

Almost at the same time, Agarwal and Biswas [25] also considered the entanglement detection for bipartite non-Gaussian states. They choose the same quadratic operators and apply partial transposition to the product form of Heisenberg uncertainty relation. The obtained separability conditions are

$$(\Delta L_1)_{\rho^{\rm PT}} (\Delta L_2)_{\rho^{\rm PT}} \leq \left| \langle N_a - N_b \rangle_{\rho^{\rm PT}} \right|, \tag{17.22}$$

$$(\Delta K_1)_{\rho^{\rm PT}} (\Delta K_2)_{\rho^{\rm PT}} \leq \langle N_a + N_b + 1 \rangle_{\rho^{\rm PT}}. \tag{17.23}$$

In this way, they showed the entanglement of a continuous variable Bell state

$$\psi(x_a, x_b) = \sqrt{\frac{2}{\pi}} (\alpha x_a + \beta x_b) \mathrm{e}^{-(x_a^2 + x_b^2)/2}, \quad |\alpha|^2 + |\beta|^2 = 1. \tag{17.24}$$

Nha and Kim [26] partially transposed the sum form of Heisenberg uncertainty relation:

$$\alpha (\Delta K_1)^2_{\rho^{\rm PT}} + \beta (\Delta K_2)^2_{\rho^{\rm PT}} \leq \sqrt{\alpha \beta} \langle N_a + N_b + 1 \rangle_{\rho^{\rm PT}}. \tag{17.25}$$

Optimal choice of α and β provides the same criterion as in [25]. They show that this optimal criterion can detect entanglement for the su(2) minimum-uncertainty states.

Later, Nha pointed out that stricter conditions can be obtained from Schrödinger–Robertson uncertainty relation (SRUR) instead of Heisenberg uncertainty relation (HUR) [27]:

$$\langle (\Delta A)^2 \rangle \langle (\Delta B)^2 \rangle \geq \tfrac{1}{4} |\langle [A, B] \rangle|^2 + \tfrac{1}{4} \langle \Delta A \Delta B + \Delta B \Delta A \rangle^2. \tag{17.26}$$

These conditions are invariant to local phase shifts.

17.3.6 Shchukin–Vogel Criterion

Alternatively but equivalently, Shchukin and Vogel [28] used the matrix forms to obtain a hierarchy of inequalities in terms of all orders of moments, also based on the positivity of partial transposition. They first derived a condition satisfied by any legitimate density matrix ρ, that is, for any operator $\hat{f}$, $\langle \hat{f}^\dagger \hat{f} \rangle_\rho \geq 0$. Here, we show it in a much simpler way

$$\langle \hat{f}^\dagger \hat{f} \rangle_\rho = \sum_\psi p_\psi \langle \psi | \hat{f}^\dagger \hat{f} | \psi \rangle = \sum_\psi p_\psi \langle \psi | \hat{f}^\dagger \left(\sum_n |n\rangle\langle n| \right) \hat{f} | \psi \rangle$$
$$= \sum_{\psi,n} p_\psi \left| \langle n | \hat{f} | \psi \rangle \right|^2 \geq 0, \quad (17.27)$$

where we use the identity $\sum_n |n\rangle\langle n| = 1$.

For a separable state, the density matrix after partial transposition should still be legitimate and therefore satisfy $\langle \hat{f}^\dagger \hat{f} \rangle_{\rho^{\mathrm{PT}}} \geq 0$. A general operator can be written in the normally ordered form

$$\hat{f} = \sum_{n,m,k,l=0}^{+\infty} c_{nmkl} \hat{a}^{\dagger n} \hat{a}^m \hat{b}^{\dagger k} \hat{b}^l. \quad (17.28)$$

Upon substituting, we have the necessary condition for separability

$$\langle \hat{f}^\dagger \hat{f} \rangle_{\rho^{\mathrm{PT}}} = \sum_{n,m,k,l,p,q,r,s=0}^{+\infty} c^*_{pqrs} c_{nmkl} M_{pqrs,nmkl} \geq 0, \quad (17.29)$$

where the moments of the partial transposition

$$M_{pqrs,nmkl} = \langle \hat{a}^{\dagger q} \hat{a}^p \hat{a}^{\dagger n} \hat{a}^m \hat{b}^{\dagger s} \hat{b}^r \hat{b}^{\dagger k} \hat{b}^l \rangle_{\rho^{\mathrm{PT}}} = \langle \hat{a}^{\dagger q} \hat{a}^p \hat{a}^{\dagger n} \hat{a}^m \hat{b}^{\dagger l} \hat{b}^k \hat{b}^{\dagger r} \hat{b}^s \rangle_\rho. \quad (17.30)$$

The nonnegativity of the quadratic form in (17.29) is equivalent to the nonnegativity of all its principal minors, i.e., if and only if for all $N = 1, 2, \ldots, \infty$ and $1 \leq r_1 < \cdots < r_N$,

$$\begin{vmatrix} M_{r_1 r_1} & M_{r_1 r_2} & \cdots & M_{r_1 r_N} \\ M_{r_2 r_1} & M_{r_2 r_2} & \cdots & M_{r_2 r_N} \\ \cdots & \cdots & \cdots & \cdots \\ M_{r_N r_1} & M_{r_N r_2} & \cdots & M_{r_N r_N} \end{vmatrix} \geq 0, \quad (17.31)$$

where the value of r_i, $i = 1, \ldots, N$ is the ordinal number of a multi-index in the ordered sequence of multi-indices. Thus, we have an infinite series of the inequalities (17.31) with $N = 1, 2, 3, \ldots$, the satisfaction of all of them is equivalent to PPT. Violation of any inequality is a sufficient condition for entanglement. They showed the Simon condition as a special example in this infinite series.

17.4 Conclusion

Continuous-variable systems are common in the real world, such as the position and momentum for particles, and the annihilation and creation operators for field modes. The detection of their entanglement remains an important question. For Gaussian states, after local phase shifts only the second-order moments are nonzero. So there exist both sufficient and necessary conditions for various systems. However, non-Gaussian states have higher order moments, and in general have an infinite series of inequalities as sufficient conditions for entanglement.

It is interesting to compare how these criteria are obtained. Remember that the uncertainty relations come from the Cauchy–Schwarz inequality, which itself is a consequence of the positive inner product matrix. Then all these criteria are derived based on the physical requirement of positivity, and the separability assumption. Following this principle, one could use other physical requirement to derive new criteria.

References

1. Einstein A, Podolsky B, Rosen N (1935) Phys Rev 47:777
2. Bohm D (1951) Quantum theory. Prentice-Hall, Englewood Cliffs, NJ
3. Bell JS (1964) Physics 1:195
4. Freedman SJ, Clauser J (1972) Phys Rev Lett 28:938
5. Fry ES, Thompson RC (1976) Phys Rev Lett 37:465.
6. Aspect A, Grangier P, Roger G (1982) Phys Rev Lett 49:91
7. Perrie W, Duncan AJ, Beyer HJ, Kleinpoppen H (1985) Phys Rev Lett 54:1790
8. Shih YH, Alley CO (1988) Phys Rev Lett 61:2921
9. Ou ZY, Mandel L (1988) Phys Rev Lett 61:50
10. Rarity JG, Tapster PR (1990) Phys Rev Lett 64:2495
11. Hasegawa Y, Loidl R, Badurek G, Baron M, Rauch H (2003) Nature 425:45
12. Clauser JF, Horne MA, Shimony A, Holt RA (1969) Phys Rev Lett 23:880
13. Schrödinger E (1935) Naturwissenschaften 23:807
14. Bennett CH, Brassard G, Crepeau C, Jozsa R, Peres A, Wootters WK (1993) Phys Rev Lett 70:1895
15. Bennett CH, Wiesner SJ (1992) Phys Rev Lett 69:2881
16. Ekert AK (1991) Phys Rev Lett 67:6961
17. Barenco A, Deutsch D, Ekert A, Jozsa R (1995) Phys Rev Lett 74:4083
18. Schmidt E (1907) Math Ann 63:433
19. Peres A (1996) Phys Rev Lett 77:1413
20. Horodecki M, Horodecki P, Horodecki R (1996) Phys Lett A 223:1
21. Simon R (2000) Phys Rev Lett 84:2726
22. Duan LM, Giedke G, Cirac JI, Zoller P (2000) Phys Rev Lett 84:2722
23. Giedke G, Kraus B, Lewenstein M, Cirac JI (2001) Phys Rev Lett 87:167904
24. Hillery M, Zubairy MS (2006) Phys Rev Lett 96:050503 (Hillery M, Zubairy MS (2006) Phys Rev A 74:032333)
25. Agarwal GS, Biswas A (2005) New J Phys 7:211
26. Nha H, Kim J (2006) Phys Rev A 74:012317
27. Nha H (2007) Phys Rev A 76:014305

28. Shchukin E, Vogel W (2005) Phys Rev Lett 95:230502 (2005) (Shchukin E, Vogel W (2006) Phys Rev A 74:030302(R))
29. Adesso G, Illuminati F (2007) J Phys A 40:7821
30. Li F-L, Zubairy MS (2007) In: Chen G, Kauffman L, Lomonaco SJ (eds) Mathematics of quantum computation and quantum technology, pp 349–385
31. Werner RF (1989) Phys Rev A 40:4277
32. Horodecki R, Horodecki P, Horodecki M (1996) Phys Lett A 210:377
33. Horodecki P (1997) Phys Lett A232:333
34. Horodecki M, Horodecki P, Horodecki R (1998) Phys Rev Lett 80:5239
35. Simon R, Sudarshan ECG, Mukunda N (1987) Phys Rev A 6:3868

Chapter 18
Quantum Carpets: Factorization with Degeneracies

Sabine Wölk and Wolfgang P. Schleich

18.1 Introduction

In 1789, Antoine Laurent Lavoisier (1743–1794) wrote [1] in his *Traité Élémentaire de Chimie*:

> "It is impossible to dissociate language from science or science from language, because every natural science always involves three things: the sequence of phenomena on which the science is based; the abstract concepts which call these phenomena to mind; and the words in which the concepts are expressed. To call forth a concept, a word is needed, to portray a phenomenon, a concept is needed. All three mirror one and the same reality."

The work of David Middleton illustrates this intimate relation between science and language in the most vivid way. Noise in transmission lines, biological systems, or quantum mechanics constitutes the phenomenon on which his science is based. Correlation functions are the abstract concepts, which he uses to call these phenomena to mind. Finally, he expresses these concepts in words found in his monumental work entitled *An Introduction to Statistical Communication Theory* [2].

It is indeed a great honor and pleasure for us to join the larger community celebrating the scientific achievements of David Middleton. In the spirit of Middleton's love for interdisciplinary research, we have chosen a topic at the interface of quantum mechanics and number theory: The factorization of numbers with the help of quantum carpets.

S. Wölk (✉) • W.P. Schleich
Institut für Quantenphysik, Universität Ulm, Albert-Einstein Allee 11, D-89069 Ulm, Germany
e-mail: sabine.woelk@uni-ulm.de; wolfgang.schleich@uni-ulm.de

L. Cohen et al. (eds.), *Classical, Semi-classical and Quantum Noise*,
DOI 10.1007/978-1-4419-6624-7_18,

18.2 Quantum Carpets and Number Theory

Quantum carpets [3] such as the one shown in Fig. 18.1 are spatiotemporal patterns [4] in the probability density of a quantum mechanical wave packet [5]. They have been studied theoretically in great detail [6] and have been observed experimentally in many physical systems involving atomic [7], electronic [8], and nuclear [9] wave packets as well as Bose-Einstein condensates (BEC) [10]. For example, a beautiful quantum carpet woven by a BEC [11] appears on the cover page of the journal *Fortschritte der Physik* during the year 2009. Moreover, the Talbot effect with classical light creates impressive carpets [12, 13].

In their purest form, quantum carpets emerge in the quantum mechanical description of a particle caught in a box. In this context, Gauss sums and the Jacobi theta function play a crucial role and build a bridge to number theory. In particular, there is a close connection between Gauss sums and prime numbers. Indeed, the Riemann zeta function whose nontrivial zeros determine the prime number distribution is given [14] by the Mellin-transform of the Jacobi theta function. In an appropriate limit, the theta function reduces to a Gauss sum used in the problem of a quantum particle in a box.

18.2.1 Prime Number Decomposition

Primes play a rather special role in number theory. They are the elementary building blocks of the integers, which Leopold Kronecker (1823–1891) once characterized [1] by the words

> "God made integers, all else is the work of man."

Indeed, any integer N can be represented as a product of primes and powers of them. While it is straightforward to multiply these primes and their powers to obtain N, it is an extremely difficult task to start from N and find its unique prime number decomposition.

The complexity inherent in the factorization problem is taken advantage in many cryptographic systems. However, the advent of the Shor algorithm [15] relying on a quantum computer made such an encoding method insecure. This observation has revived the interest in new methods to factor numbers.

Recently, we have embarked on a journey to investigate the possibilities offered by the periodicity properties of Gauss sums [16, 17] to factor numbers. These techniques rely solely on interference and pursue the path of analog computers. Indeed, in our approach a physical system computes a Gauss sum. It is remarkable that these ideas have been verified in several experiments [18] even factoring a 17-digit number.

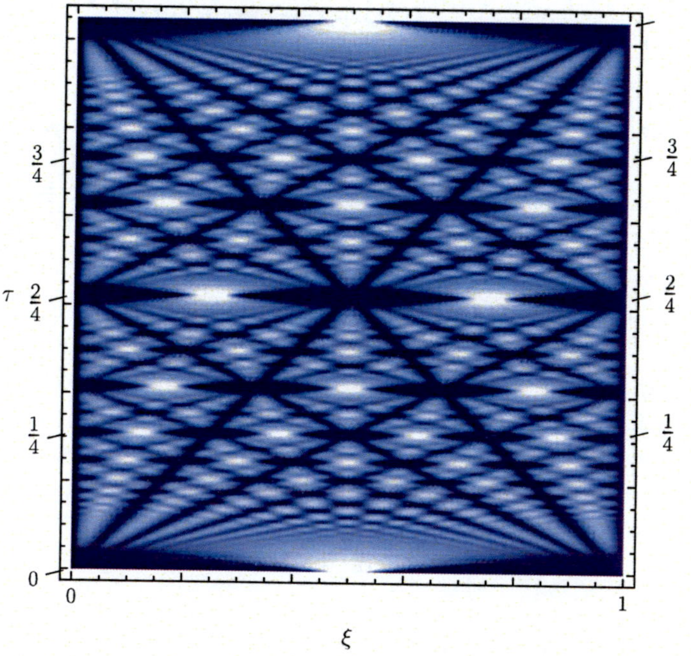

Fig. 18.1 Factorization of numbers with the help of a quantum carpet formed by the quantum mechanical probability density of a particle starting around $\xi = 0.5$ and diffusing in a box of length $L = 1$. Dark areas correspond to low probability whereas light areas indicate high probability. Our idea takes advantage of the degree of degeneracy of ratios, which manifests itself in features of the design of the carpet. Instead of a single canal or ridge approaching a fraction ℓ/N at the right edge of the carpet, several of them with different steepness cross there. We illustrate our idea of the degree of degeneracy as a tool to factor numbers for the example $N = 4$. For this purpose, we have indicated the ratios $\ell/4$ on the right- and left-hand sides of the carpet. Here, we have slightly tilted the ticks as to make most clearly visible four canals of steepness $1/4$, marked by dashed lines. The test factor lines $\tau_4^{(1/4)}$, $\tau_4^{(2/4)}$, $\tau_4^{(3/4)}$ and $\tau_4^{(4/4)}$ end at the right side at the times $\tau = 1/4, 2/4, 3/4$ and $\tau = 4/4$, respectively. At $\tau = 2/4$, an additional canal $\tau_2^{(1/2)}$ with slope $1/2$ which lies below the test factor line $\tau_4^{(2/4)}$ enters. This factor line expresses the degree 2 of degeneracy of $2/4$, which indicates that 2 and 4 share a common divisor.

In[19], we have proposed a new approach toward factorization with Gauss sums. It rests on the observation that the ratio N/s can be rewritten as p_k/s' with $s' < s$ if N and s share a common divisor. As a consequence, if the degree of degeneracy of a given ratio N/s is larger than one, then this ratio gives us information about the factors of N. Whereas in [19], the factorization with degeneracies was only a theoretical consideration, this paper demonstrates how these degeneracies manifest themselves in quantum carpets.

18.2.2 Outline

The paper is organized as follows: in Sect. 18.3 we summarize the main aspects of quantum carpets and bring out the origin of their design. In Sect. 18.4, we describe two methods to factor numbers with quantum carpets, which have been proposed previously. One of them uses truncated Gauss sums, which inspired us to this article. In Sect. 18.5, we show that the degeneracies of the ratio N/ℓ manifest themselves in the crossing of several canals or ridges of the carpet. We conclude in Sect. 18.6 by a brief summary of our results.

18.3 Quantum Carpets: Basic Facts

In this section, we provide a brief introduction [5] into the physics of quantum carpets using the most elementary example, that is the carpet woven by a particle in a box of length L. Our main goal is to obtain a representation of the probability density $W(x,t)$, which brings out most clearly the design of the carpet.

18.3.1 Probability Density

Quantum carpets emerge from the space-time representation of the probability density

$$W(x,t) \equiv |\psi(x,t)|^2 \tag{18.1}$$

defined by the time-dependent wave function

$$\psi(x,t) = \sum_{m=0}^{\infty} \psi_m u_m(x) \exp\left[-\frac{\mathrm{i}}{\hbar} E_m t\right]. \tag{18.2}$$

Here, we have used the energy eigenfunctions

$$u_m(x) \equiv \sqrt{\frac{1}{2L}} \frac{1}{i} \left(\mathrm{e}^{\mathrm{i}k_m x} - \mathrm{e}^{-\mathrm{i}k_m x}\right) \tag{18.3}$$

of a particle restricted in its motion along the x-axis by infinitely high potential walls located at $x = 0$ and $x = L$. The wave vector

$$k_m \equiv m\frac{\pi}{L} \tag{18.4}$$

is linear in the quantum number m, which gives rise to a quadratic energy spectrum

$$E_m \equiv \frac{\hbar^2}{2M} k_m^2 = m^2 \frac{\hbar^2 \pi^2}{2ML^2} \equiv m^2 \hbar \, \frac{2\pi}{T} \tag{18.5}$$

for a particle of mass M. At the revival time

$$T \equiv \frac{4ML^2}{\hbar\pi} \tag{18.6}$$

the wave function is identical to its initial form, that is $\psi(x, t = T) = \psi(x, t = 0)$.

When we substitute the explicit form (18.3) of the energy wave function u_m together with the eigen energies E_m given by (18.5) into the wave function ψ defined by (18.2) we find

$$\psi(x,t)=\frac{1}{\mathrm{i}\sqrt{2L}}\sum_{m=0}^{\infty}\left\{\psi_m \exp\left[\mathrm{i}\pi m\left(\frac{x}{L}-m\frac{2t}{T}\right)\right]-\psi_m \exp\left[-\mathrm{i}\pi m\left(\frac{x}{L}+m\frac{2t}{T}\right)\right]\right\}, \tag{18.7}$$

which we can cast into the compact form

$$\psi(x,t) = \frac{1}{\mathrm{i}\sqrt{2L}}\sum_{m=-\infty}^{\infty} \psi_m \exp\left[\mathrm{i}\pi m\left(\frac{x}{L}-m\frac{2t}{T}\right)\right] \tag{18.8}$$

if we introduce the expansion coefficients

$$\psi_{-|m|} \equiv -\psi_{|m|} \tag{18.9}$$

for negative values of m and $\psi_0 = 0$.

As a result, the probability density given by (18.1) reads

$$W(x,t) = \frac{1}{2L}\sum_{m,n=-\infty}^{\infty} \psi_m^* \psi_n \exp\left\{-\mathrm{i}\pi(m-n)\left[\frac{x}{L}-(m+n)\frac{2t}{T}\right]\right\}. \tag{18.10}$$

Here, the quantum numbers m and n enter either as the difference or the sum of each other. This decomposition is made possible by the quadratic dispersion relation (18.5).

18.3.2 Lines of Constant Phase and Degeneracy

The probability distribution $W(x, t)$ given by (18.10) display a criss-cross pattern in space–time, consisting of a superposition of straight lines. These traces [4] are defined by space-time trajectories with constant phase

$$\phi_{m,n} \equiv \pi(m-n)\left[\frac{x}{L}-(m+n)\frac{2t}{T}\right] \equiv \text{const.} \tag{18.11}$$

leading to the expression

$$\frac{t}{T/2} = \frac{1}{(m+n)}\frac{x}{L} + c \tag{18.12}$$

for these trajectories. Since $(m + n)$ can be any positive or negative integer and zero, we set $(m + n) \equiv \pm k$. The constant c is given by initial and boundary conditions.

Obviously, many different values of m and n can give rise to the same value of k. Hence, we get constructive or destructive interference of many modes leading to the same straight line. It is this degeneracy, which leads to the criss-cross design of the carpets.

18.4 Factorization with Wave Packets

The phenomenon of fractional revivals [20] provides us with several ways to factor numbers. The main idea is to concentrate on the probability distribution at times $t = T \times \ell/N$, which are fractions of the revival time T. In this section, we briefly review two approaches: (a) the first technique searches for common divisors of ℓ and N by counting the positions with high probability at time $t = T \times \ell/N$, (b) the second strategy compares the wave function at time $t = T \times \ell/N$ with the initial wave function leading us to Gauss sums.

18.4.1 Counting Maxima

William Harter [21] analyzed the behavior of rotor-like systems, which due to their quadratic spectra display fractional revivals. The corresponding probability distribution in the angle variable shows at times $t = T \times n/D$ a total of D peaks if n and D have no common divisor and if D is odd. This aspect helps us to factor numbers. For this purpose, we divide the time period of one revival into N equal parts. Then we count the peaks in the probability distribution of the angle at times $T \times \ell/N$. If there are less than N peaks, then ℓ and N share a common divisor and the fraction ℓ/N can be reduced to n/D.

As a consequence, this factorization scheme uses the degeneracies of the fractions ℓ/N. But here, only the fact, that ℓ/N can be rewritten as $\ell/N = n/D$, if ℓ and N share a common divisor, was used. The degree of degeneracy is not essential in this factorization scheme, in contrast to the method, introduced in Sect. 18.5 of this paper.

18.4.2 Locating Maxima

The connection between quantum carpets and truncated Gauss sums [16] is established with the help of the autocorrelation function

$$S(t) \equiv \langle \psi(t=0)|\psi(t)\rangle = \int_{-\infty}^{\infty} \mathrm{d}x\ \psi^*(x,t=0)\ \psi(x,t) \tag{18.13}$$

measuring the overlap between the time-evolved quantum state $|\psi(t)\rangle$ and the initial state $|\psi(t=0)\rangle$.

Due to the orthogonality of the energy wave functions u_m given by (18.3), we find substituting the expression (18.2) for $\psi(x,t)$ into the definition (18.13) of S the formula

$$S(t) = \sum_{m=0}^{\infty} |\psi_m|^2\ \mathrm{e}^{-2\pi i m^2 \frac{t}{T}}, \tag{18.14}$$

where we have used the representation (18.5) for the energy eigenvalues E_m.

This sum (18.14) is equivalent to the truncated Gauss sum

$$\mathcal{A}_N^{(M)}(\ell) \equiv \frac{1}{M+1} \sum_{m=0}^{M} \mathrm{e}^{-2\pi i m^2 \frac{N}{\ell}} \tag{18.15}$$

provided we make the following identifications

$$t \equiv N \tag{18.16}$$

and

$$T \equiv \ell \tag{18.17}$$

with

$$|\psi_m|^2 \equiv \begin{cases} \frac{1}{M+1} & \text{if } 0 \le m \le M \\ 0 & \text{else} \end{cases}. \tag{18.18}$$

The sum $\mathcal{A}_N^{(M)}(\ell)$ is equal to unity and therefore assumes a maximum if and only if ℓ is a factor of N. This property is the central idea of most of the factorization experiments [18] performed so far. However, in the mean time we could extend [19] this technique of factorization with truncated Gauss sums of integer-valued arguments ℓ to continuous arguments. Here, we take advantage of the degeneracy of the fractions N/s. If the degree of degeneracy of a given ratio N/s is larger than unity, then this ratio gives us information about the factors of N.

Unfortunately, in [19] we could only give an "idea for an idea" how to implement the technique of factoring with the help of degeneracies. However, when we recall from Sect. 18.3 that the criss-cross design of the quantum carpet originates from degeneracies , we are led to the conjecture that the degeneracies of N/ℓ are visible in quantum carpets as well. This fact will be demonstrated in the next section.

18.5 Factorization with Quantum Carpets

As shown in the previous section, there exist several ways of factoring numbers with the help of wave packets. We emphasize that all of them use properties of the probability density at a specific time. In this section we propose a method to factor with the help of the lines of maximal and minimal probability, that is canals and ridges in space–time. As a consequence, we use the whole structure of a quantum carpet. In this way we connect the degree of degeneracy of the ratio ℓ/N, which is the same as of N/ℓ in the continuous truncated Gauss sum [19] to quantum carpets.

18.5.1 Special Initial Conditions

For a better understanding of our factorization scheme, we choose special initial conditions for the particle in the box. In our approach, a Gaussian wave packet is initially localized at $L/2$ with a vanishing mean momentum. As a consequence, the resulting quantum carpet is symmetric with respect to the line $x = L/2$ and the lines of low probability follow from

$$\frac{t}{T/2} = \frac{1}{k}\frac{x}{L} + \frac{n}{k}, \qquad \text{with } n, k \in \mathbb{Z}. \tag{18.19}$$

It suffices to say that it is also possible to factor numbers using quantum carpets originating from other initial conditions, or to concentrate on lines with high probability. However, such an approach is slightly more complicated and might not yield all factors.

Moreover, it is convenient to introduce the dimensionless variables $\tau \equiv 2t/T$ and $\xi \equiv x/L$, which cast the traces (18.19) in the form

$$\tau(\xi) = \frac{1}{k}\xi + \frac{n}{k}. \tag{18.20}$$

Hence, the integer k and n determine these straight lines: their slopes are given by $1/k$ and the starting points at $\xi = 0$, that is at the left edge of the carpet, by the ratio n/k.

18.5.2 Visualization of Degeneracies

Since we are interested in the degeneracy of the fraction ℓ/N, we concentrate on the family of lines

$$\tau_N^{(\ell/N)}(\xi) = \frac{1}{N}(\xi - 1) + \frac{\ell}{N}. \tag{18.21}$$

Here, we have chosen a notation in which the subscript on τ refers to the steepness $1/N$ of the slope of the line and the superscript ℓ/N describes the time at which the line reaches the right edge of the carpet, that is at $\xi = 1$. These lines serve as our test factor lines and correspond in Fig. 18.1 to low probability.

However, also the lines

$$\tau_{rN}^{(r\ell/rN)}(\xi) = \tau_{rN}^{(\ell/N)}(\xi) \tag{18.22}$$

with r integer reach the right edge at the time ℓ/N. As a consequence of their smaller gradient $1/(rN)$, the lines are flatter and lie for $\xi < 1$ above the test factor lines defined by (18.21). Since these lines do not help us to factor the number N, we refer to them as nonfactor lines.

But if and only if $\ell = n \times p$ and $N = D \times p$ share a common divisor p, then there exist lines

$$\tau_D^{(\ell/N)}(\xi) = \tau_D^{(np/Dp)}(\xi) = \tau_D^{(n/D)}(\xi) \tag{18.23}$$

with a gradient $1/D$, which is larger than $1/N$, and which end also at the time $\ell/N = n/D$. These factor lines lie below the test factor line given by (18.21).

In summary: if ℓ and N share a common divisor, then there exist factor lines with a gradient smaller than $1/N$ which also end at $\xi = 1$ at the time $\tau = \ell/N$. The number of these lines is the degree of degeneracy of the fraction ℓ/N. If ℓ and N do not share a common divisor, then there exists only a single line.

This criterion is clearly visible in Fig. 18.1. Indeed, at $\tau = 2/4$ we detect at the right edge of the carpet a factor line $\tau_2^{(1/2)}$, which lies below the test factor line $\tau_4^{(2/4)}$ with gradient $1/4$. The factor line has the gradient $1/2$. As a consequence, 4 and 2 must share a common factor. In contrast, at the time $\tau = 3/4$, there is no such factor line because 3 and 4 do not share a common divisor.

Due to the symmetry with respect to $x = L$, our considerations are not restricted to the right edge of the carpet but also hold true for the left one.

18.6 Summary

In this paper, we have introduced a new factorization scheme, based on the observation that the degree of degeneracy of ratios ℓ/N manifests itself in the crossing of several canals or ridges of a quantum carpet in a single point on the right or left edge of the carpet.

However, our main interest was not to establish a new factorization scheme, but to complete a circle starting from quantum carpets via truncated Gauss sums of discrete and continuous arguments to factorization with the degree of degeneracy back to quantum carpets.

We are confident that these connections will shine some new light on the many facets of Gauss sums and may lead to a faster factorization algorithm.

Acknowledgment We are grateful to M. Jakob, K.A.H. van Leuwwen, M. Štefaňák, and M.S. Zubairy for many fruitful discussions on this topic. In this context, one of us (WPS) appreciates the inspiring discussions at the University of Vienna in the summer of 2009 with W. Case and M. Tomandl. This research was partially supported by the Max Planck Prize of WPS awarded by the Humboldt Foundation and the Max Planck Society. Moreover, WPS expresses his sincere thanks to the organizers L. Cohen and M.O. Scully of the Middleton Festival in Princeton 2007 for a most stimulating conference.

References

1. Bartlett J, Kaplan J (2002) Bartlett's familiar quotations. Little, Brown and Company, Boston
2. Middleton D (1960) An introduction to statistical communication theory. McGraw-Hill, New York
3. Berry M, Marzoli I, Schleich WP (2001) Quantum carpets, carpets of light. Phys World 14:39
4. Kaplan AE, Marzoli I, Lamb WE Jr., Schleich WP (2000) Multimode interference: highly regular pattern formation in quantum wave packet evolution. Phys Rev A 61:032101
5. Marzoli I, Saif F, Bialynicki-Birula I, Friesch OM, Kaplan AE, Schleich WP (1998) Quantum Carpets made simple. Acta Physica Slovaca 48:323
6. For a recent overview see for example Marzoli I, Kaplan AE, Saif F, Schleich WP (2008) Quantum carpets of a slightly relativistic particle. Fortschr. d. Physik 56:967
7. Nowak S, Kurtsiefer Ch., Pfau T, David C (1997) High-order Talbot fringes for atomic matter waves. Opt Lett 22:1430
8. Ahn J, Hutchinson DN, Rangan C, Bucksbaum PH (2001) Quantum phase retrivial of a Rydberg wave packet using a half-cycle pulse. Phys Rev Lett 86:1179
9. Katsuki H, Chiba H, Meier Ch., Girard B, Ohmori K (2009) Actively tailored spatiotemporal images of quantum interference on the picometer and femtosecond scales. Phys Rev Lett 102:103602; see also the excellent review by Ohmori K (2009); Wave-packet and coherent dynamics. Annu Rev Phys Chem 60:487
10. Deng L, Hagley EW, Denschlag J, Simsarian JE, Edwards M, Clark CW, Helmerson K, Roston SL, Philipps WD (1999) Temporal, matter-wave-dispersion Talbot effect. Phys Rev Lett 83:5407
11. Gustavsson M, Haller E, Mark MJ, Danzl JG, Hart R, Daley AJ, Nägerl H-C (2010) Interference of interacting matter waves. New J Phys 12:065029
12. Berry MV, Klein S (1996) Integer, fractional and fractal Talbot effect. J Mod Opt 43:2139
13. Case WB, Tomandl M, Deachapunya S, Arndt M (2009) Realization of optica carpets in the Talbot and Talbot-Lau configuration. Opt Express 17:20966
14. See for example Maier H, Schleich WP (2012) Prime numbers 101: A primer on number theory. Wiley-VCH, New York
15. Shor P (1994) In: Goldwasser S (ed) Proceedings of the 35th Annual Symposium on Foundations of Computer Science, Santa Fe, NM, IEEE Computer Society Press, New York, pp. 124–134, ; for an elementary introduction to the Shor algorithm see for example Mack R, Schleich WP, Haase D, Maier H (2009) In: Arendt W, Schleich WP (eds.) Mathematical Analysis of Evolution, Information and Complexity. Wiley-VCH, Berlin
16. Mack H, Bienert M, Haug F, Straub FS, Freyberger M, Schleich WP (2002) In: Mataloni P, De Martini F (eds.) Experimental quantum computation and information. Elsevier, Amsterdam
17. Wölk S, Merkel W, Schleich WP, Averbukh ISh, Girard B (2011) Factorization of numbers with Gauss sums: I. mathematical background. New J Phys 13:103007
18. See for example Mehring M, Müller K, Averbukh ISh, Merkel W, Schleich WP (2007) NMR experiment factors numbers with Gauss sums. Phys Rev Lett 98:120502; Mahesh T, Rajendran N, Peng X, Suter D (2007) Factoring numbers with the Gauss sum technique:

NMR implementations. Phys Rev A 75:062303; Gilowski M, Wendrich T, Müller T, Jentsch C, Ertmer W, Rasel EM, Schleich WP (2008) Gauss sum Factorization with cold atoms. Phys Rev Lett 100:030201; Sadgrove M, Kumar S, Nakagawa K (2008) Enhanced factoring with a Bose-Einstein condensate. Phys Rev Lett 101:180502; Tamma V, Zhang H, He X, Garruccio A, Schleich WP, Shih Y (2011) Factoring numbers with a single interferogram. Phys Rev A 83:020304
19. Wölk S, Feiler C, Schleich WP (2009) Factorization of numbers with truncated Gauss sums at rational arguments. J Mod Opt 56:2118
20. Schleich WP (2001) Quantum optics in phase space. Wiley-VCH, Weinheim
21. Harter W (2001) Quantum-fractal revival structure in C_N quadratic spectra: Base-N quantum computer registers. Phys Rev A 64:012312

Chapter 19
Co-channel Interference Modeling and Analysis in a Poisson Field of Interferers in Wireless Communications

Xueshi Yang and Athina P. Petropulu

19.1 Introduction

In wireless communication networks, signal reception is often corrupted by interference from other sources, or users, that share the same propagation medium. Knowledge of the statistics of interference is important in achieving optimum signal detection and estimation.

Existing models for interference can be divided into two groups: empirical models and statistical–physical models. Empirical models, e.g., the hyperbolic distribution and the K-distribution [1], fit a mathematical model to the practically measured data, regardless of their physical generation mechanism. On the other hand, statistical–physical models are grounded upon the physical noise generation process. Such models include the Class A noise, proposed by Middleton [2], and the α-stable model initially proposed by Furutsu and Ishida [3], and later advanced by Giordano [4], Sousa [5], Nikias [6], Ilow [7] and others. A common feature in these interference models is that the rate of decay of their density function is much slower than that of the Gaussian model. Such noise is often referred to as impulsive noise.

Impulsivenoise has been observed in several indoor (cf. [8,9]) and outdoor [2,10] wireless communication environments. In [11], measurements of interference in mobile communication channels suggest that in some frequency ranges, impulsive noise dominates over thermal noise. Impulsive noise attains large values (outliers) more frequently than the Gaussian noise. Such noise behavior has significant consequences in optimum receiver design [6, 12]. Moreover, as optimum signal

X. Yang (✉)
Marvell Semiconductor Inc., 5488 Marvell Lane, Santa Clara, CA 94087, USA
e-mail: xueshiy@marvell.com

A.P. Petropulu
ECE Department, Rutgers The State University of New Jersey, Piscataway, NJ 08854, USA
e-mail: athinap@rutgers.edu

L. Cohen et al. (eds.), *Classical, Semi-classical and Quantum Noise*,
DOI 10.1007/978-1-4419-6624-7_19, © Springer Science+Business Media, LLC 2012

detection relies on complete knowledge of the noise instantaneous and second-order statistics [13], it is important to study the spatial and/or temporal dependence structure of the noise as well as its instantaneous statistics. In the last decade, many efforts have been devoted in this direction (see, for example, [14–16]). In [14], the authors consider a physical statistical noise model originated from antenna observations that are spatially dependent. It is shown [14] that the resulted interference can be characterized by a correlated multivariate Class A noise model. In [17], it is shown that the total downlink interference in heterogeneous wireless DS-CDMA networks follows an asymptotically self-similar process. The implied long-range dependence between interference samples can be exploited in adaptive–predictive resource control for optimizing performance at different time scales. A general framework for designing wireless communication systems using an interference model which accounts for both the spatial distribution of the interferers and the propagation effect is outlined in [18]. For mathematical simplicity, the temporal dependence structure has been traditionally modeled by moving average (MA) [19] or Markov models [20]. However, it is not clear whether these models have assumed a temporal dependence structure that is consistent with the physical generation mechanism of the noise.

Here, we consider statistical–physical modeling for co-channel interference. In particular, we are interested in the temporal dependence structure of the interference. We adopt and extend the statistical–physical model investigated in [5–7]. The model considers a receiver, surrounded by interfering sources. The receiver picks up the superposition of all the pulses that originate from the interferers, after they have experienced power loss according to a power-law function of the distance traveled. We assume a communication network with basic waveform period T (time slot). We here focus on the interference sampled at rate $1/T$. As assumed in [5–7], at any time slot, the set of interferers forms a Poisson field in space. Assuming that from slot to slot these sets of interferers correspond to independent point processes, it was shown in [5–7] that the sampled interference constitutes an independent identically distributed (i.i.d.) α-stable process. However, the independence assumption is often violated in a practical system. To see why this would be the case, consider an interferer who starts interfering at some slot n, and remains active for a random number of slots. That interferer will still be active at time slots $n+i, i \in \mathbb{N}^+$, with certain probability, and will be one of the interferers at $n+i, i \in \mathbb{N}^+$. A direct consequence of this is that the interferers at n and $n+i$ are not independent of each other. Independence can only be valid if the interferer remains active for at most one time slot.

In this chapter, we assume that the interferer's holding time, or *session life*, is a random variable with some known distribution. We obtain the first- and second-order characteristic function of the sampled interference and show that, under certain assumptions, the interference becomes jointly α-stable. Further, if the session life is heavy-tail distributed, the interference constitutes a long-range dependent process in a generalized sense (defined in Sect. 19.2.3).

19.2 Mathematical Background

19.2.1 Heavy-Tailed Distributions and α-Stable Distributions

A random variable X is *regularly varying* with index α, if

$$P(|X| \geq x) \sim \frac{L(x)}{x^\alpha} \quad \text{as } x \to \infty. \tag{19.1}$$

Here, $L(x)$ is a slowly varying function, i.e., for all positive x, $\lim_{\tau\to\infty} L(\tau x)/L(\tau) = 1$ (typically such slowly varying functions are constant, or ratios of two polynomials with identical degree). The variable X is said to be *heavy-tailed with infinite variance* if it is regularly varying with index $0 < \alpha < 2$. In those cases, the variance of X is infinite. If $\alpha < 1$ the mean and moments of order greater than or equal to α are infinite.

A particular class of heavy-tail distributions with infinite variance is the α-stable distribution. The α-stable distribution often classified as non-Gaussian, although it reduces to the Gaussian when $\alpha = 2$. The difference between α-stable and Gaussian densities is that the tails of the former are heavier than those of the latter. Due to lack of closed form expression for their probability density functions, α-stable distributions are more conveniently characterized by their characteristic functions.

A random variable is said to have a stable distribution if there are parameters $0 < \alpha \leq 2, \sigma \geq 0, -1 \leq \beta \leq 1$ and μ real such that its characteristic function is of the form:

$$\Phi(\omega) = \begin{cases} \exp\left\{-\sigma^\alpha |\omega|^\alpha (1 - j\beta \operatorname{sign}(\omega) \tan \frac{\pi\alpha}{2}) + j\mu\omega\right\}, & \alpha \neq 1, \\ \exp\left\{-\sigma|\omega|(1 + j\beta \frac{2}{\pi} \operatorname{sign}(\omega) \ln|\omega|) + j\mu\omega\right\}, & \alpha = 1. \end{cases} \tag{19.2}$$

We write $X \sim S_\alpha(\sigma, \beta, \mu)$. If $\beta = 0$, then X is symmetric α-stable, or simply $S\alpha S$. Two random variables are jointly stable if any linear combination of them is stable with exponent greater than or equal to one.

19.2.2 Codifference

α-Stable distributions are known for their lack of moments of order greater than or equal to α. In particular, for $\alpha < 2$, the second-order statistics do not exist. In such case, the role of covariance is played by the covariation or, the codifference [21].

The codifference of two jointly $S\alpha S$, $0 < \alpha \leq 2$, random variables $\mathbf{x}_1$ and $\mathbf{x}_2$ equals

$$R_{\mathbf{x}_1,\mathbf{x}_2} = \sigma^\alpha_{\mathbf{x}_1} + \sigma^\alpha_{\mathbf{x}_2} - \sigma^\alpha_{\mathbf{x}_1-\mathbf{x}_2}, \tag{19.3}$$

where $\sigma_\mathbf{x}$ is the scale parameter of the $S\alpha S$ variable $\mathbf{x}$.

A quantity that is closely related to the codifference $R_{\mathbf{x}(t+\tau),\mathbf{x}(t)}$ is [21]

$$I(\rho_1, \rho_2; \tau) = -\ln E\left\{e^{j(\rho_1 \mathbf{x}(t+\tau)+\rho_2 \mathbf{x}(t))}\right\} + \ln E\left\{e^{j\rho_1 \mathbf{x}(t+\tau)}\right\} + \ln E\left\{e^{j\rho_2 \mathbf{x}(t)}\right\}. \tag{19.4}$$

This quantity is referred to as generalized codifference [22]. It reduces to the codifference for the case of jointly $S\alpha S$ processes, i.e.,

$$R_{\mathbf{x}(t+\tau),\mathbf{x}(t)} = -I(1, -1; \tau). \tag{19.5}$$

$I(\rho_1, \rho_2; \tau)$ is defined for any stationary heavy-tailed random process.

19.2.3 *Long-Range Dependence*

A second-order process $x(t)$ is called a (wide-sense) stationary with long memory, or long-range dependence, if its autocorrelation function decays hyperbolically as its lag increases [23]. This is in contrast to the exponential decay corresponding to short-memory processes, e.g., auto-regressive moving average (ARMA) processes. This definition does not apply to processes that do not have a finite autocorrelation. The following generalization, based on the generalized codifference, can be useful for such cases [22, 24].

Let $x(t)$ be a stationary process. We say that $x(t)$ is a long-memory process in a generalized sense, if $I(1, -1; \tau)$, as defined in (19.4), satisfies

$$\lim_{\tau \to \infty} -I(1, -1; \tau)/\tau^{\beta-1} = c, \tag{19.6}$$

where c is some real positive constant and $0 < \beta < 1$.

19.3 The Interference Model

Consider a wireless communication scenario without power control, where a receiver receives the signal of interest in the presence of other interfering signals. For the sake of simplicity, we will assume that the users, which are the potential interferers, and the receiver are all on the same plane and concentrated in a disk R_b of radius b. The modeling is performed first for a finite b, and at the final step, the limit for $b \to \infty$ is taken. The three-dimensional scenario is a straightforward extension of the two-dimensional one. The receiver is placed at the origin of the coordinate system, and the users are distributed within the disk according to the two-dimensional Poisson point process.

Let us define the term *emerging interferers at time interval m*, to describe the interfering sources whose contribution arrives for the first time at the receiver in the beginning of time interval m. The interferers that emerged at any time interval are spatially located according to a Poisson point process (Poisson field) with density λ.[1] It is of course reasonable to assume that the interferers that emerged at two different time slots are independent, or more precisely, correspond to two independent Poisson point processes.

One issue that the most existing models (e.g., [5–7]) do not take into account is that, at time n, in addition to interferers that emerge at n, there could be interferers that emerged at some slot $m < n$, and still stay active at n. The latter group would exist if the holding times of the users were longer than one time slot T. The combination of these two groups would make the interferers at slots m and n dependent. Let us assume once started transmission, the user continuously emits pulses for a duration of L time slots, where L is a random variable with some known distribution.

At time n, the signal transmitted from the i-th interfering user, i.e., $p_i(t)$ propagates through the transmission medium and the receiver filters, and as a result gets attenuated and distorted. For simplicity, we assume that distortion and attenuation can be separated. Let us first consider the filtering effect. For short time intervals, the propagation channel and the receiver can be represented by a time-invariant filter with impulse response $h(t)$. Due to filtering, the contribution of the i-th interfering source at the receiver is of the form $x_i(t) = p_i(t) * h(t)$, where the asterisk denotes convolution. In wireless communications, the power attenuation increases logarithmically with the distance r_i between the transmitter and the receiver (cf. [2]). The power loss function can be expressed in terms of signal amplitude loss function $a(r_i)$, i.e.,

$$a(r_i) = \frac{1}{r_i^{\gamma/2}}, \tag{19.7}$$

where γ is the path loss exponent; γ is a function of the antenna height and the signal propagation environment, and it may vary from slightly less than 2, for hallways within buildings, to larger than 5, in dense urban environments and hard partitioned office buildings [11]. Thus, the total signal at the receiver is

$$x(t) = s(t) + \sum_{i \in \mathcal{N}} a(r_i) x_i(t), \tag{19.8}$$

where $s(t)$ is the signal of interest and $\mathcal{N}$ denotes the set of interferers at time t. Note that the transmitting power of user i has been implicitly incorporated into $x_i(t)$.

[1] λ may be function of time and locations of the unit area/volume, which forms a non-homogeneous Poisson point process. A non-homogeneous Poisson process can be mapped to a homogeneous one through transformations, cf. [5,7]. In this paper, we only consider the homogeneous case, i.e., λ is a constant.

The receiver consists of a signal demodulator followed by the detector. The signal is expanded into a series of orthonormal basis functions. Let $Z(n)$ denote the projection of $x(t)$ onto one of the basis functions it time slot n. It holds

$$Z(n) = S(n) + \sum_{i \in \mathcal{N}_n} a(r_i) X_i(n), \tag{19.9}$$

or,

$$Z(n) \triangleq S(n) + Y(n), \tag{19.10}$$

where $X_i(n)$ and $S(n)$ denote, respectively, the result of the correlations of $x_i(t)$ and $s(t)$ with the basis function, $Y(n)$ denotes interference, and $\mathcal{N}_n$ denotes the set of interferers that are active at time n.

In the following, we assume that $X_i(n)$'s are spatially independent (e.g., $X_i(n)$ is independent of $X_j(n)$ for $i \neq j$) and identically distributed.

19.4 Statistical Analysis of the Interference

Let us assume that the mean of the interferers' session life is finite, denoted by μ, and the density of the emerging interferers at each time slot is λ. At time n, the number of active interferers equals

$$C(n) = \sum_k 1_{[\Gamma_k \leq n < \Gamma_k + L_k]}, \tag{19.11}$$

where Γ_k is the time when the sources emerged, and L_k's correspond to their session life (in multiples of T). $1_{[.]}$ is the indicator function.

Let $\bar{F}_L(\cdot)$ be the survival function of the session life. Let $\Phi_{m,n}(\omega_1, \omega_2)$ denote the joint characteristic function of interference samples $Y(m)$ and $Y(n)$. It can be shown that [25]

$$\Phi_{m,n}(\omega_1, \omega_2) = \exp\{-\sigma^\alpha H_1(\tau)(|\omega_1|^\alpha + |\omega_2|^\alpha) + H_2(\tau)\Theta_{m,n}(\omega_1, \omega_2)\}, \tag{19.12}$$

where

$$\alpha = 4/\gamma, \tag{19.13}$$

$$\sigma = \left(-\lambda\pi \int_0^\infty x^{-\alpha}\, \mathrm{d}\Psi(x)\right)^{1/\alpha}, \tag{19.14}$$

$$H_1(\tau) = \sum_{l=1}^{n-m} \bar{F}_L(l), \tag{19.15}$$

$$H_2(\tau) = \sum_{l=n-m+1}^{\infty} \bar{F}_L(l), \tag{19.16}$$

$$\Theta_{m,n}(\omega_1,\omega_2) = \lim_{b\to\infty} \lambda\pi b^2 \left[\int_0^b \Psi_{m,n}(a(r)\omega_1, a(r)\omega_2)\frac{2r}{b^2}\,\mathrm{d}r - 1 \right]. \quad (19.17)$$

Here, $\Psi(x)$ is the characteristic function of $X_i(\cdot)$, and $\Psi_{m,n}(\omega_1,\omega_2)$ is the second-order characteristic function of $X_i(\cdot)$, i.e.,

$$\Psi_{m,n}(\omega_1,\omega_2) = \mathrm{E}\left[\mathrm{e}^{\mathrm{j}\omega_1 X_i(m)+j\omega_2 X_i(n)}\right]. \quad (19.18)$$

The proof is long and can found in [25]. Some hints of it are included in Appendix A.

19.4.1 Remarks

1. On set $\omega_2 = 0$, we obtain the first-order characteristic function of the interference process, i.e.,
$$\Phi(\omega_1) = \mathrm{e}^{-\sigma^\alpha \sum_{l=1}^{\infty} \bar{F}_L(l)|\omega_1|^\alpha}. \quad (19.19)$$
Recognizing that $\sum_{l=1}^{\infty} \bar{F}_L(l) = \mu$, which is the mean of the session life, we get a result consistent prior results stating that the marginal statistics of interference are $S\alpha S$ with characteristic exponent α, and scale parameter $-\lambda\mu\pi \int_0^\infty$ on setting $x^{-\alpha}\,\mathrm{d}\Psi(x)$, where $\Psi(x)$ is the characteristic function of $X_i(\cdot)$.
2. In the special case
$$\bar{F}_L(l) = \begin{cases} 1, \, l = 1, \\ 0, \, l > 1, \end{cases} \quad (19.20)$$
we have that $H_2(\tau) = 0$ for $\tau = 1, 2, \ldots$, and $H_1(\tau) = 1$. Hence,
$$\ln \Phi_{m,n}(\omega_1,\omega_2) = -\sigma^\alpha \left(|\omega_1|^\alpha + |\omega_2|^\alpha\right). \quad (19.21)$$
Equation (19.21) implies that the interference samples obtained at different time slots are independent and jointly α-stable distributed. Indeed, this is consistent with the findings in [5, 7].
3. If $H_2(\tau)$ tends to zero as τ tends to infinity, and $H_1(\tau)$ approaches the mean of session life, μ, the joint characteristic function may be simplified as
$$\lim_{\tau\to\infty} \Phi_{m,n}(\omega_1,\omega_2) = \mathrm{e}^{-\sigma^\alpha \mu(|\omega_1|^\alpha + |\omega_2|^\alpha)}. \quad (19.22)$$
Equation (19.22) implies that when the distance between two samples becomes asymptotically large, the samples are becoming independently α-stable distributed.

19.4.2 Special Cases: Jointly α-Stable Interference

It is interesting to note that although the sampled interference is marginally α-stable, it is not necessarily α-stable. The latter can be verified by checking the characteristic function of (19.12) for generally distributed $X_i(m)$. However, for certain cases of $X_i(m)$ the interference becomes jointly α-stable. Next, we discuss one such case that is of interest in communications.

In spread spectrum communication networks, most of the interference is due to co-channel users who are transmitting data during their session life. Let us assume that the data symbols are either 1 or -1 with equal probability, and are independent from slot to slot. For that case, the contribution from interferer i is

$$X_i(m) = B(m)1_{[\Gamma_i \le m < \Gamma_i + L_i]}, \tag{19.23}$$

where $B(m)$ is i.i.d. Bernoulli distributed for different m, taking 1 or -1 with equal probability of $1/2$. It can be shown that [25]

$$\begin{aligned} \ln \Phi_{m,n}(\omega_1, \omega_2) = &-\sigma^\alpha H_1(\tau)(|\omega_1|^\alpha + |\omega_2|^\alpha) \\ &-\sigma^\alpha \frac{H_2(\tau)}{2}(|\omega_1 + \omega_2|^\alpha + |\omega_1 - \omega_2|^\alpha), \end{aligned} \tag{19.24}$$

where α, $H_1(\tau)$, $H_2(\tau)$ are defined as before, and

$$\sigma = \frac{1}{2}\left(\frac{\lambda\pi^{3/2}\Gamma(1-\alpha/2)}{\Gamma(1/2+\alpha/2)}\right)^{1/\alpha}. \tag{19.25}$$

Again, (19.24) implies that the interference at m and n is jointly α-stable.

19.5 Dependence Structure of Co-channel Interference

Equation (19.12) implies that the dependence structure of the co-channel interference is determined by the session life distribution $P[L \ge k]$, $k = 1, 2, \ldots$. In this section, we provide the conditions under which the resulted interference is long-range dependent. For convenience, we will focus on the case of Bernoulli distributed $X_i(m)$ for which the interference is jointly α-stable distributed; it is not difficult to verify that the results presented also hold for generally distributed $X_i(m)$, provided that $\{X_i(m)\}_{m=-\infty}^{\infty}$ has finite second-order statistics for all i such that (19.17) is well defined.

The motivation behind considering heavy-tail distributed session life is that such a distribution can well characterize many current and future communication systems. For example, in spread spectrum packet radio networks, multiple access terminals utilize the same frequency channel. The signals received at the receiver consist of superposition of the signals from all the users in the network. Assuming that multi-user detection and power-control are not implemented, the interference from other users, or otherwise referred to as self-interference, can be characterized by our interference model. As more and more wireless users are equipped with Internet-enabled cell phones, their resource-request holding times (session life) are distributed with much fatter tail than that of voice only network users (cf. [26]). Extensive statistical analysis of high-definition network traffic measurement has shown that the holding times of data network users are heavy-tailed distributed [24, 27]. In particular, they can be modeled by Pareto distributions [24, 27].

For a discrete-time communication system, we here assume that the session life is Zipf distributed (a discrete version of the Pareto distribution). A random variable X has a Zipf distribution [28] if

$$P\{X \geq k\} = \left[1 + \left(\frac{k - k_0}{\sigma}\right)\right]^{-\alpha}, \quad k = k_0, k_0 + 1, k_0 + 2, \ldots, \tag{19.26}$$

where k_0 is an integer denoting the location parameter, $\sigma > 0$ denotes the scale parameter, and $\alpha > 0$, is the tail index. In this paper, for simplicity, we set $\sigma = k_0 = 1$, and $\alpha > 1$, which implies that $\mathrm{E}\{X\} = \zeta(\alpha)$, where $\zeta(\cdot)$ is the Riemann Zeta function. We shall denote the tail index of the session life by α_L to avoid confusion with the α defined in (19.13).

Since the interference is marginally heavy-tail distributed, conventional tools for capturing dependence structure, such as the covariance, are not applicable. Next, we use the codifference (see (19.4)) to explore the dependence structure.

It can be shown that if session life of the interferers are the Zipf distributed, with tail index $1 < \alpha_L < 2$, and $X_i(m)$ are i.i.d. Bernoulli random variables taking possible values 1 and -1 with equal probability $1/2$, then, the resulted interference is long-range dependent in the generalized sense, i.e.,

$$\lim_{\tau \to \infty} \frac{-I(1, -1; \tau)}{\tau^{-(\alpha_L - 1)}} = \frac{(2 - 2^{\alpha - 1})\sigma^{\alpha}}{\alpha_L - 1}, \tag{19.27}$$

where τ is the time lag between time intervals, α_L is the tail index of the session life distribution, and σ was as defined in (19.25).

Proof. See Appendix B. □

19.6 Conclusions

We have investigated the statistics of the interference resulted from the Poisson field of interferers. Key assumptions were that individual interferers have certain random session life, whose distribution is a priori known, and the signal propagation attenuation is a power-law function of distance. We have obtained the instantaneous and second-order distributions of the interference. We showed that although the interference process is marginally $S\alpha S$, it is in general not jointly α-stable distributed, except in some special cases, for example, when interferers send BPSK signals or constant amplitude signals.

The dependence structure of the interference must be taken into account to attain optimum signal detection, as the performance of traditional detectors may deteriorate significantly under long-range dependent noise. Some recent results on this subject can be found in [29].

Appendix A

An outline of the proof is as follows. The set of interferers that are active at time m, $\mathcal{N}_m$, can be expressed in terms of the set of interferers that are active at m and emerged at time slot t as

$$\mathcal{N}_m = \bigcup_{t=-\infty}^{m} \mathcal{N}_{m,t} \left(\sum_{i \in \mathcal{N}_m} \Longrightarrow \sum_{t=-\infty}^{m} \sum_{i \in \mathcal{N}_{m,t}} \right).$$

By assumption, the interferers in $\mathcal{N}_{m,t}$, $t = 1, 2, \ldots$ are independent of each other for different t. Thus, the joint characteristic function becomes

$$\begin{aligned}\Phi_{m,n}(\omega_1, \omega_2) = & \prod_{t=-\infty}^{m} \mathrm{E}\left\{ \exp\left[j\omega_1 \sum_{i \in \mathcal{N}_{m,t}} a(r_i) X_i(m) + j\omega_2 \sum_{i \in \mathcal{N}_{n,t}} a(r_i) X_i(n) \right] \right\} \\ & \times \prod_{t=m+1}^{n} \mathrm{E}\left\{ \exp\left[j\omega_2 \sum_{i \in \mathcal{N}_{n,t}} a(r_i) X_i(n) \right] \right\}.\end{aligned}$$

Assuming there are in total k sources that started emission at time t, due to the Poisson assumption their locations are independent and uniform distributed on the disk R_b. The probability that an interferer that emerged at time t will remain active until n is $P_1(n,t) = \bar{F}_L(n-t+1)$. The probability that an interferer that emerged at t will survive until m but will die out at n is $P_2(m,n,t) = \bar{F}_L(m-t+1) - \bar{F}_L(n-t+1)$. If there are k interferers beginning their emission at t, the probability that q of them are active until time n, and p of them are active at m but not n is

$$\begin{aligned}P\{q,p\} = & \binom{k}{q} \binom{k-q}{p} P_1(n,t)^q P_2(m,n,t)^p (1 - P_1(n,t) \\ & - P_2(m,n,t))^{(k-q-p)}.\end{aligned} \qquad (19.28)$$

Appendix B

Proof. The codifference of the interference separated by τ can be calculated as

$$\begin{aligned} -I(1,-1;\tau) &= \ln \Phi_{t,t+\tau}(-1,1) - \ln \Phi_{t,t+\tau}(0,1) - \ln \Phi_{t,t+\tau}(-1,0) \\ &= (2-2^{\alpha-1})\sigma^{\alpha} H_2(\tau) \\ &= (2-2^{\alpha-1})\sigma^{\alpha} \sum_{l=\tau+1}^{\infty} l^{-\alpha_L}, \end{aligned} \tag{19.29}$$

where (19.26) was used at the last step.

Note that

$$\begin{aligned} \frac{(\tau+1)^{-\alpha_L+1}}{\alpha_L-1} &= \int_{\tau+1}^{\infty} t^{-\alpha_L}\,\mathrm{d}t \le \sum_{l=\tau+1}^{\infty} l^{-\alpha_L} \\ &\le \int_{\tau+1}^{\infty} (t-1)^{-\alpha_L}\,\mathrm{d}t = \frac{\tau^{-\alpha_L+1}}{\alpha_L-1}, \end{aligned} \tag{19.30}$$

and since both the upper and lower bounds converge as $\tau \to \infty$, we conclude that

$$\lim_{\tau\to\infty} \frac{-I(1,-1;\tau)}{\tau^{-\alpha_L+1}} = \frac{(2-2^{\alpha-1})\sigma^{\alpha}}{\alpha_L-1}, \tag{19.31}$$

where σ is as given as in (19.25). Note that for $\alpha \in (0,2)$, $2-2^{\alpha-1}$, and σ are positive. Hence, for $\alpha_L > 1$, $(2-2^{\alpha-1})\sigma^{\alpha}/(\alpha_L-1)$ is positive, and the interference process is long-range dependent in the generalized sense. □

References

1. Wegman EJ, Schwartz SC, Thomas JB (eds) (1989) Topics in non-Gaussian signal processing. Springer, New York
2. Middleton D (1977) Statistical-physical models of electromagnetic interference. IEEE Trans Electromagn C EMC-19(3):106–127
3. Furutsu K, Ishida T (1961) On the theory of amplitude distribution of impulsive random noise. J Appl Phys 32(7):1206–1221
4. Giordano A, Haber F (1972) Modeling of atmosphere noise. Radio Sci 7(11):1011–1023
5. Sousa ES (1992) Performance of a spread spectrum packet radio network link in a Poisson field of interferers. IEEE Trans Inf Theory 38(6):1743–1754
6. Nikias CL, Shao M (1995) Signal processing with alpha-stable distributions and applications. Wiley, New York
7. Ilow J, Hatzinakos D (1998) Analytic alpha-stable noise modeling in a Poisson field of interferers or scatters. IEEE Trans Signal Process 46(6):1601–1611

8. Blackard KL, Rappaport TS, Bostian CW (1993) Measurements and models of radio frequency impulsive noise for indoor wireless communications. IEEE J Sel Areas Comm 11(7):991–1001
9. Blankenship TK, Krizman DM, Rappaport TS (1997) Measurements and simulation of radio frequency impulsive noise in hospitals and clinics. In: Proceedings of the 47th IEEE vehicular technology conference, Phoenix, AZ, May 1997, vol. 3, pp 1942–1946
10. Kogon SM, Manolakis DG (1996) Signal modeling with self-similar α-stable processes: the fractional levy stable motion model. IEEE Trans Signal Process 44:1006–1010
11. Parsons JD (1996) The mobile radio propagation channel. Wiley, New York
12. Kassam SA (1987) Signal detection in non-Gaussian noise. Springer, New York
13. Poor HV, Thomas JB (1993) Signal detection in dependent non-Gaussian noise. In: Poor HV, Thomas JB (eds) Advances in statistical signal processing, vol. 2, JAI Press, Greenwich, CT
14. McDonald KF, Blum RS (2000) A statistical and physical mechanisms-based interference and noise model for array observations. IEEE Trans Signal Process 48(7):2044–2056
15. Middleton D (1995) Threshold detection in correlated non-Gaussian noise fields. IEEE Trans Inf Theory 41(4):976–1000
16. Yang X, Petropulu AP (2002) Interference modeling in radio communication networks. In: Proakis J (ed) Trends in wireless indoor networks, Wiley encyclopedia of telecommunications, Wiley, New York
17. Navaie K, Valaee S, Sharafat AR, Sousa ES (2006) IEEE Trans Wireless Commun 5(2): 384–393
18. Chong C-C, Pinto PC, Win MZ, Watanabe F, Inamura H (2008) Method and system for wireless design subject to interference contraints. US Patent Application US2008/0188253
19. Maras AM (1988) Locally optimum detection in moving average non-Gaussian noise. IEEE Trans Commun 36(8):907–912
20. Maras AM (1994) Locally optimum Bayes detection in ergotic Markov noise. IEEE Trans Inf Theory 40(1):41–55
21. Samorodnitsky G, Taqqu MS (1994) Stable non-Gaussian random processes: stochastic models with infinite variance, Chapman and Hall, New York
22. Petropulu AP, Pesquet J-C, Yang X (2000) Power-law shot noise and relationship to long-memory processes. IEEE Trans Signal Process 48(7):1883–1892
23. Beran J (1994) Statistics for long-memory processes, Chapman & Hall, New York
24. Yang X, Petropulu AP (2001) The extended alternating fractal renewal process for modeling traffic in high-speed communication networks. IEEE Trans Signal Process 49(7):1349–1363
25. Yang X, Petropulu AP (2003) Co-channel interference modeling and analysis in a Poisson field of interferers in wireless communications. IEEE Trans Signal Process 51(1):64–76
26. Kunz T, Barry T, Zhou X et al (2000) WAP traffic: description and comparison to WWW traffic. In: Proceedings of the 3rd ACM international workshop on modeling, analysis and simulation of wireless and mobile systems, Boston, USA, August 2000.
27. Willinger W, Taqqu MS, Sherman R, Wilson DV (1997) Self-similarity through high-variability: statistical analysis of Ethernet LAN traffic at the source level. IEEE/ACM Trans Network 5(1):71–86
28. Arnold BC (1983) Pareto distributions, International Co. Publishing House, Maryland
29. Yang X, Poor HV, Petropulu AP (2004) Memoryless discrete-time signal detection in long-range dependent noise. IEEE Trans Signal Process 52(6):1607–1619

Chapter 20
Introduction to: "Elements of Non-Gaussian Space-Time Statistical Communication Theory, Propagation, Noise, and Signal Processing in the Canonical Channel"

David Middleton

This Introduction explains my purpose of writing this book and its earlier companion [1]. It is based on the observation that *communication* is the central operation of discovery in all branches of science. In its "active mode," we use it to "interrogate" the physical world, sending appropriate "signals" and receiving nature's "reply." In the "passive mode," we receive nature's signals directly. Since we never know *a priori* what particular return signal will be forthcoming, we must necessarily adopt a probabilistic model of communication. This has developed over the approximately seventy years since its beginning, into a Statistical Communication Theory (SCT). Here, it is the *set* or *ensemble* of possible results, which is meaningful. From this ensemble, we attempt to construct in the appropriate model format, based on our understanding of the observed physical data and on the associated statistical mechanism, analytically represented by suitable probability measures.

Since its inception in the late '30s of the last century, and in particular subsequent to World War II, SCT has grown into a major field of study. As we have noted above, SCT is applicable to all branches of science. The latter itself is inherently and ultimately probabilistic at all levels. Moreover, in the natural world there is always a random background "noise" as well as an inherent *a priori* uncertainty in the presentation of deterministic observations, i.e. those which are specifically obtained, *a posteriori*.

20.1 The Book's Title

Let me now begin with a brief explanation of the Book's title:

> Elements of Non-Gaussian Space-Time Statistical Communication Theory, Propagation, Noise, and Signal Processing in the Canonical Channel.

D. Middleton‡

D. Middleton, *Non-Gaussian Statistical Communication Theory*, IEEE Press,
DOI 10.1007/978-0470948477_20,

My choice of "Elements" is intended to signify a connected development of fundamental material, but with an exposition which is inevitably incomplete, with many important topics necessarily omitted, primarily for reasons of space. "Elements," however, includes the propagation physics of the channel, the rôle of spatial coupling (e.g., apertures and arrays), and noise models, both physically founded. The analyses also treat deterministic and random scatter, Doppler effects, and, of course, four-dimensional (i.e., space and time) signal processing, with particular attention to threshold reception in arbitrary noise environments. *Non-Gaussian noise* receives special analysis, since it is a phenomenon of increasing practical importance. Moreover, it is a topic which presents much greater complexities than the familiar Gaussian noise model, which has dominated so much of recent as well as earlier studies.

In addition, the class of signals considered here is entirely general or "canonical,"[1] so that the coding results of parallel studies in *Information Theory*[2] [2] can be readily applied in specific cases. This Book (Book 2) may also be considered an extension of Book 1 (*An Introduction to Statistical Communication Theory*, [1]). Book 1 considers primarily random processes and continuously sampled noise and signals. Here, on the other hand, Book 2 deals with many of Book 1's earlier features. These which require a *four-dimensional, space-time formulation*, now involve *random fields*. Particular attention is also given here to the physics of propagation. In this context, another portion of this book is then devoted to physical problems of signal detection and extraction in a Bayesian formulation, with particular attention to threshold (or weak-signal) operation.

Finally, both homogeneous and inhomogeneous media are considered here. Such media are *linear* provided their equations of propagation are themselves linear where the requirement, of course, is that the superposition principle holds: If α is a typical field in such media, we have, symbolically for two fields $\alpha = (\alpha_1, \alpha_2) \rightarrow \alpha_{12} \equiv \alpha_1 + \alpha_2$. On the other hand, for nonlinear media $(\alpha_1, \alpha_2) \rightarrow \alpha_{12} \neq \alpha_1 + \alpha_2$: superposition is violated. Furthermore, stationarity or nonstationarity itself does not invalidate linearity, or nonlinearity. Note, however, that the presence of a (finite) boundary is itself an inhomogeneity of the medium, and thus is a component of nonlinearity by the above definition. The presence of scattering elements (inhomogeneities) is a major topic of interest as well, as are the probability distributions generated by such scattering elements.

[1] Here, the usage of "canonical" is to indicate a form independent of a particular choice in specific applications or branch of physics.

[2] See Fig. 20.1 below.

20.2 Communication Theory, the Scientific Method, and the Detailed Rôle of SCT

As I have noted above, this book is primarily an analytical presentation. For numerical results, it may be regarded as providing a set of macro-algorithms, to direct the computation of the desired numerical quantities in specific cases. Because of the availability of cheap and powerful computing today, and modest software costs, such numerical results should be readily and quickly available, once the needed programs (software) are obtained from the aforementioned macro-algorithms. These macro-algorithms are the consequence of a well-known scientific methodology whose basic principles are stated in somewhat simplified terms below:

$$\textit{Hypothesis} + \textit{Experiment} = \textit{Theory}^{*} \tag{20.1}$$

Loosely stated, "hypothesis" is a conjecture or proposition – experiment is the procedure required to verify or to disprove the proposition. Verification here implies replication by any qualified observer any number of times. "Theory" is the result of successful verification. A theory is thus potentially acceptable (or not) as long as it is empirically verified by appropriate observation. A fruitful theory is one which not only accounts for the observed phenomena but also leads to further discoveries. For science, the arena of application is the physical world, where it is the ontological "what and how" of the universe, not its "why".

Here I employ two familiar types of theory. These I call *a posteriori* and *a priori* theories. In the former, everything is already known, for example, equations of state, boundary conditions, propagation relations, etc. It remains only to calculate a numerical result, where all of its elements collectively constitute a unique representation, i.e. a deterministic result. For the latter, the same structure exists, but it is not now known which representation is *a priori* present, only that a particular result will have a certain probability of actually occurring, in a universe of possible outcomes. It is this fundamental uncertainty which is the key added feature, defining the field of *statistical* communication theory. It is this probabilistic nature, combined with a set of deterministic representations and associated probability measures, which in turn defines the subject. This concept, which was introduced systematically in the 1930s, was accelerated by WWII (1939–1945), and has continued into the peace time explosion of the new science and its corresponding technology.

Apart from the broad and fundamental impact, SCT is the science indicated at the beginning of my Introduction. SCT also has a microstructure which has the specifically detailed rôle of including the physical sciences. Figure 20.1 illustrates other interdisciplinary relationships.

Figure 20.1 represents my subjective view of where Statistical Communication Theory (SCT) fits into the scientific enterprise and specifically where the signal processing and channel physics lie in this hierarchy. The double set of arrows (↑↓, ↓↑) indicates the interrelationships of the various fields of study and emphasizes their interdependence. Intimately involved in all of this is the rôle of technology,

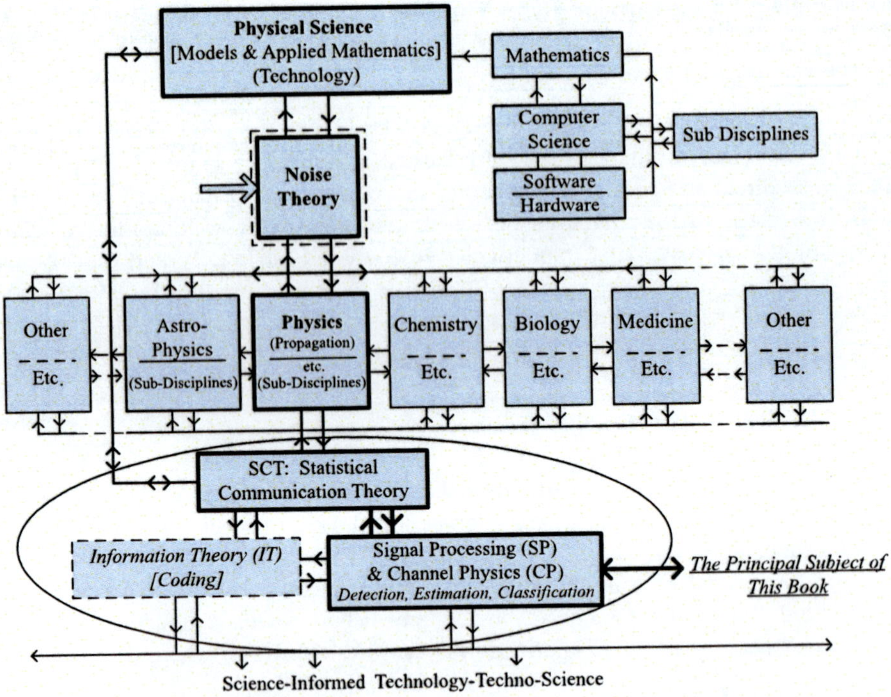

Fig. 20.1 The Rôle of Statistical Signal Processing (SSP) in the Physical Sciences

which provides the instruments of discovery, investigation, and study. The direct arrow in the diagram between SCT and the physical science themselves further emphasizes the aforementioned fact that *communication* is the link connecting the natural world with the methodology used to study it. In fact, communication in one form or another, appropriate to the scale of the phenomenon under study, is the necessary vehicle here. The progression is from the Micro to the Macro, i.e. the very small at the quantum level (e.g., the quantum fluctuations of the vacuum [2]) to the astronomical dimensions of the fluctuating gravity fields of galaxies, for example [3].

As we have noted at the beginning, physical science is also based on model building. This in turn requires both the interrogation for, and the reception of, data from the physical world. The probabilistic nature of the physical world is inherent according to modern understanding (as exemplified by Heisenberg's uncertainty principle and the behavior of subatomic particles), along with the ultimate uncertainty of measurement itself. The former is a fundamental property of matter and energy; the latter is independently a problem of technique. These remarks may be concisely summarized by the following three observations:

I. Physical Science is a *model-building* operation;*
II. Physical Science is fundamentally a *probabilistic endeavor*, and;
III. *Communication* is the process by which *hypothesis* is converted *into theory*, cf. (20.1). → (20.2)

Thus, the communication process either actively interrogates the real world or passively receives its "messages." This in either case embodies the rôle of experimental verification (or rejection) cf. Eq. 20.1. Although there is nothing really original about these remarks I-III above, they may serve as useful reminders of the scientific process. The quantitative language of science is, of course, mathematics.

Note the key place of "Noise Theory" in the hierarchy of Fig. 20.1. Whereas the rôle of noise is mostly a negative one from the point of view of treatment here in "Signal Processing etc.," it has proved to be a highly productive field in the broader context of modern physics since the beginning of the twentieth century [5]. For example, Einstein used it to prove the existence of atoms in 1905, which was a highly controversial topic up to that time. A host of other eminent scientists, among them Boltzmann and Langevin (see [5]), also advanced its theory. See, for example, [3–6]. Since "noise," a basic random phenomenon, pervades every field of science, its study since then has yielded a host of discoveries and new methodologies. Its history has indeed proved "glorious," as Prof. Cohen has so aptly described it in his recent, enlightening Review article [6].

Although I have focused here on the deleterious and ubiquities effects of noise on signal reception (particularly in Parts 1, 3 of the present book), its physical and analytic description, especially for *nongaussian noise*, is a necessary and significant major subject for discussion as well (cf. Part 2, Chapters 11 – 13). The dominant relation of noise to the physical problems inherent in Signal Processing is well known. Its important companion discipline in SCT, *Information Theory*,[3] is emphasized by the direct arrow in *Fig. 20.1*, as well as its connection to the physics of propagation. The close mutual relationship of the ensuing technology is also noted and is a major part of the advances discussed here and appearing in all fields of science.

20.3 The Scope of the Analysis and New General Results

Before summarizing the contents of this book, let me describe the physical domain of most of the book's contents. Topics not treated here are also noted. The principal areas of application are mainly acoustical and electromagnetic. Exceptions are

[3]In the *Information Theory* community, the space-time formulation represents the simultaneous use of separate multiple channels, which may be statistically related and possibly coupled for simultaneous new versions, of one (or more) signals received together. The details of the spatial environment in the immediate neighborhood of the separate receivers, however, are not directly considered from a physical point of view. The effects of the different receiver locations are subsumed in the different received waveforms. These are usually suitably combined in the receiver to take advantage of their common structure at the transmitter. The noise background at the receivers on the other hand can have noticeable statistical correlations, which can enhance the received signal(s) in reception, as is typical of MIMO (multiple input – multiple output) reception [30].

quantum mechanical (the very small) and astrophysical (the very large), where elements of Statistical Communication Theory (SCT) are also specifically, although briefly, considered. In all cases from the SCT-viewpoint, we have to deal with noise, and signals in such noise, propagating in space–time. From the SCT viewpoint, these are determined by the physical properties of the channel. Throughout, the appropriate language is statistical, specifically for the inherently random character of these channels and for reception, namely detection and estimation in the face of uncertainty, i.e. in an *a priori* theory, cf. II above.

We are also dealing here mostly with media, which are regarded as continuous. These may be described by the following simple hierarchy:

(1) *Vacuum*: empty space, no matter present at all. This is typically the usual assumption made here for most electromagnetic propagation. Such media, of course, do not support conventional acoustic propagation.
(2) *Gas*: a low-density continuum, e.g., earth's atmosphere, and other low-density environments. These media clearly do not support a *shear*: $\nabla \times \boldsymbol{\alpha} = 0$, ($\alpha$ = displacement field. Here, α is also said to be *irrotational*.)
(3) *Liquid*: usually of greater density than gas, for example, the ocean. These also do not support a shear, e.g. $\nabla \times \boldsymbol{\alpha} = \mathbf{0}$. Here, we consider the propagation of acoustic waves primarily.
(4) *Solids*: these are dense enough to maintain their shape, or at most suffer minor distortions from a rest condition, i.e. an undergo and restore shape modifications. Such media are comparatively elastic, deformable, and restorable. These bodies can support a shear, e.g. $\nabla \times \boldsymbol{\alpha} \neq \mathbf{0}$. When subject to stresses alone, these deformable media, produce a displacement (vector) $\Delta = \alpha s$ which is *solenoidal*, i.e. $\nabla - \boldsymbol{\alpha} = 0$. Electromagnetic and acoustic radiation is also possible in all of the above media (except the latter in a vacuum). The magnitude of the results in the acoustic and electromagnetic cases depends, of course, on the physical properties of the media involved. For example, EM radiation in fresh and salt water is significantly weaker than for acoustic radiation, but may be enhanced in certain conducting solids vis-à-vis vacuum or liquids, etc. In all cases, we must pay attention to the appropriate conductive properties of the medium in question.

We define the domains of the analysis here in terms of the range of frequencies used for the signals and hence for the accompanying propagation whether ambient or signal-generated (i.e., scattering). For acoustic applications, we have (for an average speed of sound in water $c_0 = 1.5 \cdot 10^3$ m/sec), a frequency range of $O(1\text{–}10^7$ Hz), or in wavelengths[4] $\lambda = 1.5 \cdot 10^3$ m to $\lambda = 1.5 \cdot 10^{-4}$ m. For the EM cases, we consider frequencies in the range $O(10^0\text{–}10^{12}$ Hz) with the speed of light in space

[4]I use $f\lambda = c_0$, a dispersionless approximation in both acoustic and EM cases, e.g. $\lambda = c_0/f$, etc.

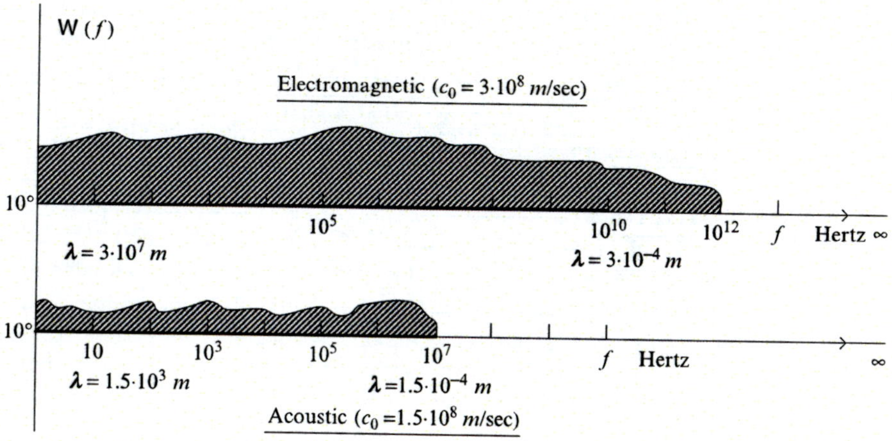

Fig. 20.2 Current frequency range of signals used in general Electromagnetic and Acoustic Communications, e.g., radar, sonar, and telecommunications generally (from the relation $\lambda f = c_0$)

$c_0 = 3\cdot10^8\,\text{m/sec}$, which is in wavelengths $O(3\cdot10^7\,\text{m}–3\cdot10^{-4}\,\text{m})$. (These frequency or equivalent wavelength intervals, are, of course, somewhat loose, and exceeded in any optical and quantum mechanical examples which are discussed here.)

Figure 20.2 illustrates the ranges of frequencies to be used in (i) the electromagnetic and (ii) acoustic applications:

20.4 Some Specific New Results

With the above in mind, let me note now what I believe to be the major *new* material presented in this book. These are discussed in detail in subsequent chapters, some of which have been published in recent journal papers as well. Note the following (mostly new) SCT topics specifically:

(1) A *Space–Time Formulation* throughout, largely *with discrete sampling* of continuous noise and signal fields. This includes *space–time matched filters* for optimum performance in reception. This includes quantification of *system degradation*, in particular for ultra- and ultra wide-band (UWB, UW^{2}B) signals when optimal space–time (*ST*) processing is replaced by the usual *separate* space and time ($S \otimes T$) processing in conventional receivers;

(2) A theory of *Jointly Coupled Detection and Estimation,* as developed by the author and R. Esposito, and generalized in Chapters 6 and 7, with additional new references;

(3) *Extensions of Classical Noise Theory* (Langevin Equations, Fokker-Planck approximations, and classical scatter methods) to include random spatial phenomena (Part 2, Chapter 9);

(4) *New Methods in Scattering Theory* (1997-), from which first- and second-order probability densities are directly obtained, cf. Chapter 10, Part 2 for this physical statistics (PS) approach vis-à-vis the limited classical Statistical Physics (SP) techniques discussed in Chapter 9;
(5) *Physically based Non-Gaussian Noise Models* (Class A and Class B canonical noise distributions), developed since 1974 by the author and presented in Chapters 11 and 12, Part 2;
(6) A systematic program of *Threshold* (or Weak-Signal) *Detection and Estimation* for general noise fields, particularly for non-Gaussian noise, and including as a special case, the Gaussian noise of earlier treatments (Part 3).
(7) The physics of propagation, in the context of the channel description and its space-time coupling, in transmission and reception for a variety of media, with general boundary conditions.

Our list here also includes:

(8) An emphasis on the *Interdisciplinary Character* of the presentation throughout; (see the diagram of *Fig. 20.1*), with specific examples, e.g., astrophysics, computer traffic probabilities. This also leads to the development of *canonical expressions* independent of particular physical disciplines, which permits their treatment by a common methodology;
(9) An extension of the scalar field analysis of Part 2 (Chapters 8–14) to *vector fields* in Chapter, Part 4, specifically illustrated by the full electromagnetic field, with the typical reception applications; and finally
(10) A very concise treatment of quantum detection and estimation is presented in Chapter, Part 4.

20.5 Overview: Parts 1–4

The following provides a brief summary of the book's contents. We begin with

Part 1. An Introduction to Non-Gaussian Statistical Communication Theory: A Bayesian Formulation

Part 1 presents most of the basic statistical apparatus needed to apply Statistical Decision Theory from the Bayesian probabilistic point of view. These are the ultimate problems of Statistical Communication Theory, namely optimal and sub-optimal detection (D) and extraction (E) of desired signal and noise. Chapters 1–3 introduce the subjects of space–time optimality of detection. These employ space–time Bayes matched filters and Gaussian fields, as well as coupling of the canonical channel[5] to receiver and transmitter. Inhomogeneous and nonstationary (non-Hom-Stat) conditions are considered as well as the more usual Hom-Stat situations, along with an introductory discussion of one- to three-dimensional arrays for explicit channel coupling. Discrete sampling is also introduced in these chapters, in contrast to the earlier treatment of Book 1, [1], in accordance with the usual digitalized handling of most data today. These first three chapters provide an introduction to the chapters which follow. Chapters 4 and 5 are reviews of much of the material in Chapters 21–23 of book 1, [1] extended to space as well as time. On the other hand, Chapters 6 and 7 present mostly new material, namely a theory of jointly coupled signal detection *and* extraction [cf. (1) above], for both binary and M-ary ($M > 2$) signals in noise. Part 1 concludes with a brief summary of the results.

Part 2. The Canonical Channel: Deterministic and Random Fields, Non-Gaussian Noise Models and Doppler

Part 2 introduces some essential elements of the classical theory of propagation. This is needed in our effort to apply these elements to the quantitative description of the channel itself, which is the origin of much of the noise, which interferes and limits reception. The aim here is to go beyond the "black box" labeled "noise" and postulated *ad hoc* statistics to the physically observed and analytically derived statistical distributions, in both time *and* space. Thus, propagation is a complex operation, involving the structure of the medium, boundaries, boundary and initial conditions. This is especially the case for inhomogeneous scattering media, and which in addition may be absorptive (i.e., dissipative). Chapter 8 discusses with deterministic cases, examples of *a posteriori* formulations. These in turn may next be regarded as "representations," which in turn form a statistical ensemble when suitable probability measures are assigned to them. This randomizing feature is characteristic of the *a priori* theory mentioned above (I), and which is our main concern here.

Two classes of problems are considered in Chapter 9: (1) where the ensemble of dynamical equations are deterministic and the driving source is itself a random field $G_T(\mathbf{R}, t)$, so that the resulting field $a(\mathbf{R}, t)$ is the probabilistic solution to $L^{(0)}\alpha_H = -G_T$ (or $\alpha_H = \hat{M}^{(0)}G_T$); $\hat{M}$ = integral Green's function. The second class of problem is the more difficult one of scattering, represented by the ensemble of equations of the form $\alpha^{(Q)} = \alpha_H + \hat{h}(Q)\alpha^{(Q)}$, which is nonlinear in the scattering elements ($\sim \hat{Q}$), (Chapters 8, 9). The *a priori* approach is next introduced for the classical treatment of scattering. It is noted that again here scattering is a *nonlinear* property (with respect to the scattering elements) of such random media. (It is also a similar type of nonlinearity for the deterministic inhomogeneous

[5] i.e., in analytic forms applicable to a variety of different specific physical cases, i.e., acoustic, or electromagnetic, etc.

media of Chapter 8.) The principal results here are the low-order moments of the governing probability distributions. Only in special cases (involving linear Langevin equations, Gaussian statistics, and Markoff assumptions) of the first class are analytic solutions for a full treatment generally available. Chapter 10 following, however, remedies this situation with a new, purely probabilistic approach. Its advantages, and limitations with respect to the classical treatment of Chapter 9, are also discussed in Chapter 10.

Chapters 11 and 12 turn next to canonical forms,[6] as well as physical, derivations of three major classes of non-Gaussian noise, namely Poisson, *Class A, and Class B noise*,[7] which represent most physical random noise processes, including Thermal, Shot, and Impulse Noise. Chapter 11 presents the first-order pdfs of these three general classes of noise, Chapter 12 extends the treatment to the second-order cases. Physical non-Gaussian noise models are also considerably more analytically complex than Gaussian noise models are often encountered in practical receiving systems. Moreover, Poisson Class A and B noise (of interest to us here in the frequency ranges of Fig. 20.2, cf. Chapters 10 and 12 of [1]) are now usually a dominant component of interference in the channel. As a possible alternative to these relating complex, physical models we use Chapter 13 to present a brief treatment of various common, so-called *ad hoc* noise pdfs. These have comparatively simple analytic forms but whose relationship to the underlying physical mechanisms is relatively tenuous. Chapter 14 concludes Part 2 with formulations for deterministic and random Doppler, which arise in many practical situations when the transmitting and/or receiving sensor platforms are in relative nation to one another, and or to a fixed frame of reference.

Part 3: Threshold Theory: Detection and Estimation – Structure and Performance

Part 3 is devoted to optimum and near optimum. Threshold signal processing in general noise environments, in particular non-Gaussian noise whose explicit pdfs are obtained in Part 2. Here, we are concerned not only with the structures of the detector and estimator, which are themselves generally sufficient statistics, but also with their performance. Chapter 15 develops canonical forms of such signal processors from the general arbitrary pdfs of the noise. This is done for both additive and multiplicative signals and noise, such as these produced in inhomogeneous media with scattering elements, c.f. Chapter 16. We then apply these results to both optimum and suboptimum signal detection, while Chapter 17 considers analogous results for signal estimation. Chapters 18 and 19 are devoted, respectively, to

---

[6] i.e. in analytic forms applicable to a variety of different specific physical cases, i.e. acoustic, or electromagnetic, etc.

[7] The author's designation.

examples from fluids, i.e., underwater acoustics, from elastic solids, and analogous electromagnetic vector field formulations, all in weak-signal régimes, which permit a general treatment.

Part 4: Special Topics

Part 4 concludes our general treatment and consists of a variety of special problems, based on the results of Parts 1–2 and selected papers. Chapter 20 describes acoustic problems connected with the reception of sonar signals in the ocean, where wave surface, volume, and bottom scatter are the principal interference mechanisms. Chapter 21 extends the analysis for radar in the full electromagnetic formulation. Chapter 22 considers next a variety of special problems listed below, which address briefly various additional features of the preceding analyses of Parts 1–3.

(i) Effect on performance of the separation of space (S) and time in reception, $[S \otimes T]$ vs. $[ST]$; See (6), **IV** above and (xi) below;
(ii) Path integrals, for Class A and B noise;
(iii) Optical communications and quantum effects; lasers as technical enablers;
(iv) (Introductory) Exposition of *Quantum Mechanics for Communications* (Lasers, Optics, etc.);
(v) Matched Field Processing (*MFP*);
(vi) Noise Signals in Noise – the (Non-Singular) Gaussian Case for Space–time Fields;
(vii) Soliton Models, Spectrum; Wind-Wave surface structures;
(viii) Astrophysics and Computer LAN Traffic Applications;
(ix) Signal Fading and *Multipath*;
(x) Ocean Wave Surface Models – Surface Scatter, etc.
(xi) Ultra Wide-Band Systems (see (i) above)
(xii) Propagations for Time-Reversal or Reciprocal Media.

The book concludes with a series of appendixes, mathematical relations needed above, and additional references.

20.6 Level of Treatment

The analytic requirements are comparable to that required in texts on theoretical physics or engineering. Morse and Feshbach [15] have noted, in particular, with the addition of the special probability methods guided by SCT (as used for example in [15]). From the physical point of view here, the slogan "more vigor than rigor" is to be expected. The (ε, δ), etc. of pure mathematics is implied, and is presented elsewhere. For example, the Dirac delta function, $\delta(x{-}x_0)$ and its generalizations have been shown to belong to the class of "generalized functions." These are described in Lighthill's book [vi], based on the concepts of Lebesque, Stieltjes, and L. Schwartz, [cf. Chapters 2, 3 of [16]), and the various extensions of the limit concept in integration. In addition, to facilitate the handling of the mathematical details in the propagation models, an operator formalism is frequently employed here. This provides a certain measure of compactness to the analytic treatment in many cases. (See, in particular, Part 2.)

The class of functions representing the physical models used here and in similar problems can, and has been shown to give correct answers for these physical models, in addition to satisfying one's intuitive expectations. I have thus tried to avoid "cluttering up" the physical arguments represented by these macro-algorithms [cf. I, p. 1 above] by avoiding the full rigor of the associated "pure" mathematics, with its often arcane (to us) symbolisms and operations.

Finally, I have also included examples of the so-called "*Engineering Approach*" to these problems. Here, the (linear) canonical channel is represented by a *linear time-variable filter*. It is shown that this representation is valid, i.e. is equivalent to the general physical description of the channel in the cases treated here, *only in the far-field (Frauenhofer) régimes*. Moreover, it does not explicitly indicate range effects, namely, the attenuation of the propagating field due to "spreading." Additional conditions on the channel itself are that the signal applied to each sensor of the transmitting array must be the same and that the receiving portion of coupling to the channel must be an all-pass-network [see Chapter 8, Sec. 8.2, and Eqs. (8.2–8.8), (8.2–8.14). These conditions can often be met in practice, since they are usually acceptable in many applications.

20.7 Referencing, and Some Topics not Covered

In addition to the above, let me add a few remarks about the referencing, with respect to the selection of book and journal articles: These selections are based on a number of criteria and personal observations:

(1) Obviously important and pertinent books;
(2) The need for an inevitably limited number of sources, chosen now from thousands of possibilities, which are available via various search engines (Google, AOL, etc.);
(3) Recognition that there *is* an historical record, which is both informative and needed. This must necessarily involve a relatively small, finite number of books and papers to be manageable and thereby readily useful. Many of the references which I have used here since 1942 (including those cited in later editions of [1]) are from the formative period (1940–1970) in the development of SCI. These are still pertinent today.
(4) That any finite selection inevitably reflects the subjective choices of the author. However, I feel that these choices deserve attention, although they are now ignored in much of the current literature, which appears to have a "corporate memory" of only a decade and a half [8]. The "new" is not necessarily better.

(5) During the seventy years of my activity in Statistical Communication Theory (SCT), as it developed during World War II and subsequently [cf. Fig. 20.1 above], I have encountered much relevant material, cf. (4) above. From these, I have chosen the references used throughout. Undoubtedly, I have missed many others, for which I ask the readers' indulgence.

In addition, there are also many important topics in SCT which of necessity I have had to omit, in order to keep the sheer size of the presentation under some control. Clearly, a major field of equal importance is *Information Theory,* essentially the theory of Coding [2], [9], to be applied to the canonical signals postulated here [cf. *Fig. 20.1*] above. The purpose of such limits is: (1) to preserve an acceptable combination of probabilistic methods (Part 1); (2) to present not an entirely trivial account of the relevant physics (Part 2); (3) to give the development of threshold theory (Part 3), which extends the applicability of SCT to the important (and more difficult) cases involving non-Gaussian noise, in addition to the usual Gaussian treatment; and (4) to illustrate (in Part 4) the scope of SCT through its applicability to a variety of diverse special topics.

Thus, Book 1 [ref. [1]] may be regarded as a treatment of SCT involving temporal *processes* only, whereas Book (2) here provides an extension of SCT to space–time *fields*. Both books require the same level of capability; both are primarily research monographs, at the doctoral level, and both require a measure of familiarity with mathematical physics, as well as theoretical engineering, cf. References below. Of course, suitably prepared graduate students can also expect to find the book useful. In this connection, the rôle of the problems here is the same: to provide useful and special results, in addition to the text itself.

Acknowledgements I thank my many colleagues who critiqued various chapters and who in some instances cast critical eyes on the entire manuscript. They have been most helpful indeed. Their number include H. Vincent Poor, Professor and Dean of the Electrical Engineering and Computer Department at Princeton University, Prof. Leon Cohen, Hunter College Graduate Physics Department, Prof. Steven S. Kay, Engineering & Computer Science Department, University of Rhode Island, Prof. J. W. Goodman (Optical Science, Stanford Univ.), Prof. John Proakis (EE & Comp, Northeastern University), Dr. Julian J. Bussgang (Founder of *Signitatron*, MA,), Prof. V. Bhargava (Univ. of Victoria, Victoria, B.C. (EE & Comp.), Prof. Eric Jakeman, Royal Society, (the Univ. of Manchester, G.B.).

My grateful thanks to my colleague Dr. William A. Von Winkle, Assoc. Director of the Naval Under Sea Warfare Center (NUWC) of the Navy [alas, now recently deceased (1930 – Aug. 12, 2007)], who supported and encouraged my work during the period of 1960–1980. In addition, I am also particularly indebted to the US Navy's Office of Naval Research, which through the years (1949–2005) has also supported much of my original research on the above topics, in addition to various private Industrial Research departments during this period. (See DM, *Who's Who in America*.)

For my errors of commission and omission, I take full responsibility. My reviewers have added positive comments and corrective support, for which I am most grateful (See VIII, Acknowledgements).

In addition, I thank the two young ladies, Ms. Dawn M. Goldstein (Centreville, VA) and Ms. Nancy S. Russell (Niantic, CT), who have been my most accomplished technical typists. They have had a very difficult (hand-written) manuscript to deal with. It is because of their efforts over quite a few years, that they have produced a beautiful text.

Finally, and not by no means the least, I am deeply grateful to my wife, Joan Bartlett Middleton, for her consideration and encouragement, particularly over many years (1992 – 2010) during which I have been creating this book.

References

1. Middleton D (1965) An introduction to statistical communication theory (ISCT), McGraw-Hill, New York, International series in pure and applied physics, 1960–1972, 1st edn, 2nd edn, Peninsula Publishing, Los Altos, 1987–1995; 3rd edn (Classic edn.), IEEE Press, Piscataway, 08855 and Wiley, New York, 1996; also, Topics in communication theory, McGraw-Hill, New York
2. Shannon CE (1949) The mathematical theory of communication, with Warren Weaver. University of Illinois Press, Urbana
3. Chandrasekhar S (1943) Stochastic problems in physics and astronomy. Rev Mod Phys 15(1):1–91
4. Wax N (ed) (1954) Noise and stochastic processes (Selected Papers). Dover, New York, p 19
5. Lax M (1960) Fluctuations from the non-equilibrium steady state (I). Rev Mod Phys 32 (1):25–64; (1966) Classical noise (III): nonlinear Markoff processes. Rev Mod Phys 38(2):359–379; (1966) Classical noise (IV): Langevin methods. Rev Mod Phys 38(3):544–566; (1960) Influence of trapping, diffusion, and recombination on carrier concentration fluctuations. J Phys Chem Solid (II), 14:248–267 (Printed in Great Britain). (See also Supplementary Refs (1996) to 3rd ed. of Ref. [1] above, Chapter 10, pp 1113, and Ref. [3], Chapter 11, Rev Mod Phys). These papers contain many additional references to related work for 1966 and earlier
6. Cohen L (2005) The history of noise. IEEE Signal Process Mag 20–45
7. Middleton D, Esposito R (1968) Simultaneous optimum detection and estimation of signals in noise. IEEE Trans Inform Theor IT-14(3):434–444; (1970) New results in the theory of simultaneous optimum detection and estimation of signals in noise. Problemy Peredachi Informatii, Moscow, USSR, 6(2):3–20, April–June, 1970, English Translation Consultants Bureau (New York), Plenum, New York, 1972
8. Middleton D (2005) Information Theory Society Newsletters. Reflections Reminiscences 55(3):3, 5–7
9. Bykovsky M (2005) Pioneers of the information era.In: History of the development of communication theory, 192006. Technosphera, Moscow, p 375
10. Helstrom CW (1976) Quantum detection and estimation theory. In: Mathematics in science and engineering, vol 123. Academic, New York

Some additional references are noted, which are also used in much of the Book following, with many others mainly specific to individual Chapters. These are:

11. Margenau H, Murphy GM (1943) The mathematics of physics and chemistry. D. Van Nostrand, New York
12. Watson GN (1944) Theory of Bessel functions, 2nd edn. Cambridge University Press, New York
13. Campbell, GA, Forster RM (1948) Fourier integrals. D. Van Nostrand, New York
14. Stratton JA (1941) Electromagnetic theory. McGraw-Hill, New York

15. Morse, PM, Feshbach H (1953) Methods of theoretical physics. In: International series in pure and applied physics, vols 1, 2. McGraw-Hill, New York (A comprehensive non-random physical account of propagation, etc.)
16. Lighthill MJ (1958) Fourier series and generalized functions, Chap. 2. Cambridge University Press, Cambridge
17. Kotelnikov VA (1959) The theory of optimum noise immunity (Translated from the Russian by R. A. Silverman), originally Kotelnikov's doctoral dissertation in 1947, published in the Soviet Union in 1956. McGraw-Hill, New York
18. Lindsey RB (1960) Mechanical radiation. In: International series in pure and applied physics. McGraw-Hill, New York
19. Brillouin L (1962) Science and information theory, 2nd edn. Academic, New York; (1946) Wave propagation in periodic structures (Electric filters and crystal lattices), International series in pure and applied physics. McGraw-Hill, New York
20. Tolstoy I, Clay CS (1966) Ocean acoustics, theory and experiment in underwater sound. McGraw-Hill, New York
21. Rytov SM, Kravtsov YA, Tatarskii VI (1976) Vvedenie v. Statisticheskuyu Radiofiziku, **I**, Sluchainue Protscesui, Nauka, Moscow; Vvedenie v. Statisticheskuyu Radiofiziku, **II**, Sluchainuie Polya, 1978, originally in Russian; Principles of Statistical Radiophysics, **I**, Elements of Random Process Theory 1987; **II**, Correlation Theory of Random Processes 1988; **III**, Elements of Random Fields, 1989; **IV**, Wave Propagation Through Random Media, 1989. Springer, New York (in English). Basically, classical theory: Gaussian random processes; continuous media "microstructure" and boundary conditions; finite, low order moments with structure, boundaries, and then nonlinear media, cf. Chapter 9, Sec. 9.________, of present Book; classical theory.
22. Ishimaru A (1978) Wave propagation and scattering in random media. In: Single scatter and transport theory,vol I; Multiple scatter, turbulence, rough surfaces, and remote sensing, vol II. Academic, New York (classical theory, cf. Refs. [20], [21].)
23. Gradshteyn IS, Ryzhrik IN (1980) Table of integrals, series and products (Corrected and Enlarged edn, by Alan Jeffrey)
24. Goodman JW (1985) Statistical optics. Wiley, New York; (2007) Speckle phenomenon in optics. Roberts and Co., Englewood
25. Gardiner GW (1985) Handbook of stochastic methods, 2nd edn. Springer, New York
26. Chew WC (1995) Waves and fields in inhomogeneous media. IEEE, New York; Series on electromagnetic waves. IEEE, New York (originally pub. By D. Van Nostrand-Reinhold, 1990; reprinted by IEEE Press, 1995 (Classical, non-random, a posteriori theory)
27. Helstrom CW (1995) Elements of signal detection and estimation. Prentice-Hall, New Jersey (SCT with Gaussian noise)
28. Poor HV (1994) An introduction to signal detection and estimation, 2nd edn. Springer, New York
29. VanKampen NG (1992) Stochastic processes in physics and chemistry (Revised and enlarged edn.), North Holland, New York
30. Hochwald BM, Hassik B, Marzetta TL, Guest (eds) (2003) The academic and industrial embrace of space-time methods. IEEE Trans Info Theor 49:2329
31. Candy JV (2006) Model-based signal processing. IEEE Press, and Wiley, Hoboken, New Jersey
32. Griffiths RB (2002) Consistent quantum mechanics. Cambridge University Press, Cambridge (in paperback, 2003)

For further references to Information Theory itself, besides [2] above, there is an extensive bibliography. We list a short introductory samples below, of fundamental work (from 1953 – 1979):

(i) Goldman S (1953) Information theory. Prentice-Hall, New York
(ii) Feinstein A (1958) Foundations of information theory. McGraw-Hill, New York
(iii) Fano RM (1961) Transmission of information (MIT). Wiley, New York
(iv) Slepian D (ed) (1973) Key papers in the development of information theory. IEEE, New York (Selected Reprint Series)
(v) Berlekamp ER (ed) (1973) Key papers in the development of coding theory. IEEE Press, New York (Selected Reprint Series)